CAMBRIDGE STUDIES IN
ADVANCED MATHEMATICS 7

Introduction to higher order categorical logic

Introduction to higher order categorical logic

J. LAMBEK
McGill University
P.J. SCOTT
University of Ottawa

The right of the University of Cambridge to print and sell all manner of books was granted by Henry VIII in 1534. The University has printed and published continuously since 1584.

CAMBRIDGE UNIVERSITY PRESS
Cambridge
London New York New Rochelle
Melbourne Sydney

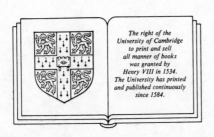

Published by the Press Syndicate of the University of Cambridge
The Pitt Building, Trumpington Street, Cambridge CB2 1RP
32 East 57th Street, New York, NY 10022, USA
10 Stamford Road, Oakleigh, Melbourne 3166, Australia

First published 1986

Printed in Great Britain at the University Press, Cambridge

Library of Congress catalogue card number: 85–5941

British Library cataloguing in publication data

Lambek, J.
Introduction to higher order categorical logic.
–(Cambridge studies in advanced mathematics; 7)
1. Categories (Mathematics)
I. Title II. Scott, P.J.
512′.55 QA169

ISBN 0 521 24665 2

Contents

Preface

This book makes an effort to reconcile two different attempts to come to grips with the foundations of mathematics. One is mathematical logic, which traditionally consists of proof theory, model theory and the theory of recursive functions; the other is category theory. It has been our experience that, when lecturing on the applications of logic to category theory, we met with approval from logicians and with disapproval from categorists, while the opposite was the case when we mentioned applications of category theory to logic. Unfortunately, to show that the logicians' viewpoint is essentially equivalent to the categorists' one, we have to slightly distort both. For example, categorists may be unhappy when we treat categories as special kinds of deductive systems and logicians may be unhappy when we insist that deductive systems need not be freely generated from axioms and rules of inference. The situation becomes even worse when we take the point of view of universal algebra. For example, combinatory logics are for us certain kinds of algebras, which goes against the grain for those logicians who have spent a life-time studying what we call the free such algebra. On the other hand, cartesian closed categories and even toposes are for us also certain kinds of algebras, although not over sets but over graphs, and this goes against the grain of those categorists who like to think of products and the like as being given only up to isomorphism. To make matters worse, universal algebraists may not be happy when we stress the logical or the categorical aspects, and even graph theorists may feel offended because we have had to choose a definition of graph which is by no means standard.

This is not the first book on categorical logic, as there already exists a classical monograph on first order categorical logic by Makkai and Reyes, not to mention a book on toposes written by a categorist (Johnstone) and a book on topoi written by a logician (Goldblatt), both of whom mention the internal language of toposes*. Our point is rather this: logicians have made

* Let us also draw attention to the important recent book by Barr and Wells, which manages to cover an amazing amount of material without explicit use of logical tools, relying on embedding theorems instead.

three attempts to formulate higher order logic, in increasing power: typed λ-calculus, Martin–Löf type theory and the usual (let us say intuitionistic) type theory. Categorists quite independently, though later, have developed cartesian closed categories, locally cartesian closed categories and toposes. We claim here that typed λ-calculi and cartesian closed categories are essentially the same, in the sense that there is an equivalence of categories (even untyped λ-calculi are essentially the same as certain algebras we call C-monoids). All this will be found in Part I. We also claim that intuitionistic type theories and toposes are closely related, in as much as there is a pair of adjoint functors between their respective categories. This is worked out in Part II. The relationship between Martin–Löf type theories and locally cartesian closed categories was established too recently (by Robert Seely) to be treated here. Logicians will find applications of proof theory in Part I, while many possible applications of proof theory in Part II have been replaced by categorical techniques. They will find some mention of model theory in Part I and more in Part II, but with emphasis on a categorical presentation: models are functors. All discussion of recursive functions is relegated to Part III.

We deliberately excluded certain topics from consideration, such as geometric logic and geometric morphisms. There are other topics which we omitted with some regret, because of limitations of time and space. These include the results of Robert Seely already mentioned, Gödel's *Dialectica interpretation* (1958), which greatly influenced much of this book, the relation between Gödel's double negation translation and double negation sheaves noted by Peter Freyd, Joyal's proof of Brouwer's principle that arrows from R to R in the free topos necessarily represent continuous functions (and related results), the proof that N is projective in the free topos and the important work on graphical algebras by Burroni.

Of course, like other authors, we have some axes to grind. Aside from what some people may consider to be undue emphasis on category theory, logic, universal algebra or graph theory, we stress the following views:

> We decry overzealous applications of Occam's razor.
>
> We believe that type theory is the proper foundation for mathematics.
>
> We believe that the free topos, constructed linguistically but determined uniquely (up to isomorphism) by its universal property, is an acceptable universe of mathematics for a moderate intuitionist and, therefore, that Platonism, formalism and intuitionism are reconcilable philosophies of mathematics.

This may be the place for discussing very briefly who did what. Many results in categorical logic were in the air and were discovered by a number of people simultaneously. Many results were discussed at the Séminaire Bénabou in Paris and published only in preprint form if at all. (Since we are referring to a number of preprints in our bibliography, we should point out that preliminary versions of portions of this book had also been circulated in preprint form, namely Part I in 1982, Part II in 1983 and Part 0 in 1983.) If we are allowed to say to whom we owe the principal ideas exposed in this monograph, we single out Bill Lawvere, Peter Freyd, André Joyal and Dana Scott, and hope that no one whose name has been omitted will be offended.

Let us also take this opportunity to thank all those who have provided us with some feedback on preliminary versions of Parts 0 and I. Again, hoping not to give offence to others, we single out for special thanks (in alphabetic order) Alan Adamson, Bill Anglin, John Gray, Bill Hatcher, Denis Higgs, Bill Lawvere, Fred Linton, Adam Obtułowicz and Birge Zimmermann-Huysgen. We also thank Peter Johnstone for his astute comments on our seminar presentation of Part II. Of course, we take full responsibility for all errors and oversights that still remain.

Finally let us thank Marcia Rodríguez for her conscientious handling of the bibliography, Pat Ferguson for her excellent and patient typing of successive versions of our manuscript and David Tranah for initiating the whole project.

The authors wish to acknowledge support from the Natural Sciences and Engineering Research Council of Canada and the Quebec Department of Education.

Montreal, July, 1984

Remark on notation: throughout this book, we frequently, though not exclusively, use the symbol ≡ for definitional equality.

0

Introduction to category theory

Introduction to Part 0

In Part 0 we recall the basic background in category theory which may be required in later portions of this book. The reader who is familiar with category theory should certainly skip Part 0, but even the reader who is not is advised to consult it only in addition to standard texts.

Most of the material in Part 0 is standard and may also be found in other books. Therefore, on the whole we shall refrain from making historical remarks. However, our exposition differs from treatments elsewhere in several respects.

Firstly, our exposition is slanted towards readers with some acquaintance with logic. Quite early we introduce the notion of a 'deductive system'. For us, this is just a category without the usual equations between arrows. In particular, we do not insist that a deductive system is freely generated from certain axioms, as is customary in logic. In fact, we really believe that logicians should turn attention to categories, which are deductive systems with suitable equations between proofs.

Secondly, we have summarized some of the main thrusts of category theory in the form of succinct slogans. Most of these are due to Bill Lawvere (whose influence on the development of category theory is difficult to overestimate), even if we do not use his exact words. Slogan V represents the point of view of a series of papers by one of the authors in collaboration with Basil Rattray.

Thirdly, we have emphasized the algebraic or equational nature of many of the systems studied in category theory. Just as groups or rings are algebraic over sets, it has been known for a long time that categories with finite products are equational over graphs. More recently, Albert Burroni made the surprising discovery that categories with equalizers are also algebraic over graphs. We have included this result, without going into his more technical concept of 'graphical algebra'.

In Part 0, as in the rest of this book, we have been rather cavalier about set theoretical foundations. Essentially, we are using Gödel–Bernays, as do most mathematicians, but occasionally we refer to universes in the sense of Grothendieck. The reason for our lack of enthusiasm in presenting the foundations properly is our belief that mathematics should be based on a version of type theory, a variant of which adequate for arithmetic and analysis is developed in Part II. For a detailed discussion of these foundational questions see Hatcher (1982, Chapter 8.)

1 Categories and functors

In this section we present what our reader is expected to know about category theory. We begin with a rather informal definition.

Definition 1.1. A *concrete category* is a collection of two kinds of entities, called *objects* and *morphisms*. The former are sets which are endowed with some kind of structure, and the latter are mappings, that is, functions from one object to another, in some sense preserving that structure. Among the morphisms, there is attached to each object A the *identity mapping* 1_A: $A \to A$ such that $1_A(a) = a$ for all $a \in A$. Moreover, morphisms $f: A \to B$ and $g: B \to C$ may be *composed* to produce a morphism $gf: A \to C$ such that $(gf)(a) = g(f(a))$ for all $a \in A$. (See also Exercise 2 below.)

Examples of concrete categories abound in mathematics; here are just three:

Example C1. The category of *sets*. Its objects are arbitrary sets and its morphisms are arbitrary mappings. We call this category '**Sets**'.

Example C2. The category of *monoids*. Its objects are monoids, that is, semigroups with unity element, and its morphisms are homomorphisms, that is, mappings which preserve multiplication (the semigroup operation) and the unity element.

Example C3. The category of *preordered sets*. Its objects are preordered sets, that is, sets with a transitive and reflexive relation on them, and its morphisms are monotone mappings, that is, mappings which preserve this relation.

The reader will be able to think of many other examples: the categories of rings, topological spaces and Banach algebras, to name just a few. In fact, one is tempted to make a generalization, which may be summed up as follows, provided we understand 'object' to mean 'structured set'.

Slogan I. Many objects of interest in mathematics congregate in concrete categories.

We shall now progress from concrete categories to abstract ones, in three easy stages.

Definition 1.2. A *graph* (usually called a *directed graph*) consists of two classes: the class of *arrows* (or *oriented edges*) and the class of *objects* (usually called *nodes* or *vertices*) and two mappings from the class of arrows to the class of objects, called *source* and *target* (often also *domain* and *codomain*).

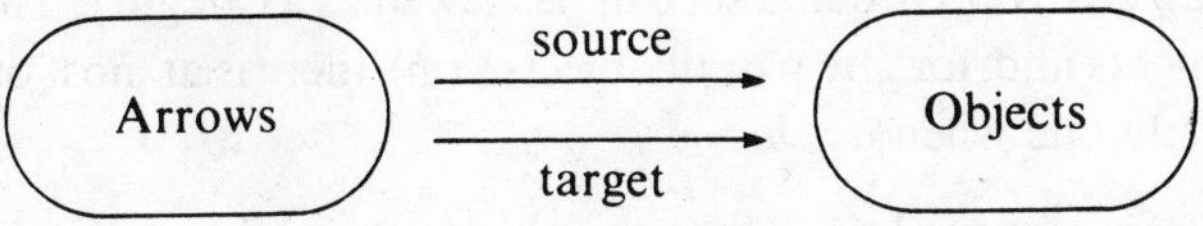

One writes '$f: A \to B$' for 'source $f = A$ and target $f = B$'. A graph is said to be *small* if the classes of objects and arrows are sets.

Example C4. The category of small *graphs* is another concrete category. Its objects are small graphs and its morphisms are functions F which send arrows to arrows and vertices to vertices so that, whenever $f: A \to B$, then $F(f): F(A) \to F(B)$.

A *deductive system* is a graph in which to each object A there is associated an arrow $1_A: A \to A$, the *identity* arrow, and to each pair of arrows $f: A \to B$ and $g: B \to C$ there is associated an arrow $gf: A \to C$, the *composition* of f with g. A logician may think of the objects as *formulas* and of the arrows as *deductions* or *proofs*, hence of

$$\frac{f: A \to B \quad g: B \to C}{gf: A \to C}$$

as a *rule of inference*. (Deductive systems will be discussed further in Part I.)

A *category* is a deductive system in which the following equations hold, for all $f: A \to B$, $g: B \to C$ and $h: C \to D$:

$$f 1_A = f = 1_B f, \quad (hg)f = h(gf).$$

Of course, all concrete categories are categories. A category is said to be *small* if the classes of arrows and objects are sets. While the concrete categories described in examples 1 to 4 are not small, a somewhat surprising observation is summarized as follows:

Slogan II. Many objects of interest to mathematicians are themselves small categories.

Example C1′. Any set can be viewed as a category: a small *discrete* category. The objects are its elements and there are no arrows except the obligatory identity arrows.

Example C2′. Any monoid can be viewed as a category. There is only one object, which may remain nameless, and the arrows of the monoid are its elements. In particular, the identity arrow is the unity element. Composition is the binary operation of the monoid.

Example C3′. Any preordered set can be viewed as a category. The objects are its elements and, for any pair of objects (a, b), there is at most one arrow $a \to b$, exactly one when $a \leqslant b$.

It follows from slogans I and II that small categories themselves should be the objects of a category worthy of study.

Example C5. The category **Cat** has as objects small categories and as morphisms functors, which we shall now define.

Definition 1.3. A *functor* $F: \mathscr{A} \to \mathscr{B}$ is first of all a morphism of graphs (see Example C4), that is, it sends objects of $\mathscr{A}$ to objects of $\mathscr{B}$ and arrows of $\mathscr{A}$ to arrows of $\mathscr{B}$ such that, if $f: A \to A'$, then $F(f): F(A) \to F(A')$. Moreover, a functor preserves identities and composition; thus

$$F(1_A) = 1_{F(A)}, \quad F(gf) = F(g)F(f).$$

In particular, the identity functor $1_{\mathscr{A}}: \mathscr{A} \to \mathscr{A}$ leaves objects and arrows unchanged and the composition of functors $F: \mathscr{A} \to \mathscr{B}$ and $G: \mathscr{B} \to \mathscr{C}$ is given by

$$(GF)(A) = G(F(A)), \quad (GF)(f) = G(F(f)),$$

for all objects A of $\mathscr{A}$ and all arrows $f: A \to A'$ in $\mathscr{A}$.

The reader will now easily check the following assertion.

Proposition 1.4. When sets, monoids and preordered sets are regarded as small categories, the morphisms between them are the same as the functors between them.

The above definition of a functor $F: \mathscr{A} \to \mathscr{B}$ applies equally well when $\mathscr{A}$ and $\mathscr{B}$ are not necessarily small, provided we allow mappings between classes. Of special interest is the situation when $\mathscr{B} = \mathbf{Sets}$ and $\mathscr{A}$ is small.

Slogan III. Many objects of interest to mathematicians may be viewed as functors from small categories to **Sets**.

Example F1. A set may be viewed as a functor from a discrete one-object category to **Sets**.

Example F2. A small graph may be viewed as a functor from the small category $\cdot \rightrightarrows \cdot$ (with identity arrows not shown) to **Sets**.

Example F3. If $\mathcal{M} = (M, 1, \cdot)$ is a monoid viewed as a one-object category, an $\mathcal{M}$-set may be regarded as a functor from $\mathcal{M}$ to **Sets**. (An *$\mathcal{M}$-set* is a set A together with a mapping $M \times A \to A$, usually denoted by $(m, a) \mapsto ma$, such that $1a = a$ and $(m \cdot m')a = m(m'a)$ for all $a \in A$, m and $m' \in M$.)

Once we admit that functors $\mathcal{A} \to \mathcal{B}$ are interesting objects to study, we should see in them the objects of yet another category. We shall study such functor categories in the next section. For the present, let us mention two other ways of forming new categories from old.

Example C6. From any category (or graph) $\mathcal{A}$ one forms a new category (respectively graph) $\mathcal{A}^{\mathrm{op}}$ with the same objects but with arrows reversed, that is, with the two mappings 'source' and 'target' interchanged. $\mathcal{A}^{\mathrm{op}}$ is called the *opposite* or *dual* of $\mathcal{A}$. A functor from $\mathcal{A}^{\mathrm{op}}$ to $\mathcal{B}$ is often called a *contravariant* functor from $\mathcal{A}$ to $\mathcal{B}$, but we shall avoid this terminology except for occasional emphasis.

Example C7. Given two categories $\mathcal{A}$ and $\mathcal{B}$, one forms a new category $\mathcal{A} \times \mathcal{B}$ whose objects are pairs (A, B), A in $\mathcal{A}$ and B in $\mathcal{B}$, and whose arrows are pairs $(f, g): (A, B) \to (A', B')$, where $f: A \to A'$ in $\mathcal{A}$ and $g: B \to B'$ in $\mathcal{B}$. Composition of arrows is defined componentwise.

Definition 1.5. An arrow $f: A \to B$ in a category is called an *isomorphism* if there is an arrow $g: B \to A$ such that $gf = 1_A$ and $fg = 1_B$. One writes $A \cong B$ to mean that such an isomorphism exists and says that A is *isomorphic* with B.

In particular, a functor $F: \mathcal{A} \to \mathcal{B}$ between two categories is an isomorphism if there is a functor $G: \mathcal{B} \to \mathcal{A}$ such that $GF = 1_{\mathcal{A}}$ and $FG = 1_{\mathcal{B}}$. We also remark that a group is a one-object category in which all arrows are isomorphisms.

To end this section, we shall record three basic isomorphisms. Here **1** is the category with one object and one arrow.

Proposition 1.6. For any categories $\mathcal{A}$, $\mathcal{B}$ and $\mathcal{C}$,

$$\mathcal{A} \times \mathbf{1} \cong \mathcal{A}, \quad (\mathcal{A} \times \mathcal{B}) \times \mathcal{C} \cong \mathcal{A} \times (\mathcal{B} \times \mathcal{C}), \quad \mathcal{A} \times \mathcal{B} \cong \mathcal{B} \times \mathcal{A}.$$

Exercises

1. Prove Propositions 1.4 and 1.6.

2. Show that for any concrete category $\mathcal{A}$ there is a functor $U: \mathcal{A} \to \mathbf{Sets}$

which 'forgets' the structure, often called the *forgetful* functor. Clearly U is *faithful* in the sense that, for all $f, g\colon A \rightrightarrows B$, if $U(f) = U(g)$ then $f = g$. (A more formal version of Definition 1.1 describes a *concrete* category as a pair $(\mathscr{A}, U)$, where $\mathscr{A}$ is a category and $U\colon \mathscr{A} \to \mathbf{Sets}$ is a faithful functor.)

Show that for any category $\mathscr{A}$ there are functors $\Delta\colon \mathscr{A} \to \mathscr{A} \times \mathscr{A}$ and $\bigcirc_{\mathscr{A}}\colon \mathscr{A} \to \mathbf{1}$ given on objects A of $\mathscr{A}$ by $\Delta(A) = (A, A)$ and $\bigcirc_{\mathscr{A}}(A) =$ the object of $\mathbf{1}$.

2 Natural transformations

In this section we shall investigate morphisms between functors.

Definition 2.1. Given functors $F, G\colon \mathscr{A} \rightrightarrows \mathscr{B}$, a *natural transformation* $t\colon F \to G$ is a family of arrows $t(A)\colon F(A) \to G(A)$ in $\mathscr{B}$, one arrow for each object A of $\mathscr{A}$, such that the following square commutes for all arrows $f\colon A \to B$ in $\mathscr{A}$:

$$\begin{array}{ccc} F(A) & \xrightarrow{\quad t(A) \quad} & G(A) \\ {\scriptstyle F(f)}\big\downarrow & & \big\downarrow{\scriptstyle G(f)} \\ F(B) & \xrightarrow[\quad t(B) \quad]{} & G(B) \end{array}$$

that is to say, such that

$$G(f)t(A) = t(B)F(f).$$

It is this concept about which it has been said that it necessitated the invention of category theory. We shall give examples of natural transformations later. For the moment, we are interested in another example of a category.

Example C8. Given categories $\mathscr{A}$ and $\mathscr{B}$, the *functor category* $\mathscr{B}^{\mathscr{A}}$ has as objects functors $F\colon \mathscr{A} \to \mathscr{B}$ and as arrows natural transformations. The *identity* natural transformation $1_F\colon F \to F$ is of course given by stipulating that $1_F(A) = 1_{F(A)}$ for each object A of $\mathscr{A}$. If $t\colon F \to G$ and $u\colon G \to H$ are natural transformations, their *composition* $u \circ t$ is given by stipulating that $(u \circ t)(A) = u(A)t(A)$ for each object A of $\mathscr{A}$.

To appreciate the usefulness of natural transformations, the reader should prove for himself the following, which supports Slogan III.

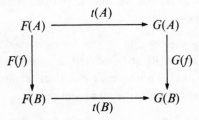

Proposition 2.2. When objects such as sets, small graphs and $\mathscr{M}$-sets are

viewed as functors into **Sets** (see Examples F1 to F3 in Section 1), the morphisms between two objects are precisely the natural transformations. Thus, the categories of sets, small graphs and $\mathcal{M}$-sets may be identified with the functor categories $\mathbf{Sets}^{\mathbf{1}}$, $\mathbf{Sets}^{\cdot\rightrightarrows}$ and $\mathbf{Sets}^{\mathcal{M}}$ respectively.

Of course, morphisms between sets are mappings, morphisms between graphs were described in Definition 1.3 and morphisms between $\mathcal{M}$-sets are $\mathcal{M}$-homomorphisms. (An *$\mathcal{M}$-homomorphism* $f: A \to B$ between $\mathcal{M}$-sets is a mapping such that $f(ma) = mf(a)$ for all $m \in M$ and $a \in A$.)

We record three more basic isomorphisms in the spirit of Proposition 1.6.

Proposition 2.3. For any categories $\mathcal{A}$, $\mathcal{B}$ and $\mathcal{C}$,

$$\mathcal{A}^1 \cong \mathcal{A}, \quad \mathcal{C}^{\mathcal{A} \times \mathcal{B}} \cong (\mathcal{C}^{\mathcal{B}})^{\mathcal{A}}, \quad (\mathcal{A} \times \mathcal{B})^{\mathcal{C}} \cong \mathcal{A}^{\mathcal{C}} \times \mathcal{B}^{\mathcal{C}}.$$

We shall leave the lengthy proof of this to the reader. We only mention here the functor $\mathcal{C}^{\mathcal{A} \times \mathcal{B}} \to (\mathcal{C}^{\mathcal{B}})^{\mathcal{A}}$, which will be used later. We describe its action on objects by stipulating that it assigns to a functor $F: \mathcal{A} \times \mathcal{B} \to \mathcal{C}$ the functor $F^*: \mathcal{A} \to \mathcal{C}^{\mathcal{B}}$ which is defined as follows:

For any object A of $\mathcal{A}$, the functor $F^*(A): \mathcal{B} \to \mathcal{C}$ is given by $F^*(A)(B) = F(A, B)$ and $F^*(A)(g) = F(1_A, g)$, for any object B of $\mathcal{B}$ and any arrow $g: B \to B'$ in $\mathcal{B}$.

For any arrow $f: A \to A'$, $F^*(f): F^*(A) \to F^*(A')$ is the natural transformation given by $F^*(f)(B) = F(f, 1_B)$, for all objects B of $\mathcal{B}$.

Finally, to any natural transformation $t: F \to G$ between functors F, G: $\mathcal{A} \times \mathcal{B} \rightrightarrows \mathcal{C}$ we assign the natural transformation $t^*: F^* \to G^*$ which is given by $t^*(A)(B) = t(A, B)$ for all objects A of $\mathcal{A}$ and B of $\mathcal{B}$.

This may be as good a place as any to mention that natural transformations may also be composed with functors.

Definition 2.4. In the situation

$$\mathcal{D} \xrightarrow{L} \mathcal{A} \underset{G}{\overset{F}{\rightrightarrows}} \mathcal{B} \xrightarrow{K} \mathcal{C},$$

if $t: F \to G$ is a natural transformation, one obtains natural transformations $Kt: KF \to KG$ between functors from $\mathcal{A}$ to $\mathcal{C}$ and $tL: FL \to GL$ between functors from $\mathcal{D}$ to $\mathcal{B}$ defined as follows:

$$(Kt)(A) = K(t(A)), \quad (tL)(D) = t(L(D)),$$

for all objects A of $\mathcal{A}$ and D of $\mathcal{D}$.

If $H: \mathcal{A} \to \mathcal{B}$ is another functor and $u: G \to H$ another natural transformation, then the reader will easily check the following 'distributive laws':

$$K(u \circ t) = (Ku) \circ (Kt), \quad (u \circ t)L = (uL) \circ (tL).$$

If we compare Slogans I and III, we are led to ask: which categories may be viewed as categories of functors into **Sets**? In preparation for an answer to that question we need another definition.

***Definition* 2.5.** If A and B are objects of a category $\mathscr{A}$, we denote by $\mathrm{Hom}_{\mathscr{A}}(A, B)$ the class of arrows $A \to B$. (Later, the subscript $\mathscr{A}$ will often be omitted.) If it so happens that $\mathrm{Hom}_{\mathscr{A}}(A, B)$ is a set for all objects A and B, $\mathscr{A}$ is said to be *locally small*.

One purpose of this definition is to describe the following functor.

***Example* F4.** If $\mathscr{A}$ is a locally small category, then there is a functor $\mathrm{Hom}_{\mathscr{A}}: \mathscr{A}^{\mathrm{op}} \times \mathscr{A} \to \mathrm{Sets}$. For an object (A, B) of $\mathscr{A}^{\mathrm{op}} \times \mathscr{A}$, the value of this functor is $\mathrm{Hom}_{\mathscr{A}}(A, B)$, as suggested by the notation. For an arrow $(g, h): (A, B) \to (A', B')$ of $\mathscr{A}^{\mathrm{op}} \times \mathscr{A}$, where $g: A' \to A$ and $h: B \to B'$ in $\mathscr{A}$, $\mathrm{Hom}_{\mathscr{A}}(g, h)$ sends $f \in \mathrm{Hom}_{\mathscr{A}}(A, B)$ to $hfg \in \mathrm{Hom}_{\mathscr{A}}(A', B')$.

Applying the isomorphism $\mathrm{Sets}^{\mathscr{A}^{\mathrm{op}} \times \mathscr{A}} \to (\mathrm{Sets}^{\mathscr{A}})^{\mathscr{A}^{\mathrm{op}}}$ of Proposition 2.3, we obtain a functor $\mathrm{Hom}^*_{\mathscr{A}}: \mathscr{A}^{\mathrm{op}} \to \mathrm{Sets}^{\mathscr{A}}$ and, dually, a functor $\mathrm{Hom}^*_{\mathscr{A}^{\mathrm{op}}}$: $\mathscr{A} \to \mathrm{Set}^{\mathscr{A}^{\mathrm{op}}}$. We shall see that the latter functor allows us to assert that $\mathscr{A}$ is isomorphic to a 'full' subcategory of $\mathrm{Sets}^{\mathscr{A}^{\mathrm{op}}}$.

***Definition* 2.6.** A *subcategory* $\mathscr{C}$ of a category $\mathscr{B}$ is any category whose class of objects and arrows is contained in the class of objects and arrows of $\mathscr{C}$ respectively and which is closed under the 'operations' source, target, identity and composition. By saying that a subcategory $\mathscr{C}$ of $\mathscr{B}$ is *full* we mean that, for any objects C, C' of $\mathscr{C}$, $\mathrm{Hom}_{\mathscr{C}}(C, C') = \mathrm{Hom}_{\mathscr{B}}(C, C')$.

For example, a proper subgroup of a group is a subcategory which is not full, but the category of Abelian groups is a full subcategory of the category of all groups.

The arrows $F \to G$ in $\mathbf{Sets}^{\mathscr{A}^{\mathrm{op}}}$ are natural transformations. We therefore write $\mathrm{Nat}(F, G)$ in place of $\mathrm{Hom}(F, G)$ in $\mathbf{Sets}^{\mathscr{A}^{\mathrm{op}}}$.

Objects of the latter category are sometimes called 'contravariant' functors from $\mathscr{A}$ to **Sets**. Among them is the functor $h_A \equiv \mathrm{Hom}_{\mathscr{A}}(-, A)$ which sends the object A' of $\mathscr{A}$ onto the set $\mathrm{Hom}_{\mathscr{A}}(A', A)$ and the arrow $f: A' \to A''$ onto the mapping $\mathrm{Hom}_{\mathscr{A}}(f, 1_A)$: $\mathrm{Hom}_{\mathscr{A}}(A'', A) \to \mathrm{Hom}_{\mathscr{A}}(A', A)$.

The following is known as Yoneda's Lemma.

***Proposition* 2.7.** If $\mathscr{A}$ is locally small and $F: \mathscr{A}^{\mathrm{op}} \to \mathbf{Sets}$, then $\mathrm{Nat}(h_A, F)$ is in one-to-one correspondence with $F(A)$.

Proof. If $a \in F(A)$, we obtain a natural transformation $\breve{a}: h_A \to F$ by stipulating that $\check{a}(B)\colon \mathrm{Hom}_{\mathscr{A}}(B, A) \to F(B)$ sends $g\colon B \to A$ onto $F(g)(a)$. (Note that F is contravariant, so $F(g)\colon F(A) \to F(B)$.)

Conversely, if $t\colon h_A \to F$ is a natural transformation, we obtain the element $t(A)(1_A) \in F(A)$. It is a routine exercise to check that the mappings $a \mapsto \check{a}$ and $t \mapsto t(A)(1_A)$ are inverse to one another.

Definition 2.8. A functor $H\colon \mathscr{A} \to \mathscr{B}$ is said to be *faithful* if the induced mappings $\mathrm{Hom}_{\mathscr{A}}(A, A') \to \mathrm{Hom}_{\mathscr{B}}(H(A), H(A'))$ sending $f\colon A \to A'$ onto $H(f)\colon H(A) \to H(A')$ for all $A', A \in \mathscr{A}$ are injective and *full* if they are surjective. A *full embedding* is a full and faithful functor which is also injective on objects, that is, for which $H(A) = H(A')$ implies $A = A'$.

Corollary 2.9. If $\mathscr{A}$ is locally small, the Yoneda functor $\mathrm{Hom}^*_{\mathscr{A}^{\mathrm{op}}}\colon \mathscr{A} \to \mathbf{Sets}^{\mathscr{A}^{\mathrm{op}}}$ is a full embedding.

Proof. Writing $H \equiv \mathrm{Hom}^*_{\mathscr{A}^{\mathrm{op}}}$, we see that the induced mapping $\mathrm{Hom}(A, A') \to \mathrm{Nat}(H(A), H(A'))$ sends $f\colon A \to A'$ onto the natural transformation $H(f)\colon H(A) \to H(A')$ which, for all objects B of $\mathscr{A}$, gives rise to the mapping $H(f)(B) = \mathrm{Hom}(1_B, f)\colon \quad \mathrm{Hom}(B, A) \to \mathrm{Hom}(B, A')$. Now $f \in H(A')(A)$, hence $\check{f}\colon H(A) \to H(A')$, as defined in the proof of Proposition 2.7, is given by

$$\begin{aligned}\check{f}(B)(g) &= H(A')(g)(f) = \mathrm{Hom}_{\mathscr{A}}(g, 1_{A'})(f)\\ &= fg = \mathrm{Hom}_{\mathscr{A}}(1_B, f)(g) = H(f)(B)(g),\end{aligned}$$

hence $\check{f} = H(f)$. Thus the mapping $f \mapsto H(f)$ is a bijection and so H is full and faithful.

Finally, to show that H is injective on objects, assume $H(A) = H(A')$, then $\mathrm{Hom}(A, A) = \mathrm{Hom}(A, A')$, so A' must be the target of the identity arrow 1_A, thus $A' = A$.

Exercises

1. Prove propositions 2.2 and 2.3.
2. If $\mathbf{2}$ is the category $\cdot \to \cdot$ (with identity arrows not shown), show that the objects of $\mathscr{A}^{\mathbf{2}}$ are essentially the arrows of $\mathscr{A}$ and that 'source' and 'target' may be viewed as functors $\delta, \delta'\colon \mathscr{A}^{\mathbf{2}} \rightrightarrows \mathscr{A}$.
3. If $F, G\colon \mathscr{A} \rightrightarrows \mathscr{B}$ are given functors, show that a natural transformation $t\colon F \to G$ is essentially the same as a functor $t\colon \mathscr{A} \to \mathscr{A}^{\mathbf{2}}$ such that $\delta t = F$ and $\delta' t = G$.
4. Show that the isomorphism in Yoneda's Lemma (Proposition 2.7) is natural in both A and F, that is, if $f\colon B \to A$ and $t\colon F \to G$ then the relevant diagrams commute.

3 Adjoint functors

Perhaps the most important concept which category theory has helped to formulate is that of adjoint functors. Aspects of this idea were known even before the advent of category theory and we shall begin by looking at one such.

We recall from Proposition 1.4 that a functor $\mathscr{A} \to \mathscr{B}$ between two preordered sets $\mathscr{A} = (A, \leqslant)$ and $\mathscr{B} = (B, \leqslant)$ regarded as categories is an order preserving mapping $F: A \to B$, that is, such that, for all elements a, a' of A, if $a \leqslant a'$ then $F(a) \leqslant F(a')$. A functor $G: \mathscr{B} \to \mathscr{A}$ in the opposite direction is said to be *right adjoint* to F provided, for all $a \in A$ and $b \in B$,

$$F(a) \leqslant b \quad \text{if and only if} \quad a \leqslant G(b).$$

Classically, a pair of order preserving mappings (F, G) is called a covariant *Galois correspondence* if it satisfies this condition.

Once we have such a Galois correspondence, we see immediately that $GF: \mathscr{A} \to \mathscr{A}$ is a *closure operation*, that is, for all $a, a' \in A$,

$$a \leqslant GF(a),$$
$$GFGF(a) \leqslant GF(a),$$
$$\text{if } a \leqslant a' \quad \text{then} \quad GF(a) \leqslant GF(a').$$

Similarly, $FG: \mathscr{B} \to \mathscr{B}$ may be called an *interior operation*: it satisfies the conditions dual to the above.

In a preordered set an isomorphism $a \cong a'$ just means that $a \leqslant a'$ and $a' \leqslant a$. (In a *poset*, or *partially ordered* set, one has the antisymmetry law: if $a \cong a'$ then $a = a'$.) We note that it follows from the above that $GFGF(a) \cong GF(a)$ and, dually, $FGFG(b) \cong FG(b)$, for all $a \in A$ and $b \in B$.

The most interesting consequence of a Galois correspondence is this: the functors F and G set up a one-to-one correspondence between isomorphism classes of 'closed' elements a of A such that $GF(a) \cong a$ and isomorphism classes of 'open' elements b of B such that $FG(b) \cong b$. We also say that F and G determine an *equivalence* between the preordered set $\mathscr{A}_0$ of closed elements of $\mathscr{A}$ and the preordered set $\mathscr{B}_0$ of open elements of $\mathscr{B}$. The following picture illustrates this principle of 'unity of opposites', which will be generalized later in this section.

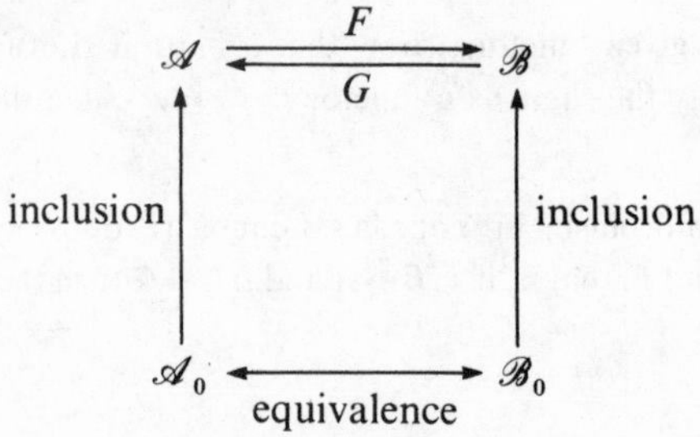

Before carrying out the promised generalization, let us look at a couple of examples of Galois correspondence; others will be found in the exercises.

Example G1. Take both $\mathscr{A}$ and $\mathscr{B}$ to be $(\mathbb{N}, \leqslant)$, the set of natural numbers with the usual ordering, and let

$$F(0) = 0, F(a) = p_a = \text{the } a\text{th prime number when } a > 0,$$

$$G(b) = \pi(b) = \text{the number of primes} \leqslant b.$$

Then F and G form a pair of adjoint functors and the 'unity of opposites' describes the biunique correspondence between positive integers and prime numbers.

Many examples arise from a binary relation $R \subseteq X \times Y$ between two sets X and Y. Take $\mathscr{A} = (\mathscr{P}(X), \subseteq)$, the set of subsets of X ordered by inclusion, and $\mathscr{B} = (\mathscr{P}(Y), \supseteq)$, ordered by inverse inclusion, and put

$$F(A) = \{y \in Y \mid \forall_{x \in A}(x, y) \in R\},$$

$$G(B) = \{x \in X \mid \forall_{y \in B}(x, y) \in R\},$$

for all $A \subseteq X$ and $B \subseteq Y$. This situation is called a *polarity*; it gives rise to an isomorphism between the lattice $\mathscr{A}_0$ of 'closed' subsets of X and the lattice $\mathscr{B}_0$ of 'closed' subsets of Y. (Note that the open elements of $\mathscr{B}$ are closed subsets of Y.)

Example G2. Take X to be the set of points of a plane, Y the set of half-planes, and write $(x, y) \in R$ for $x \in y$. Then, for any set A of points, $GF(A)$ is the intersection of all halfplanes containing A, in other words, the *convex hull* of A. The 'unity of opposites' here asserts that there are two equivalent ways of describing a convex set: by the points on it or by the halfplanes containing it.

We shall now generalize the notion of adjoint functor from preordered sets to arbitrary categories. In so doing, we shall bow to a notational prejudice of many categorists and replace the letter 'G' by the letter 'U'. ('U' is for 'underlying', 'F' for 'free'.)

Definition 3.1. An *adjointness* between categories $\mathscr{A}$ and $\mathscr{B}$ is given by a quadruple $(F, U, \eta, \varepsilon)$, where $F: \mathscr{A} \to \mathscr{B}$ and $U: \mathscr{B} \to \mathscr{A}$ are functors and $\eta: 1_{\mathscr{A}} \to UF$ and $\varepsilon: FU \to 1_{\mathscr{B}}$ are natural transformations such that

$$(U\varepsilon) \circ (\eta U) = 1_U, \quad (\varepsilon F) \circ (F\eta) = 1_F.$$

One says that U is *right adjoint* to F or that F is *left adjoint* to U and one calls η and ε the two *adjunctions*.

Before going into examples, let us give another formulation of what will turn out to be an equivalent concept (in Proposition 3.3 below).

Definition 3.2. A solution to the *universal mapping problem* for a functor $U:\mathscr{B}\to\mathscr{A}$ is given by the following data: for each object A of $\mathscr{A}$ an object $F(A)$ of $\mathscr{B}$ and an arrow $\eta(A): A\to UF(A)$ such that, for each object B of $\mathscr{B}$ and each arrow $f: A\to U(B)$ in $\mathscr{A}$, there exists a unique arrow $f^*: F(A)\to B$ in $\mathscr{B}$ such that $U(f^*)\eta(A)=f$.

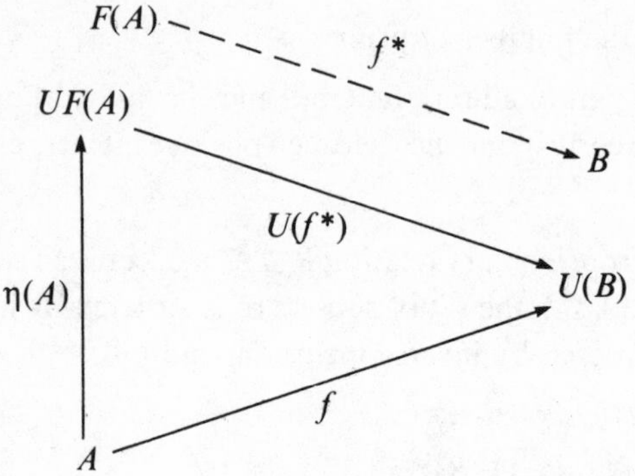

Example U1. Let $\mathscr{B}$ be the category of monoids, $\mathscr{A}$ the category of sets, $U:\mathscr{B}\to\mathscr{A}$ the forgetful (= underlying) functor, $F(A)$ the free monoid generated by the set A and $\eta(A)$ the obvious mapping of A into the underlying set of the monoid $F(A)$.

Definition 3.2′. Of special interest is the case of Definition 3.2 in which $\mathscr{B}$ is a full subcategory of $\mathscr{A}$ and $U:\mathscr{B}\to\mathscr{A}$ is the inclusion. Then $\eta(A): A\to F(A)$ may be called the *best approximation* of A by an object of $\mathscr{B}$ in the sense that, for each arrow $f: A\to B$ with B in $\mathscr{B}$, there is a unique arrow $f^*: F(A)\to B$ such that $f^*\eta(A)=f$. One then says that $\mathscr{B}$ is a full *reflective* subcategory of $\mathscr{A}$ with *reflector* F and *reflection* η.

Example U2. Let $\mathscr{A}$ be the category of Abelian groups, $\mathscr{B}$ the full subcategory of torsion free Abelian groups and $F(A)=A/T(A)$, where $T(A)$ is the torsion subgroup of A.

Proposition 3.3. Given two categories $\mathscr{A}$ and $\mathscr{B}$, there is a one-to-one correspondence between adjointnesses $(F, U, \eta, \varepsilon)$ and solutions $(F, \eta, *)$ of the universal mapping problem for $U:\mathscr{B}\to\mathscr{A}$.

Proof. If $(F, U, \eta, \varepsilon)$ is given, put $f^* = \varepsilon(B)F(f)$. Conversely, if U and $(F, \eta, *)$ are given, for each $f: A\to A'$, put $F(f)=(\eta(A')f)^*$ and check that this makes F a functor and η a natural transformation; moreover define $\varepsilon(B)=(1_{U(B)})^*$.

It follows from symmetry considerations that an adjointness is also equivalent to a 'co-universal mapping problem', obtained by dualizing

Definition 3.2. (A left adjoint to $\mathscr{B} \to \mathscr{A}$ is a right adjoint to $\mathscr{B}^{\text{op}} \to \mathscr{A}^{\text{op}}$.)

In view of Proposition 3.3, Examples U1 and U2 are examples of adjoint functors. We shall give other examples later.

There is yet another way of looking at adjoint functors, at least when $\mathscr{A}$ and $\mathscr{B}$ are locally small.

Proposition 3.4. An adjointness $(F, U, \eta, \varepsilon)$ between locally small categories $\mathscr{A}$ and $\mathscr{B}$ gives rise to and is determined by a natural isomorphism $\text{Hom}_{\mathscr{B}}(F(-),-) \cong \text{Hom}_{\mathscr{A}}(-, U(-))$ between functors $\mathscr{A}^{\text{op}} \times \mathscr{B} \rightrightarrows$ **Sets.**

We leave the proof of this to the reader.

Even if $\mathscr{A}$ is not locally small, there is a natural bijection between arrows $FA \to B$ in $\mathscr{B}$ and arrows $A \to UB$ in $\mathscr{A}$. Logicians may think of such a bijection as comprising two rules of inference; and this point of view has been quite influential in the development of categorical logic. An analogous situation in the propositional calculus would be the bijection between proofs of the entailments $C \wedge A \vdash B$ and $A \vdash C \Rightarrow B$ (see Exercise 4 below). Inasmuch as implication is a more sophisticated notion than conjunction, the adjointness here explains the emergence of one concept from another. This point of view, due to Lawvere, may be summarized by yet another slogan, illustrations of which will be found throughout this book (see, for instance, Exercise 6 below).

Slogan IV. Many important concepts in mathematics arise as adjoints, right or left, to previously known functors.

We summarize two important properties of adjoint functors, which will be useful later.

Proposition 3.5. (i) Adjoint functors determine each other uniquely up to natural isomorphisms.

(ii) If (U, F) and (U', F') are pairs of adjoint functors, as in the diagram

$$\mathscr{C} \underset{F'}{\overset{U'}{\rightleftarrows}} \mathscr{B} \underset{F}{\overset{U}{\rightleftarrows}} \mathscr{A},$$

then $(UU', F'F)$ is also an adjoint pair.

Exercise

1. If (F, G) is a Galois correspondence between posets $\mathscr{A}$ and $\mathscr{B}$, show that F preserves supremums and G preserves infimums. If $\mathscr{A}$ has and F preserves supremums, show that its right adjoint $G: \mathscr{B} \to \mathscr{A}$ can be calculated by the

formula

$G(b) = \sup\{a \in \mathscr{A} \mid F(a) \leqslant b\}$.

2. In Example G1, show that the two sets $\{F(a) + a \mid a \in \mathbb{N}\}$ and $\{G(b) + b + 1 \mid b \in \mathbb{N}\}$ are complementary sets.

3. Given a commutative ring C, take X to be the set of elements of C, Y the set of prime ideals of C and define $R \subseteq X \times Y$ by writing $(x, y) \in R$ for $x \in y$. If F and G are defined as for any polarity, show that, for any subset A of X, $GF(A) = \{x \in X \mid \exists_{n \in \mathbb{N}} x^n \in A\}$, the so-called *radical* of A. Also show that the closure operation FG on the set of subsets of Y makes Y into a compact topological space called the *spectrum* of C. The 'unity of opposites' here describes a one-to-one correspondence between closed subspaces of the spectrum and ideals which are equal to their radical.

4. Take $\mathscr{A}$ and $\mathscr{B}$ to be the preordered sets of formulas of the propositional calculus, the order being entailment. For a fixed formula C, show that $F\colon \mathscr{A} \to \mathscr{B}$ and $G\colon \mathscr{B} \to \mathscr{A}$ defined by $F(A) \equiv C \wedge A$ and $G(B) \equiv C \Rightarrow B$ are a pair of adjoint functors. What is the 'unity of opposites' in this case?

5. Prove propositions 3.4 and 3.5.

6. If $\mathscr{A} = \mathscr{B} =$ **Sets**, C a given set, let $F(A) = C \times A$ and $U(B) = B^C$ for any sets A and B. Extend U and F to functors and show that U is right adjoint to F.

7. Show that the forgetful functor from **Cat** to **Sets** which sends every small category onto its set of objects has both a left and a right adjoint.

8. Show that the forgetful functor from **Cat** to the category of graphs has a left adjoint, which assigns to each graph the category 'generated by it'.

4 Equivalence of categories

We shall extend the 'unity of opposites' to general categories, but first we need to extend the notion of 'equivalence'.

Definition 4.1. An adjointness $(F, U, \eta, \varepsilon)$ is an *adjoint equivalence* if η and ε are natural isomorphisms. More generally, an *equivalence* between categories $\mathscr{A}$ and $\mathscr{B}$ is given by a pair of functors $F\colon \mathscr{A} \to \mathscr{B}$ and $U\colon \mathscr{B} \to \mathscr{A}$ such that $UF \cong 1_{\mathscr{A}}$ and $FU \cong 1_{\mathscr{B}}$.

The extra generality is an illusion: given that $\eta\colon 1_{\mathscr{A}} \to UF$ and $\varepsilon'\colon FU \to 1_{\mathscr{B}}$ are isomorphisms, one obtains an adjoint equivalence by putting

$$\varepsilon(B) \equiv \varepsilon'(B)F(U\varepsilon'(B)\eta U(B))^{-1}.$$

Proposition 4.2. An adjointness $(F, U, \eta, \varepsilon)$ between categories $\mathscr{A}$ and $\mathscr{B}$ induces an adjoint equivalence between full subcategories $\mathscr{A}_0$ of $\mathscr{A}$ and $\mathscr{B}_0$

of $\mathscr{B}$, where

$$\mathscr{A}_0 \equiv \text{Fix}\,\eta \equiv \{A \in \mathscr{A} \mid \eta(A) \text{ is an isomorphism}\},$$
$$\mathscr{B}_0 \equiv \text{Fix}\,\varepsilon \equiv \{B \in \mathscr{B} \mid \varepsilon(B) \text{ is an isomorphism}\}.$$

Moreover, ηU is an isomorphism if and only if εF is.

The significance of the last statement is this: if ηU is an isomorphism, $\mathscr{B}_0$ becomes a reflective subcategory of $\mathscr{B}$; if εF is an isomorphism $\mathscr{A}_0$ becomes a coreflective subcategory of $\mathscr{A}$. (See Definition 3.2′, 'coreflective' being the dual of 'reflective'.)

Proof. Only the last statement requires proof. It is a consequence of the following.

Lemma 4.3. Given an adjointness $(F, U, \eta, \varepsilon)$ between categories $\mathscr{A}$ and $\mathscr{B}$, the following statements are equivalent:

(1) $\eta UF = UF\eta$,
(2) ηU is an isomorphism,
(3) $\varepsilon FU = FU\varepsilon$,
(4) εF is an isomorphism.

Proof. We show that (1)⇒(2)⇒(3)⇒(4)⇒(1).

(1)⇒(2). Suppose for the moment that $\eta(A)$ has a left inverse g, we claim that, in the presence of (1), g is also a right inverse. For

$$\begin{aligned}\eta(A)g &= UF(g)\eta UF(A) \text{ by naturality of } \eta\\ &= UF(g)UF\eta(A) \text{ by (1)}\\ &= UF(g\eta(A)) = UF(1_A) = 1_{UF(A)}.\end{aligned}$$

Now, by Definition 3.1, $\eta U(B)$ has a left inverse $U\varepsilon(B)$, hence $\eta U(B)$ is an isomorphism, which proves (2).

(2)⇒(3). Assume that $\eta U(B)$ is an isomorphism, then its inverse is $U\varepsilon(B)$, by Definition 3.1. Hence

$$\begin{aligned}\varepsilon FU(B) &= \varepsilon FU(B)F(1_{U(B)})\\ &= \varepsilon FU(B)F(\eta U(B)U\varepsilon(B))\\ &= \varepsilon FU(B)F\eta U(B)FU\varepsilon(B)\\ &= 1_{FU(B)}FU\varepsilon(B) \text{ by Definition 3.1}\\ &= FU\varepsilon(B).\end{aligned}$$

(3)⇒(4). This is proved exactly like (1)⇒(2). In fact, we may quote (1)⇒(2), since there is an adjointness between $\mathscr{B}^{\text{op}}$ and $\mathscr{A}^{\text{op}}$.

(4)⇒(1). This is proved like (2)⇒(3) or by duality quoting (2)⇒(3).

Examples of Proposition 4.2 abound in mathematics. The main problem is usually the identification of $\mathscr{A}_0$ and $\mathscr{B}_0$. The following examples require some knowledge of mathematics that has not been developed in this book. (The same will be true for exercises 2 and 3 below.)

Example A1. Let $\mathscr{A}$ be the category of Abelian groups and $\mathscr{B}$ the opposite of the category of topological Abelian groups. Let K be the compact group of the reals modulo the integers: $K \equiv \mathbb{R}/\mathbb{Z}$. For any abstract Abelian group A, define $F(A)$ as the group of all homomorphisms of A into K, with the topology induced by K. For any topological Abelian group B, define $U(B)$ as the group of all continuous homomorphisms of B into K. Then U and F are easily seen to be the object parts of a pair of adjoint functors. Here $\mathscr{A}_0$ is $\mathscr{A}$, while $\mathscr{B}_0$ is the opposite of the category of compact Abelian groups. The 'unity of opposites' asserts the well-known Pontrjagin duality between abstract and compact Abelian groups. The last statement of Proposition 4.2 tells us that the compact Abelian groups form a reflective subcategory of the category of all topological Abelian groups.

Example A2. Let $\mathscr{A}$ be the category of rings and $\mathscr{B}$ the opposite of the category of topological spaces. For any ring A, $F(A)$ is the topological space of homomorphisms of A into $\mathbb{Z}/(2)$, the ring of integers modulo 2, the topology being induced by the discrete topology of $\mathbb{Z}/(2)$. For any topological space B, $U(B)$ is the ring of continuous functions of B into $\mathbb{Z}/(2)$ (with the discrete topology), with the ring structure inherited by that of $\mathbb{Z}/(2)$. Here $\mathscr{A}_0$ is the category of Boolean rings and $\mathscr{B}_0$ is the opposite of the category of zero-dimensional compact Hausdorff spaces. The 'unity of opposites' asserts the well-known Stone duality. Both $\mathscr{A}_0$ and $\mathscr{B}_0^{\mathrm{op}}$ are full reflective subcategories.

We summarize the 'unity of opposites' principle in another slogan. (The reader will have noticed that a *duality* between categories $\mathscr{A}$ and $\mathscr{B}$ is nothing but an equivalence between $\mathscr{A}$ and $\mathscr{B}^{\mathrm{op}}$.)

Slogan V. Many equivalence and duality theorems in mathematics arise as an equivalence of fixed subcategories induced by a pair of adjoint functors.

Exercises

1. Prove the statement following Definition 4.1 that every equivalence gives rise to an adjoint equivalence. (Hint: first show that $\eta UF = UF\eta$.)

2. Give a presentation of the well-known Gelfand duality between commutative C*-algebras and compact Hausdorff spaces in a manner similar to Example A2. (Let $\mathscr{A}$ be the category of commutative Banach algebras.)
3. If $\mathscr{A}$ is the category of presheaves on a topological space X and $\mathscr{B}$ is the category of spaces over X, show that there is a pair of adjoint functors between $\mathscr{A}$ and $\mathscr{B}$ which induces an equivalence between sheaves and local homeomorphisms. (See also Part II, Theorem 10.3.)
4. Prove that $U: \mathscr{B} \to \mathscr{A}$ is (half of) an equivalence if and only if it is full and faithful and every object of $\mathscr{A}$ is isomorphic to one of the form $U(B)$, for some object B to $\mathscr{B}$.

5 Limits in categories

In this section we shall study limits in categories. They contain as special cases many important constructions, for example products, equalizers and pullbacks, as well as their duals. Moreover, they serve as an illustration of Slogan IV. We begin with the following special case.

Definition 5.1. An object T of a category $\mathscr{A}$ is said to be a *terminal* object if for each object A of $\mathscr{A}$ there is a unique arrow $\bigcirc_A: A \to T$. (Later, we shall usually write 1 for T.)

We note that the uniqueness of $\bigcirc_A$ may be expressed equationally by saying that, for all arrows $h: A \to T, h = \bigcirc_A$.

It is easily seen that T is unique up to isomorphism: if T' is another terminal object, then $T' \cong T$. Hence, one often speaks of *the* terminal object. For example, in the category of sets, any one element set $\{*\}$ is terminal and, in the category of groups, any one element group is terminal. A terminal object in $\mathscr{A}^{\text{op}}$ is also called an *initial* object in $\mathscr{A}$. In **Sets**, the only initial object is the empty set $\varnothing$, while, in the category of groups, any terminal object is also initial.

As an illustration of Slogan IV, we note that to say that $\mathscr{A}$ has a terminal (respectively initial) object is the same as saying that the functor $\bigcirc_{\mathscr{A}}: \mathscr{A} \to \mathbf{1}$ has a right (respectively left) adjoint.

Definition 5.2. Given a set I and a family $\{A_i | i \in I\}$ of objects in a category $\mathscr{A}$, their *product* is given by an object P and a family of *projections* $\{p_i: P \to A_i | i \in I\}$ with the following universal property: given any object Q and any family of arrows $\{q_i: Q \to A_i | i \in I\}$, there is a unique arrow $f: Q \to P$ such that $p_i f = q_i$ for all $i \in I$.

We may also say that the family $\{p_i: P \to A_i | i \in I\}$ is a terminal object in the category of all families $\{q_i: Q \to A_i | i \in I\}$ (with appropriate arrows).

It is easily seen that the object P is unique up to isomorphism. Hence, one speaks of *the* product. It is often denoted by $\prod_{i \in I} A_i$. In the category of sets, products are 'cartesian' products. In many concrete categories, products are constructed on the underlying sets with an obvious induced structure. This is true for the categories of monoids, groups, rings etc., in fact all 'algebraic' categories (that is, varieties of universal algebras), as well as for the categories of posets and topological spaces.

A product in $\mathscr{A}^{\mathrm{op}}$ is also called a *coproduct* in $\mathscr{A}$. There is no one preferred name for coproducts in the literature; in **Sets**, coproducts are disjoint unions, while, in the category of groups, they are free products.

What if I is the empty set? Then the universal property asserts that, for each object Q, there is a unique arrow $Q \to P$, in other words, that P is a terminal object.

Again we have an illustration of Slogan IV: to say that all I-indexed families in $\mathscr{A}$ have products (respectively coproducts) is the same as saying that the functor $\mathscr{A} \to \mathscr{A}^I$ which sends an object A of $\mathscr{A}$ onto the constant family $\{A | i \in I\}$ has a right (respectively left) adjoint.

It may be worth looking at the product of two objects A and B of $\mathscr{A}$ in some detail. It is given by an object $A \times B$ with projections $\pi_{A,B}: A \times B \to A$ and $\pi'_{A,B}: A \times B \to B$ such that, for all arrows $f: C \to A$ and $g: C \to B$, there is a unique arrow $\langle f, g \rangle: C \to A \times B$ satisfying the equations:

$$\pi_{A,B}\langle f, g \rangle = f, \quad \pi'_{A,B}\langle f, g \rangle = g.$$

Note that the uniqueness of $\langle f, g \rangle$ may also be expressed by an equation, namely:

$$\langle \pi_{A,B} h, \pi'_{A,B} h \rangle = h,$$

for all $h: C \to A \times B$.

Evidently, the defining property of $A \times B$ establishes a bijection between pairs of arrows $(C \to A, C \to B)$ and arrows $C \to A \times B$. To say that all such products exist is the same as saying that the diagonal functor $\Delta: \mathscr{A} \to \mathscr{A} \times \mathscr{A}$ has a right adjoint. Dually, all binary coproducts exist if and only if Δ has a left adjoint.

Definition 5.3. A pair of arrows $f, g: A \rightrightarrows B$ is said to have an *equalizer* $e: C \to A$ provided $fe = ge$ and, for all $h: D \to A$ such that $fh = gh$, there is a unique arrow $k: D \to C$ satisfying $ek = h$. Another way of expressing this is to say that $e: C \to A$ is terminal in the category of all arrows $h: D \to A$ such that $fh = gh$.

It is easily seen that the equalizing object C is unique up to isomorphism.

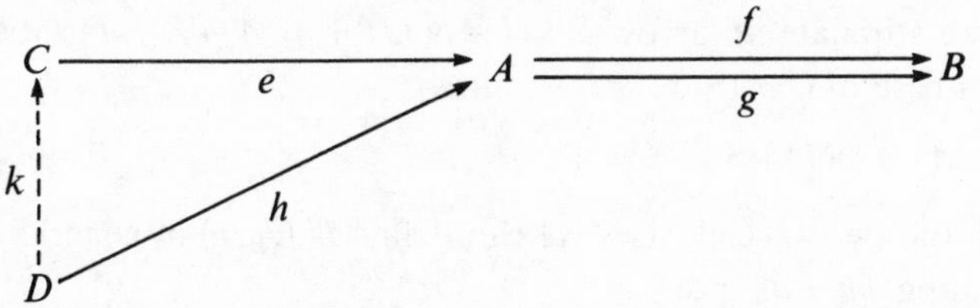

In the category of sets or groups, one may take $C \equiv \{a \in A \mid f(a) = g(a)\}$ and $e: C \to A$ as the inclusion. As is the case for products, equalizers in many concrete categories are formed on the underlying sets. An equalizer in $\mathscr{A}^{\mathrm{op}}$ is also called a *coequalizer* in $\mathscr{A}$. In **Sets**, the coequalizer of two mappings $f, g: B \rightrightarrows A$ is given by $e: A \to C$, where C is obtained from A by identifying all elements $f(b)$ and $g(b)$ with $b \in B$, and where e is the obvious surjection. (More precisely, $C = A/\equiv$, where $\equiv$ is the smallest equivalence relation on A such that $f(b) \equiv g(b)$ for all $b \in B$.) In the category of groups, the coequalizer of two homomorphisms $f, g: B \rightrightarrows A$ is obtained similarly from a suitable congruence relation on A (or normal subgroup of A).

While it was evident how finite products could be presented equationally, it is by no means obvious how this can be done for equalizers. The following discussion is our version of Burroni's pioneering ideas.

With any diagram $A \underset{g}{\overset{f}{\rightrightarrows}} B$ we associate another diagram $E(f, g) \xrightarrow{\alpha(f, g)} A$ which is to serve as its equalizer. Clearly, we must stipulate the equation

(B1) $\quad f\alpha(f, g) = g\alpha(f, g).$

Next, let us consider the universal property of $\alpha(f, g)$. Given an arrow $h: D \to A$ such that $fh = gh$, we seek a unique arrow $\beta(f, g, h): D \to E(f, g)$ such that

(*) $\quad \alpha(f, g)\beta(f, g, h) = h.$

While (*) is an equation, it depends on the condition $fh = gh$, which we would like to get rid of. We shall consider two special cases of $\beta(f, g, h)$ in which the condition $fh = gh$ is automatically satisfied.

First special case: consider any arrow $h: D \to A$, then surely

$$fh\alpha(fh, gh) = gh\alpha(fh, gh).$$

Hence we stipulate an arrow $\gamma(f, g, h)$ $(\equiv \beta(f, g, h\alpha(fh, gh)))$: $E(fh, gh) \to E(f, g)$ satisfying as a special case of (*):

(B2) $\quad \alpha(f, g)\gamma(f, g, h) = h\alpha(fh, gh).$

Second special case: consider any arrow $f: A \to B$, then surely $f1_A = f1_A$.

Hence we stipulate an arrow $\delta(f)\ (\equiv \beta(f, f, 1_A))\colon A \to E(f, f)$ satisfying as a special case of (*):

(B3) $\quad \alpha(f, f)\delta(f) = 1_A$.

From the two special cases we can define $\beta(f, g, h)$ in general: Assuming $fh = gh$, put

(**) $\quad \beta(f, g, h) \equiv \gamma(f, g, h)\delta(fh)$.

Then

$$\alpha(f, g)\beta(f, g, h) = \alpha(f, g)\gamma(f, g, h)\delta(fh) = h\alpha(fh, gh)\delta(fh),$$

by (B2). As it so happens that $fh = gh$, this becomes equal to

$$h1_D = h$$

by (B3), and so we recapture (*).

It remains to express the uniqueness of $\beta(f, g, h)$ equationally. So suppose that $\alpha(f, g)k = h$, we want this to imply that $k = \beta(f, g, h)$. This is evidently done by

(B4) $\quad \beta(f, g, \alpha(f, g)k) = k$.

Here β can be eliminated in favour of γ and δ using (**).

We summarize the preceding discussion of equalizers as follows.

Proposition 5.4. (Burroni). Equalizers for all pairs of arrows $f, g\colon A \rightrightarrows B$ are given by the following data: an arrow $\alpha(f, g)\colon E(f, g) \to A$ for each such pair, a family of arrows $\gamma(f, g, h)\colon E(fh, gh) \to E(f, g)$ one for each $h\colon D \to A$, and an arrow $\delta(f)\colon A \to E(f, f)$ satisfying (B1) to (B4) (with β eliminated from (B4) by (**)).

Definition 5.5. A *pullback* of a diagram $\begin{array}{l} A\searrow \\ \qquad C \\ B\nearrow \end{array}$ is given by a diagram $P\begin{array}{l}\nearrow A\\ \searrow B\end{array}$ which is terminal in the category of all diagrams $D\begin{array}{l}\nearrow A\\ \searrow B\end{array}$ such that

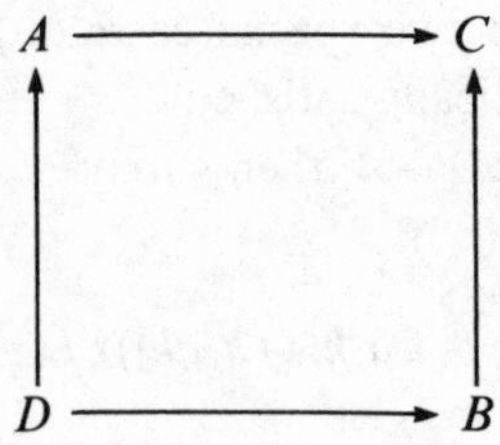

commutes. In other words,

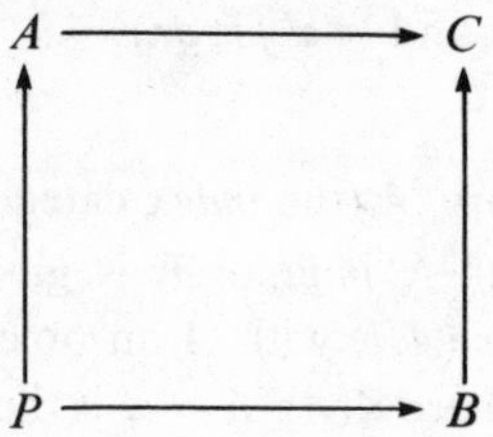

commutes and, for any other commutative square as above, there is a unique arrow $D \to P$ such that the two triangles

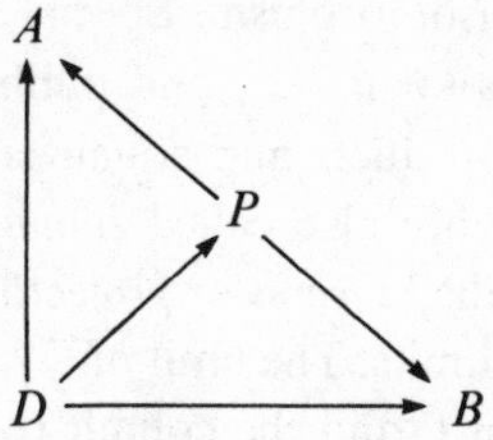

commute.

It is easily seen that P is unique up to isomorphism. A pullback in $\mathscr{A}^{\mathrm{op}}$ is called a *pushout* in $\mathscr{A}$. In a category with a terminal object T, binary products are special cases of pullbacks, namely when $C \equiv T$. Instead of describing pullbacks in other special categories, we shall show how, in general, they may be constructed from products and equalizers.

Proposition 5.6. If a category has binary products and equalizers, pullbacks may be constructed as follows:

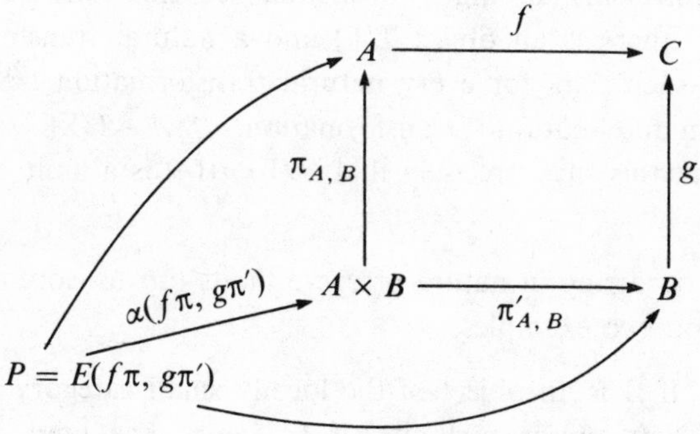

Proof. Note that $f\pi\alpha(f\pi, g\pi') = g\pi'\alpha(f\pi, g\pi')$, by (B1). Suppose $h: D \to A$ and $k: D \to B$ are such that $fh = gk$. Then there is a unique arrow $\langle h, k\rangle: D \to A \times B$ such that $\pi\langle h, k\rangle = h$ and $\pi'\langle h, k\rangle = k$, hence a unique arrow $s: D \to P$ such that $\pi\alpha(f\pi, g\pi)s = h$ and $\pi'\alpha(f\pi, g\pi)s = k$, that is, $\alpha(f\pi, g\pi)s = \langle h, k \rangle$.

***Definition* 5.7.** Let there be given a category $\mathscr{I}$ (the *index* category) and a functor $\Gamma: \mathscr{I} \to \mathscr{A}$ (called an *$\mathscr{I}$-diagram*). A *limit* of Γ is given by a terminal object in the category of all pairs (A, t) with A an object of $\mathscr{A}$ and $t: K(A) \to \Gamma$ a natural transformation, where $K(A): \mathscr{I} \to \mathscr{A}$ is the functor with *constant* value A. In other words, $(A_0, t_0: K(A_0) \to \Gamma)$ is a limit of Γ if for all $(A, t: K(A) \to \Gamma)$ there is unique $f: A \to A_0$ such that $t_0(I)f = t(I)$ for all objects I of $\mathscr{I}$.

It is easily seen that A_0 is unique up to isomorphism. Special cases of limits are products ($\mathscr{I}$ discrete), equalizers ($\mathscr{I}$ is $\cdot \rightrightarrows \cdot$) and pullbacks ($\mathscr{I}$ is $\begin{smallmatrix}\cdot & \\ & \searrow \\ & & \cdot \\ & \nearrow \\ \cdot & \end{smallmatrix}$). Limits may be constructed from products and equalizers as are pullbacks (Proposition 5.6). Limits in $\mathscr{A}^{\mathrm{op}}$ are also called *colimits* in $\mathscr{A}$. If $\mathscr{I}$ is a directed poset, limits are usually called *inverse* or *projective* limits, while colimits are called *direct* or *inductive* limits. The limit of Γ (or rather the object A_0) is sometimes denoted by $\varprojlim \Gamma$ and the colimit by $\varinjlim \Gamma$.

The following connection between limits and adjoint functors illustrates Slogan IV.

***Proposition* 5.8.** To say for given categories $\mathscr{I}$ and $\mathscr{A}$ that every $\mathscr{I}$-diagram $\Gamma: \mathscr{I} \to \mathscr{A}$ has a limit (respectively colimit) is equivalent to saying that the *constancy* functor $K: \mathscr{A} \to \mathscr{A}^{\mathscr{I}}$, which associates to every object A of $\mathscr{A}$ the functor $K(A): \mathscr{I} \to \mathscr{A}$ with constant value A, has a right adjoint (respectively left adjoint).

Proof. One way of asserting that K has a right adjoint $L: \mathscr{A}^{\mathscr{I}} \to \mathscr{A}$ is by the solution to the universal mapping problem (dualize Definition 3.2): for each object Γ of $\mathscr{A}^{\mathscr{I}}$ there is an object $L(\Gamma)$ and a natural transformation $\varepsilon(\Gamma): KL(\Gamma) \to \Gamma$ such that, for every natural transformation $t: K(A) \to \Gamma$ there is a unique natural transformation $t^*: A \to L(\Gamma)$ satisfying $\varepsilon(\Gamma)K(t^*) = t$. But this says precisely that $(L(\Gamma), \varepsilon(\Gamma))$ is a limit of Γ (see Definition 4.7).

Many functors occurring in nature preserve limits (up to isomorphism). We shall mention two examples.

***Proposition* 5.9.** If A is an object of the locally small category $\mathscr{A}$, then $\mathrm{Hom}(A, -): \mathscr{A} \to \mathbf{Sets}$ preserves limits: if $\Gamma: \mathscr{I} \to \mathscr{A}$ has limit A_0 then $\mathrm{Hom}(A, \Gamma(-)): \mathscr{I} \to \mathbf{Sets}$ has limit $\mathrm{Hom}(A, A_0)$.

Proof. Write $h^A \equiv \mathrm{Hom}(A, -)$ and assume that $(A_0, t_0: K(A_0) \to \Gamma)$ is terminal in the category of all pairs $(A, t: K(A) \to \Gamma)$. We assert that $(h^A(A_0),\ h^A t_0: h^A K(A_0) \to h^A \Gamma)$ is terminal in the category of all pairs

$(X, \tau: K(X) \to h^A\Gamma)$, X being a set. (Note that $h^A K(A_0) = K(h^A(A_0))$.) In other words, we claim that there is a unique mapping $\psi: X \to h^A(A_0)$ such that $(h^A t_0) \circ K(\psi) = \tau$. To see what this last equation means, apply it to any object I of $\mathscr{I}$, then it asserts

$$\mathrm{Hom}(1_A, t_0(I))\psi = \tau(I).$$

Again, applying this equation to any $x \in X$, we obtain

$$t_0(I)\psi(x) = \tau(I)(x).$$

If $t_x: K(A) \to \Gamma$ is defined by $t_x(I) \equiv \tau(I)(x)$, we see that this means

$$t_0 \circ K(\psi(x)) = t_0.$$

The existence of a unique $\psi(x): A_0 \to A$ with this property is assured by the fact that (A_0, t_0) was terminal.

Proposition 5.10. If $F: \mathscr{A} \to \mathscr{B}$ is left adjoint to $U: \mathscr{B} \to \mathscr{A}$, then U preserves limits and F preserves colimits.

Proof. If $\mathscr{A}$ and $\mathscr{B}$ are locally small, this is an easy corollary of Proposition 5.9. However, one may just as well prove the result directly, without assuming local smallness, and we shall do so for U.

While it is easy to give a precise argument as in the proof of Proposition 5.9, the reader may find the following sketch more intuitive.

Let $\mathscr{C}$ (respectively $\mathscr{D}$) be the full subcategory of $\mathscr{A}^{\mathscr{I}}$ (respectively $\mathscr{B}^{\mathscr{I}}$) consisting of those $\mathscr{I}$-diagrams which have limits. Evidently, $\mathscr{C}$ contains all constant $\mathscr{I}$-diagrams $K_{\mathscr{A}}(A)$, with A in $\mathscr{A}$, such that $K_{\mathscr{A}}(A)(I) = A$ for all I in $\mathscr{I}$. Hence we may factor the constancy functor $K_{\mathscr{A}}: \mathscr{A} \to \mathscr{A}^{\mathscr{I}}$ through $K'_{\mathscr{A}}: \mathscr{A} \to \mathscr{C}$. As in Proposition 5.8, we may regard $\varprojlim_{\mathscr{A}}$ as right adjoint to $K'_{\mathscr{A}}$. Now $F^{\mathscr{I}}: \mathscr{A}^{\mathscr{I}} \to \mathscr{B}^{\mathscr{I}}$ (respectively $U^{\mathscr{I}}: \mathscr{B}^{\mathscr{I}} \to \mathscr{A}^{\mathscr{I}}$) factors through $F': \mathscr{C} \to \mathscr{D}$ (respectively $U': \mathscr{D} \to \mathscr{C}$) and U' is right adjoint to F'. Then, clearly, $F'K'_{\mathscr{A}} = K'_{\mathscr{B}}F$.

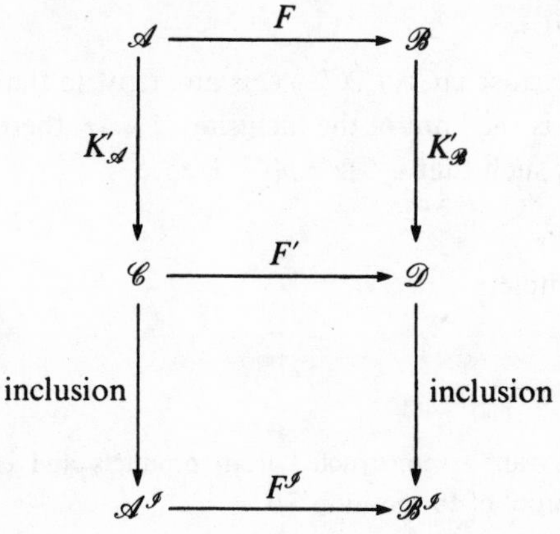

Taking right adjoints, we obtain, in view of Proposition 3.5, $\varprojlim_{\mathscr{A}} U' \cong U \varprojlim_{\mathscr{B}}$. Applying both sides to any diagram $\Delta: \mathscr{I} \to \mathscr{B}$ and noting that $U'(\Delta) = U\Delta$, we finally obtain $\varprojlim (U\Delta) \cong U(\varprojlim(\Delta))$.

Definition 5.11. A category $\mathscr{A}$ is said to be *complete* (cocomplete) if it has all limits (colimits) of diagrams $\Gamma: \mathscr{I} \to \mathscr{A}$, $\mathscr{I}$ being small. This means that products (coproducts) and equalizers (coequalizers) exist.

Assuming completeness of $\mathscr{A}$ or $\mathscr{B}$ one can prove a kind of converse of Proposition 5.9 and of 5.10. For example, if $U: \mathscr{B} \to \mathscr{A}$ preserves limits and $\mathscr{A}$ is complete, one can construct a left adjoint $F: \mathscr{A} \to \mathscr{B}$, as in Exercise 1 of Section 3, provided a certain 'solution set condition' holds; this is the content of Freyd's Adjoint Functor Theorem. These converse results will be brought out in the exercises; they depend on the following lemma, the proof of which is a bit tricky.

Lemma 5.12. If $\mathscr{A}$ is complete, then $\mathscr{A}$ has an initial object if and only if it has a small *pre-initial* full subcategory $\mathscr{C}$, that is to say, for any object A of $\mathscr{A}$ there is an object C of $\mathscr{C}$ and an arrow $f: C \to A$ in $\mathscr{A}$.

Proof. The necessity of the condition is obvious. To prove its sufficiency, let $(A_0, u: K(A_0) \to \Gamma)$ be the limit of the inclusion functor $\Gamma: \mathscr{C} \to \mathscr{A}$. In particular, for each object C of $\mathscr{C}$ there is an arrow $u(C): A_0 \to C$. Take any object A of $\mathscr{A}$, then, by assumption, we can find C in $\mathscr{C}$ and an arrow $f: C \to A$, hence an arrow $fu(C): A_0 \to A$. It remains to show that there is only one arrow $A_0 \to A$.

Suppose we have two arrows $g, h: A_0 \rightrightarrows A$ and let $k: K \to A_0$ be their equalizer. It will follow that $g = h$ if we can show that k has a right inverse. By assumption, there exists C' in $\mathscr{C}$ and an arrow $f': C' \to K$. It will suffice to show that $kf'u(C') = 1_{A_0}$.

Now, for any object C of $\mathscr{C}$,

$$u(C)kf'u(C') = u(C),$$

by naturality of u and because $u(C)kf': C' \to C$ is an arrow in the full subcategory $\mathscr{C}$. Since (A_0, u) is the limit of the inclusion $\mathscr{C} \to \mathscr{A}$, there exists a unique arrow $e: A_0 \to A_0$ such that $u(C)e = u(C)$. Hence

$$kf'u(C') = e = 1_{A_0},$$

and our argument is complete.

Exercises

1. Prove that limits can be constructed from products and equalizers, generalizing the proof of Proposition 5.6.

2. Deduce from Proposition 5.10 that, in the propositional calculus regarded as a preordered set (see Exercise 4 of Section 3), the distributive law holds: $p \wedge (a \vee b) \cong (p \wedge a) \vee (p \wedge b)$.

3. Given two functors $F, G: \mathscr{A} \rightrightarrows \mathscr{B}$, let $(F; G)$ be the category whose objects are pairs $(A, b: F(A) \to G(A))$, A any object of $\mathscr{A}$, and whose arrows $(A, b) \to (A', b')$ are arrows $a: A \to A'$ in $\mathscr{A}$, such that $G(a)b = b'F(a)$. Assuming that $\mathscr{A}$ is complete and that G preserves limits, show that $(F; G)$ has an initial object if and only if it has a small pre-initial full subcategory. (Hint: Use Proposition 5.12.)

4. If $\mathscr{A}$ is locally small, a functor $U: \mathscr{A} \to$ **Sets** is said to be *representable* if $U \cong \mathrm{Hom}(A, -)$ for some objects A of $\mathscr{A}$. Show that U is representable if and only if the category $(K(\{*\}); U)$ has a small pre-initial full subcategory. (Hint: Use Exercise 3 with $\mathscr{B} =$ **Sets**.)

5. Let $\mathscr{A}$ be a complete category. Show that a functor $U: \mathscr{A} \to \mathscr{B}$ has a left adjoint if and only if U preserves limits and, for each object B of $\mathscr{B}$, the category $(K(B); U)$ has a small pre-initial full sub-category.

6. Let $\mathscr{A}$ be a complete category. Show that a functor $\Gamma: \mathscr{I} \to \mathscr{A}$ has a colimit if and only if the category $(K(\Gamma); K)$ has a small pre-initial full subcategory. (Here the first K denotes the constancy functor $\mathscr{A}^{\mathscr{I}} \to (\mathscr{A}^{\mathscr{I}})^{\mathscr{A}}$, while the second K denotes the constancy functor $\mathscr{A} \to \mathscr{A}^{\mathscr{I}}$.)

7. Given a small category $\mathscr{A}$ and any functor $F: \mathscr{A}^{\mathrm{op}} \to$ **Sets**, show that F is a colimit of representable functors as follows. Let $\mathscr{I}_F$ be the category whose objects are pairs (A, t), A an object of $\mathscr{A}$ and $t: \mathrm{Hom}_{\mathscr{A}}(-, A) \to F$ a natural transformation, and whose arrows $(A, t) \to (A', t')$ are arrows $a: A \to A'$ in $\mathscr{A}$ such that $t' \circ \mathrm{Hom}_{\mathscr{A}}(-, a) = t$. Then F is the colimit of the functor Γ_F: $\mathscr{I}_F \to \mathbf{Sets}^{\mathscr{A}^{\mathrm{op}}}$ obtained by composing the Yoneda embedding $\mathscr{A} \to \mathbf{Sets}^{\mathscr{A}^{\mathrm{op}}}$ with the obvious forgetful functor $\mathscr{I}_F \to \mathscr{A}$. (The associated natural transformation $t_0: \Gamma_F \to K(F)$ is defined by $t_0(A, t) \equiv t$.)

6 Triples

We recall that a *closure operation* on a preordered set $\mathscr{A} = (|\mathscr{A}|, \leqslant)$ is a mapping $T: |\mathscr{A}| \to |\mathscr{A}|$ with the following properties:

$$\frac{A \leqslant B}{T(A) \leqslant T(B)}, \quad A \leqslant T(A), \quad TT(A) \leqslant T(A),$$

for all elements A, B of $|\mathscr{A}|$. The first of these says, of course, that T is order preserving. This notion has been generalized from preordered sets to arbitrary categories and is then called a 'standard construction', 'triple' or 'monad'. Reluctantly, we choose the second term, as it appears to be the most widely used.

Definition 6.1. A *triple* (T,η,μ) on a category $\mathscr{A}$ consists of a functor T: $\mathscr{A}\to\mathscr{A}$ and natural transformations $\eta\colon 1_{\mathscr{A}}\to T$ and $\mu\colon T^2\to T$ satisfying the equations

$$\mu\circ T\eta = 1_T = \mu\circ\eta T,\quad \mu\circ\mu T = \mu\circ T\mu.$$

These equations are sometimes called the *unity laws* and *associative law* respectively and are illustrated by the following commutative diagrams:

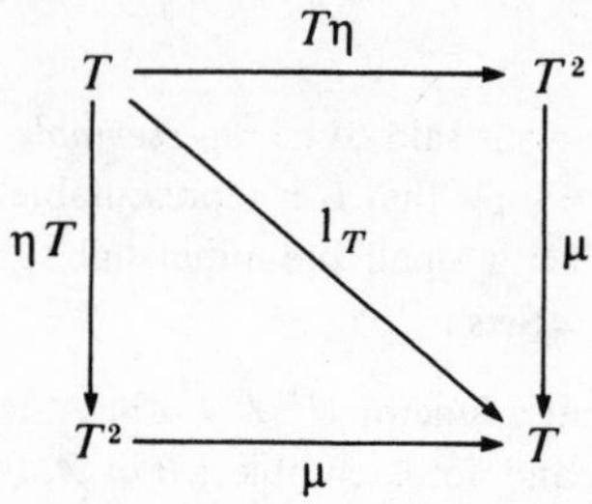

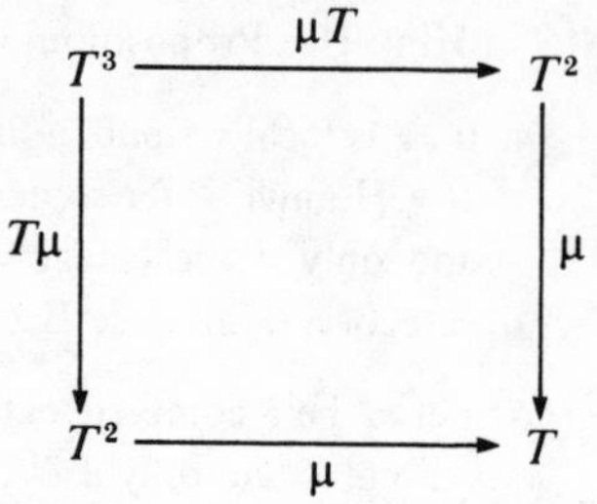

The reader will recall how natural transformations are composed (see Example C8); for example, the associative law asserts that, for every object A of $\mathscr{A}$,

$$\mu(A)\mu(T(A)) = \mu(A)T(\mu(A)).$$

Proposition 6.2. (Huber). If $F\colon\mathscr{A}\to\mathscr{B}$ is left adjoint to the functor $U\colon\mathscr{B}\to\mathscr{A}$ with adjunctions $\eta\colon 1_{\mathscr{A}}\to UF$ and $\varepsilon\colon FU\to 1_{\mathscr{B}}$, then $(UF,\eta,U\varepsilon F)$ is a triple on $\mathscr{A}$.

Proof. For example, let us prove one of the unity laws:

$$\mu\circ T\eta = U\varepsilon F\circ UF\eta = U(\varepsilon F\circ F\eta) = U1_F = 1_{UF},$$

by Definition 3.1, and since

$$(U1_F)(A) = U(1_{F(A)}) = 1_{U(F(A))} = 1_{UF}(A).$$

We leave the proofs of the other two laws to the reader.

We shall see that the converse of this proposition is also true; but first we shall look at a number of examples of triples, which, on the face of it, do not seem to arise from a pair of adjoint functors.

Example T1. Let there be given a monoid $\mathscr{M} = (M,1,\cdot)$. For each set A define the set $T(A)\equiv M\times A$ and the mappings

$$\begin{array}{ll}\eta(A)\colon A\to M\times A, & \mu(A)\colon M\times(M\times A)\to M\times A\\ \quad a\mapsto(1,a) & \quad (m,(m',a))\mapsto(m\cdot m',a).\end{array}$$

One easily makes T into a functor **Sets** $\to$ **Sets** and checks that η and μ are natural transformations. Moreover, one obtains a triple (T, η, μ) on **Sets**, the unity laws and associative law here following from the equations

$$m \cdot 1 = m = 1 \cdot m, \quad (m \cdot m') \cdot m'' = m \cdot (m' \cdot m'')$$

for all m, m' and $m'' \in M$, which will explain their names.

Example T2. Let $T \equiv P$ be the covariant power set functor **Sets** $\to$ **Sets**, that is, for any set A,

$$P(A) \equiv \{X \mid X \subseteq A\}$$

and, for any mapping $f: A \to B$, and any subset $X \subseteq A$,

$$P(f)(X) = \{f(x) \mid x \in X\}.$$

Furthermore, let the natural transformations η and μ be given by the mappings $\eta(A): A \to P(A)$ and $\mu(A): P(P(A)) \to P(A)$ defined by

$$\eta(A)(a) \equiv \{a\}, \quad \mu(A)(\mathscr{X}) \equiv \bigcup \mathscr{X} \equiv \bigcup_{X \in \mathscr{X}} X,$$

for any set A, any element $a \in A$ and any set $\mathscr{X}$ of subsets of A. The reader is invited to show that (T, η, μ) is a triple by verifying the unity and associative laws in this case.

We now return to the question: does every triple on $\mathscr{A}$ arise from a pair of adjoint functors $\mathscr{A} \xrightarrow{F} \mathscr{B} \xrightarrow{U} \mathscr{A}$ as in Proposition 6.2? The answer is 'yes', but the category $\mathscr{B}$ is not unique. In fact, we shall present two extremes for the construction of $\mathscr{B}$.

Definition 6.3. Given a triple (T, η, μ) on a category $\mathscr{A}$, the *Eilenberg–Moore* category $\mathscr{A}^T$ of the triple is defined as follows. Its objects, called *algebras*, are pairs (A, φ), where $\varphi: T(A) \to A$ is an arrow of $\mathscr{A}$ satisfying the equations

$$\varphi\eta(A) = 1_A, \quad \varphi\mu(A) = \varphi T(\varphi)$$

for all objects A of $\mathscr{A}$. Its arrows, called *homomorphisms*, $(A, \varphi) \to (A', \varphi')$ are arrows $\alpha: A \to A'$ of $\mathscr{A}$ satisfying the equation

$$\varphi' T(\alpha) = \alpha\varphi.$$

These equations are illustrated by the following commutative diagrams:

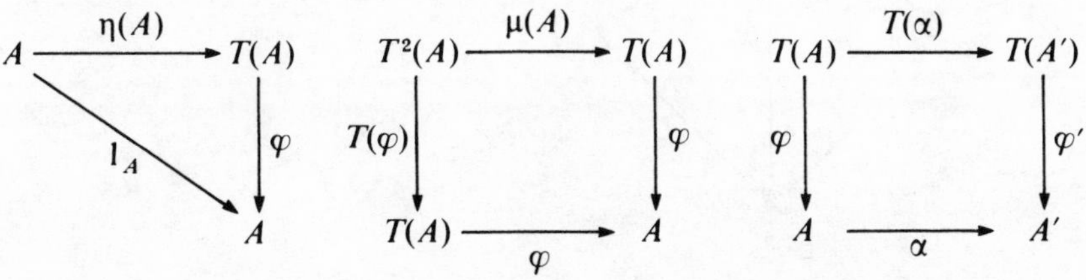

Example T1 (continued). An element of $T(A) \equiv M \times A$ is a pair (m, a) with $m \in M$ and $a \in A$. One usually writes $ma \equiv \varphi(m, a)$. The equations of an algebra then read

$$1a = a, \quad (m \cdot m')a = m(m'a),$$

for all $a \in A, m$ and $m' \in M$. In other words, an algebra is an $\mathcal{M}$-set (see Example F3 in Section 1). The equation satisfied by a homomorphism reads

$$\alpha(ma) = m\alpha(a),$$

for all $a \in A$, so we recapture the usual homomorphisms of $\mathcal{M}$-sets (see Proposition 2.2).

Example T2 (continued). The algebras of the power set triple on **Sets** are sup-complete (hence inf-complete) lattices and the homomorphisms are sup-preserving (hence also order preserving) mappings.

In view of these examples and many others like them, we enunciate our final slogan.

Slogan VI. Many categories of interest are the Eilenberg–Moore categories of triples on familiar categories.

In both examples above, the familiar category is **Sets**, but in Exercise 2 below it is **Ab**, the category of abelian groups. Categories, on the other hand, may be viewed as algebras over **Grph**, the category of graphs.

Definition 6.4. Given a triple (T, η, μ) on a category $\mathscr{A}$, by a *resolution* $(\mathscr{B}, U, F, \varepsilon)$ of this triple we mean a category $\mathscr{B}$ and a pair of adjoint functors $\mathscr{A} \xrightarrow{F} \mathscr{B} \xrightarrow{U} \mathscr{A}$ such that $UF = T$ with adjunctions η (as given) and ε such that $U\varepsilon F = \mu$ (as in Proposition 6.2). The resolutions of the given triple form a category whose arrows $\Phi: (\mathscr{B}, U, F, \varepsilon) \to (\mathscr{B}', U', F', \varepsilon')$ are functors $\Phi: \mathscr{B} \to \mathscr{B}'$ such that $\Phi F = F'$, $U'\Phi = U$ and $\Phi\varepsilon = \varepsilon'\Phi$. In particular, the following two triangles commute:

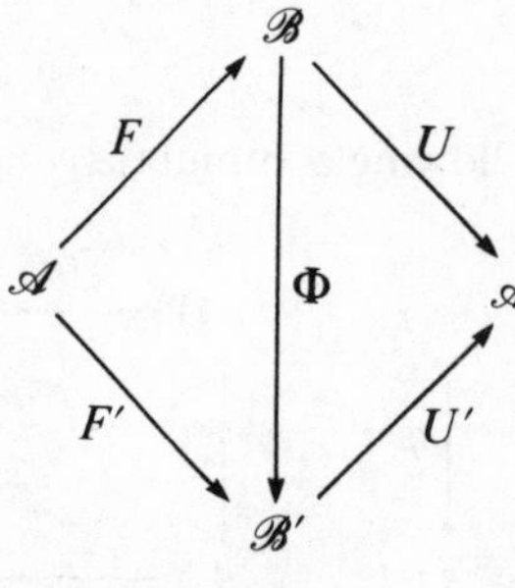

Proposition 6.5. The Eilenberg–Moore category $\mathscr{A}^T$ of the triple (T, η, μ) on $\mathscr{A}$ gives rise to a resolution $(\mathscr{A}^T, U^T, F^T, \varepsilon^T)$, which is a terminal object in the category of all resolutions. Thus, given any resolution $(\mathscr{B}, U, F, \varepsilon)$, there is a unique functor $K^T: \mathscr{B} \to \mathscr{A}^T$, called the *comparison functor*, such that $K^T F = F^T, U^T K^T = U$ and $K^T \varepsilon = \varepsilon^T K^T$. Moreover, U^T is faithful.

Proof. (1) We define $U^T: \mathscr{A}^T \to \mathscr{A}$ by

$$U^T(A, \varphi) \equiv A, \quad U^T(\alpha) \equiv \alpha,$$

for any algebra (A, φ) and any homomorphism α. Evidently, U^T is faithful.

(2) We define $F^T: \mathscr{A} \to \mathscr{A}^T$ by

$$F^T(A) \equiv (T(A), \mu(A)), \quad F^T(f) \equiv T(f),$$

for any object A and any arrow f of $\mathscr{A}$. It is easily checked that $(T(A), \mu(A))$ is an algebra, that $T(f)$ is a homomorphism and that $U^T F^T = T$.

(3) We define the natural transformation ε^T from $F^T U^T$ to the identity functor on $\mathscr{A}^T$ by its action on the algebra (A, φ) as follows: the homomorphism $\varepsilon^T(A, \varphi) \equiv \varphi$. Indeed, the square

$$\begin{array}{ccc} T^2(A) & \xrightarrow{T(\varphi)} & T(A) \\ {\scriptstyle \mu(A)}\downarrow & & \downarrow{\scriptstyle \varphi} \\ T(A) & \xrightarrow[\varphi]{} & A \end{array}$$

commutes by Definition 6.3. To see that $U^T \varepsilon^T F^T = \mu$, one calculates

$$(U^T \varepsilon^T F^T)(A) = (U^T \varepsilon^T)(T(A), \mu(A)) = U^T(\mu(A)) = \mu(A).$$

We let the reader check that

$$(\varepsilon^T F^T \circ F^T \eta)(A) = A, \quad (U^T \varepsilon^T \circ \eta U^T)(A, \varphi) = (A, \varphi),$$

for any object A of $\mathscr{A}$ and any algebra (A, φ), whence it follows that $(\mathscr{A}^T, U^T, F^T, \varepsilon^T)$ is a resolution of the given triple.

(4) Let $(\mathscr{B}, U, F, \varepsilon)$ be another resolution of the same triple, we shall construct the comparison functor $K^T: \mathscr{B} \to \mathscr{A}^T$ and show that it is the unique functor with the desired properties. For any object B and any arrow g of $\mathscr{B}$, we put

$$K^T(B) \equiv (U(B), U\varepsilon(B)), \quad K^T(g) \equiv U(g).$$

Then surely $U^T K^T = U$; in fact, this result forces the definitions of $K^T(g)$ and of the first component of $K^T(B)$. Moreover, $\varepsilon^T K^T(B) = U\varepsilon(B)$, and this forces the definition of the second component of $K^T(B)$. It remains to check that $K^TF = F^T$. Indeed, for any object A of $\mathscr{A}$,

$$K^TF(A) = (UF(A), U\varepsilon F(A)) = (T(A), \mu(A)) = F^T(A).$$

This completes the proof.

We remark that, in view of Slogan VI, it is of interest to know when the comparison functor is an equivalence of categories. Conditions for this to be the case were found by Beck. Without going into these conditions here, let us only mention that a functor $U: \mathscr{B} \to \mathscr{A}$ is called *tripleable* or *monadic* if it has a left adjoint and if the comparison functor K^T is an equivalence. Examples of tripleable concrete categories $U: \mathscr{B} \to$ **Sets** are all algebraic categories, that is, varieties of universal algebras, and the category of compact Hausdorff spaces.

The category of resolutions of a triple also has an initial object.

Definition 6.6. The *Kleisli* category $\mathscr{A}_T$ of a triple (T, η, μ) on a category $\mathscr{A}$ is defined as follows. Its objects are the same as those of $\mathscr{A}$; however, arrows $A \to A'$ in $\mathscr{A}_T$ are not the same as they would be in $\mathscr{A}$, instead they are arrows $A \to T(A')$ in $\mathscr{A}$. How do we compose arrows $f: A \to T(A')$ and $g: A' \to T(A'')$? Denoting their composition in $\mathscr{A}_T$ by $g * f: A \to T(A'')$ in $\mathscr{A}$, we define

$$g * f \equiv \mu(A'')T(g)f.$$

In particular,

$$f * \eta(A) = \mu(A')T(f)\eta(A) = \mu(A')\eta T(A')f = 1_T(A')f = f$$

and

$$\eta(A') * f = \mu(A')T\eta(A')f = 1_T(A')f = f,$$

hence $\eta(A): A \to T(A)$ serves as the identity arrow $A \to A$ in $\mathscr{A}_T$. We leave it to the reader to check the associativity of composition in $\mathscr{A}_T$.

Example T2 (continued). What is the Kleisli category of the power set triple on **Sets**? An arrow $A \to P(B)$ in **Sets** may be regarded as a multi-valued function from A to B or, equivalently, as a relation between A and B. More precisely, let $f: A \to P(B)$ correspond to $R_f \subseteq A \times B$, where $(a, b) \in R_f$ means $b \in f(a)$. What about the composition of f with $g: B \to P(C)$? According to Definition 6.6,

$$\begin{aligned}(g * f)(a) &\equiv \mu(C)(P(g)(f(a))) \\ &\equiv \bigcup \{g(b) \mid b \in f(a)\},\end{aligned}$$

hence

$$\begin{aligned}(a,c)\in R_{g*f} &\Leftrightarrow c\in(g*f)(a)\\ &\Leftrightarrow \exists_{b\in B}(b\in f(a)\wedge c\in g(b))\\ &\Leftrightarrow \exists_{b\in B}((a,b)\in R_f\wedge(b,c)\in R_g)\\ &\Leftrightarrow (a,c)\in R_gR_f,\end{aligned}$$

according to one way of defining the 'relative product'. Moreover, the identity arrow 1_A in the Kleisli category is represented by the mapping $\eta(A): A\to P(A)$ in **Sets**, which sends $a\in A$ onto $\{a\}\subseteq A$. Hence $(a,a')\in R_{\eta(A)}\Leftrightarrow a'\in\{a\}$, so $R_{\eta(A)}$ is the identity relation on A. We conclude that the Kleisli category of the power set triple on **Sets** is (isomorphic to) the category whose objects are sets and whose arrows are binary relations.

Proposition 6.7. The Kleisli category $\mathscr{A}_T$ of the triple (T,η,μ) on $\mathscr{A}$ gives rise to a resolution $(\mathscr{A}_T,U_T,F_T,\varepsilon_T)$, which is an initial object in the category of all resolutions. Thus, given any resolution $(\mathscr{B},U,F,\varepsilon)$, there is a unique functor $K_T:\mathscr{A}_T\to\mathscr{B}$ such that $K_TF_T=F$, $UK_T=U_T$ and $K_T\varepsilon_T=\varepsilon K_T$. Moreover, F_T is bijective on objects.

Proof. (1) We define $U_T:\mathscr{A}_T\to\mathscr{A}$ by

$$U_T(A)\equiv T(A),\quad U_T(f)\equiv\mu(B)T(f),$$

for any object A of $\mathscr{A}_T$, that is of $\mathscr{A}$, and for any arrow $f:A\to B$ in $\mathscr{A}_T$, that is, $f:A\to T(B)$ in $\mathscr{A}$. It is easily verified that U_T is a functor.

(2) We define $F_T:\mathscr{A}\to\mathscr{A}_T$ by

$$F_T(A)\equiv A,\quad F_T(f)\equiv\eta(B)f,$$

for any object A and any arrow $f:A\to B$ in $\mathscr{A}$. Evidently, F_T is bijective on objects and it is easily checked that $U_TF_T=T$ and that F_T is a functor.

(3) We define the natural transformation ε_T from F_TU_T to the identity functor on $\mathscr{A}_T$ by putting $\varepsilon_T(A)\equiv 1_{T(A)}$ in $\mathscr{A}$. To see that $U_T\varepsilon_TF_T=\mu$ one calculates

$$(U_T\varepsilon_TF_T)(A)=(U_T\varepsilon_T)(A)=U_T(1_{T(A)})=\mu(A).$$

We let the reader check that

$$(\varepsilon_TF_T\circ F_T\eta)(A)=A,\quad (U_T\varepsilon_T\circ\eta U_T)(A)=A,$$

for any object A of $\mathscr{A}$, hence of $\mathscr{A}_T$, whence it follows that $(\mathscr{A}_T,U_T,F_T,\varepsilon_T)$ is a resolution of the given triple.

(4) Let $(\mathscr{B},U,F,\varepsilon)$ be another resolution of the same triple. We shall construct a functor $K_T:\mathscr{A}_T\to\mathscr{B}$ and show that is the unique functor with the desired properties.

For any object A of $\mathscr{A}_T$ and any arrow $g: A \to A'$ in $\mathscr{A}_T$, that is, g: $A \to T(A')$ in $\mathscr{A}$, we put

$$K_T(A) \equiv F(A), \quad K_T(g) \equiv \varepsilon F(A')F(g).$$

Then surely $K_T F_T(A) = K_T(A) = F(A)$, and this forces the definition of K_T on objects. Moreover, for any $f: A \to B$ in $\mathscr{A}, K_T F_T(f) = K_T(\eta(B)f) = \varepsilon F(B)F\eta(B)F(f) = F(f)$. Thus $K_T F_T = F$.

Conversely, $K_T F_T = F$ implies that $K_T(\eta(B)f) = F(f)$; in particular, it implies for $g: A \to T(A')$ in $\mathscr{A}$ that $K_T(\eta T(A')g) = F(g)$. We shall see later that this forces the definition of K_T on arrows, once we know what it does to the arrow $1_{T(A')}$.

We calculate

$$K_T \varepsilon_T(A') = K_T(1_{T(A')}) = \varepsilon F(A') = \varepsilon K_T(A')$$

as required, and this forces the definition of $K_T(1_{T(A')})$. Now if $g: A \to T(A')$ in $\mathscr{A}$ is any arrow $A \to A'$ in $\mathscr{A}_T$,

$$g = \mu(A')\eta T(A')g = \mu(A')T(1_{T(A')})\eta T(A')g = 1_{T(A')} * \eta T(A')g,$$

where $*$ denotes composition in $\mathscr{A}_T$, hence

$$\begin{aligned} K_T(g) &= K_T(1_{T(A')})K_T(\eta T(A')g) \\ &= \varepsilon F(A')F(g), \end{aligned}$$

which finally establishes the uniqueness of K_T.

It remains to check that

$$\begin{aligned} UK_T(A) &= UF(A) = T(A) = U_T(A), \\ UK_T(g) &= U\varepsilon F(A')UF(g) = \mu(A')T(g) = U_T(g), \end{aligned}$$

and this completes the proof.

Corollary 6.8. Let $L_T: \mathscr{A}_T \to \mathscr{A}^T$ be the special case of the comparison functor K^T when $\mathscr{B} = \mathscr{A}_T$ (or of K_T when $\mathscr{B} = \mathscr{A}^T$), then we have functors

$$\mathscr{A} \xrightarrow{F_T} \mathscr{A}_T \xrightarrow{L_T} \mathscr{A}^T \xrightarrow{U^T} \mathscr{A}$$

with $F^T = L_T F_T$ left adjoint to U^T and $U_T = U^T L_T$ right adjoint to F_T. Moreover, F_T is bijective on objects, U^T is faithful and L_T is full and faithful.

Proof. In view of Propositions 6.5 and 6.7, it only remains to show that L_T is full and faithful. This follows from the following calculation: for any $g: A \to T(A')$ in $\mathscr{A}$,

$$L_T(g) = K^T(g) = U_T(g) = \mu(A')T(g),$$

hence

$$g = \mu(A')\eta T(A')g = \mu(A')T(g)\eta(A) = L_T(g)\eta(A).$$

Corollary 6.9. The Kleisli category of a triple is equivalent to the full subcategory of the Eilenberg–Moore category consisting of all free algebras.

Proof. The full and faithful functor L_T establishes an equivalence between $\mathscr{A}_T$ and a full subcategory of $\mathscr{A}^T$. Since, for any object A of $\mathscr{A}_T$,

$$L_T(A) = K^T(A) = (U_T(A), U_T\varepsilon_T(A)) = (T(A), \mu(A)) = F^T(A),$$

it follows that the objects of this subcategory are precisely the 'free' algebras of the triple.

Example T1 (continued). The Kleisli category of the triple associated with a monoid $\mathscr{M}$ is equivalent to the category of all free $\mathscr{M}$-sets regarded as a full subcategory of the category of all $\mathscr{M}$-sets.

Exercises.

1. Complete the proofs of Propositions 6.2 and 6.4 and the proofs in Examples T1 and T2.
2. Given a ring R (associative with unity element), construct a triple (T, η, μ) on the category **Ab** of abelian groups with $T(A) = R \otimes A$ for any abelian group A. What is the Eilenberg–Moore category of this triple?
3. Prove the associativity of composition in the Kleisli category of a triple.
4. (Linton). Show that the Eilenberg–Moore category may be constructed from the Kleisli category as a pullback:

$$\begin{array}{ccc} \mathscr{A}^T & \longrightarrow & \mathbf{Sets}^{\mathscr{A}_T^{op}} \\ {\scriptstyle U^T}\Big\downarrow & & \Big\downarrow{\scriptstyle \mathbf{Sets}^{L_T^{op}}} \\ \mathscr{A} & \xrightarrow[\text{Yoneda}]{} & \mathbf{Sets}^{\mathscr{A}^{op}} \end{array}$$

7 Examples of cartesian closed categories

In Part I we shall talk at length about 'cartesian closed categories', which will be defined equationally. In preparation, it may be useful to give a less formal definition and to present some examples.

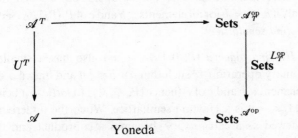

A *cartesian closed* category is a category $\mathscr{C}$ with finite products (hence having a terminal object) such that, for each object B of $\mathscr{C}$, the functor $(-) \times B: \mathscr{C} \to \mathscr{C}$ has a right adjoint, denoted by $(-)^B: \mathscr{C} \to \mathscr{C}$. This means that, for all objects A, B and C of $\mathscr{C}$, there is an isomorphism

$$(*) \qquad \mathrm{Hom}_{\mathscr{C}}(A \times B, C) \simeq \mathrm{Hom}_{\mathscr{C}}(A, C^B)$$

and, moreover, that this isomorphism is natural in A, B and C.

***Example* 7.1.** The category **Sets** is cartesian closed. Here $A \times B$ is the usual cartesian product of sets and C^B is the set of all functions $B \to C$. The bijection $(*)$ sends the function $f: A \times B \to C$ onto the function $f^*: A \to C^B$, where $f^*(a)(b) = f(a, b)$ for all $a \in A$ and $b \in B$. (See Section 3, Exercise 6.)

***Example* 7.2.** More generally, for any small category $\mathscr{X}$, the functor category $\mathbf{Sets}^{\mathscr{X}}$ is cartesian closed. Also cartesian closed is the category of sheaves on a topological space and, in fact, every so-called topos (see Part II, Sections 9 and 10, even without natural numbers object).

***Example* 7.3.** We recall from Section 1 that a poset $(P, \leqslant)$ (that is, preordered set satisfying the antisymmetry law) may be regarded as a category. As such, it has finite products if and only if it has a largest element 1 and a binary operation $\wedge$ such that $c \leqslant a \wedge b$ if and only if $c \leqslant a$ and $c \leqslant b$ for all elements a, b and c of P. In fact, $(P, 1, \wedge)$ is then a monoid satisfying the commutative and idempotent laws:

$$a \wedge b = b \wedge a, \quad a \wedge a = a.$$

Such a monoid is usually called a *semilattice*, and one may recapture the partial order by defining $a \leqslant b$ to mean $a \wedge b = a$. For $(P, 1, \wedge)$ to be cartesian closed there must be another binary operation $\Leftarrow$ such that $a \wedge b \leqslant c$ if and only if $a \leqslant c \Leftarrow b$ for all elements a, b and c of P. $(P, 1, \wedge, \Leftarrow)$ is then called a *Heyting semilattice*.

***Example* 7.4.** A *Heyting algebra* $(P, 0, 1, \wedge, \vee, \Leftarrow)$ also has a smallest element 0 and a binary operation $\vee$ such that $a \vee b \leqslant c$ if and only if $a \leqslant c$ and $b \leqslant c$ for all element a, b and c of P (hence $(P, \wedge, \vee)$ is a *lattice*), it being assumed that $(P, 1, \wedge, \Leftarrow)$ is a Heyting semilattice. When the underlying poset $(P, \leqslant)$ is viewed as a category, $\vee$ becomes a coproduct and the category is called *bicartesian closed*. Incidentally, the distributive law

$$a \wedge (b \vee c) = (a \wedge b) \vee (a \wedge c)$$

then follows from general categorical principles (see Section 5, Exercise 2).

A typical example of a Heyting algebra is the lattice of open subsets of a topological space X, with the following structure:

$$1 \equiv X, 0 \equiv \varnothing, U \wedge V \equiv U \cap V, U \vee V \equiv U \cup V,$$

$$V \Leftarrow U \equiv \mathrm{int}((X - U) \cup V),$$

for all open subsets U and V of X, where 'int' denotes the interior operation.

Another example of a Heyting algebra will be the lattice of subobjects of an object in a topos (see Part II, Section 5, Exercise 3). Many other examples are found in the literature (see the books by Balbes and Dwinger and by Rasiowa and Sikorski).

***Example* 7.5.** **Cat**, the category of small categories, is cartesian closed. For any small categories $\mathscr{A}$ and $\mathscr{B}$, $\mathscr{A} \times \mathscr{B}$ is their product and $\mathscr{B}^{\mathscr{A}}$ is the category of all functors $\mathscr{A} \to \mathscr{B}$. (See: Section 1, Example C7; Section 2, Example C8; Proposition 2.3.)

***Example* 7.6.** Although the category **top** of topological spaces and continuous mappings is not itself cartesian closed, various full subcategories of **top** are. For example, the category of Kelley spaces (that is, compactly generated Hausdorff spaces) is cartesian closed if products are defined in the usual way and Y^X is the set of all continuous functions $X \to Y$ with the compact–open topology. (See the book by MacLane for more details.)

***Example* 7.7.** The category of ω-posets is cartesian closed. An *ω-poset* is a poset in which every countable ascending chain $a_0 \leqslant a_1 \leqslant a_2 \leqslant \ldots$ of elements has a supremum. Morphisms of ω-posets are mappings which preserve supremums of countable ascending chains (such mappings necessarily preserve order). The product structure is inherited from **Sets** and B^A is $\mathrm{Hom}(A, B)$ with order and supremum being defined componentwise. (For details see Part I, Proposition 18.1. For related cartesian closed categories see the book by Gierz *et al.*)

***Example* 7.8.** The category of Kuratowski limit spaces is cartesian closed. A *limit space* is a set X with a partial ω-ary operation (that is, an operation defined on a subset of $X^{\mathbb{N}}$, the set of all countable sequences of elements of X) satisfying the following conditions:

(i) the constant sequence $(x, x, \ldots)$ has limit x;
(ii) if a sequence has limit x, then so does every subsequence;
(iii) if every subsequence of a sequence has a subsequence with limit x, then the sequence itself has limit x.

A morphism $f: X \to Y$ between limit spaces is a function such that, whenever $\{x_n \mid n \in \mathbb{N}\}$ is a sequence of elements of X with limit x, then $\{f(x_n) \mid n \in \mathbb{N}\}$ has limit $f(x)$. The product is defined as for sets, with limits given componentwise, and Y^X is the set of all morphisms $X \to Y$, where the limit of $\{f_n \mid n \in \mathbb{N}\}$ is said to be f provided the limit of $\{f_n(x_n) \mid n \in \mathbb{N}\}$ is $f(x)$

whenever the limit of $\{x_n | n \in \mathbb{N}\}$ is x. (For details see the book by Kuratowski, Chapter 2.)

Exercises

1. Carry out the detailed proof in any of the above examples.
2. Show that Heyting semilattices may be defined equationally.

I

Cartesian closed categories and λ-calculus

Introduction to Part I

λ-calculus or combinatory logic is a topic that logicians have studied since 1924. Cartesian closed categories are more recent in origin, having been invented by Lawvere (1964, see also Eilenberg and Kelly, 1966). Both are attempts to describe axiomatically the process of substitution, so it is not surprising to find that these two subjects are essentially the same. More precisely, there is an equivalence of categories between the category of cartesian closed categories and the category of typed λ-calculi with surjective pairing. This remains true if cartesian closed categories are provided with a weak natural numbers object and if typed λ-calculi are assumed to have a natural numbers type with iterator.

This result depends crucially on the *functional completeness* of cartesian closed categories, which goes back to the functional completeness of combinatory logic due to Schönfinkel and Curry. It asserts, in particular, that every arrow $\varphi(x): 1 \to B$ expressible as a polynomial in an indeterminate arrow $x: 1 \to A$ over a cartesian closed category $\mathscr{A}$ (with given objects A and B) is uniquely of the form $1 \xrightarrow{x} A \xrightarrow{f} B$, where f is an arrow in $\mathscr{A}$ not depending on x.

Functional completeness is closely related to the *deduction theorem* for positive intuitionistic propositional calculi presented as deductive systems. In our version, it associates with each proof of $\mathrm{T} \vdash \mathrm{B}$ on the assumption $\mathrm{T} \vdash \mathrm{A}$ a proof of $\mathrm{A} \vdash \mathrm{B}$ without assumptions. However, functional completeness goes beyond this; it asserts that the proof of $\mathrm{T} \vdash \mathrm{B}$ on the assumption $\mathrm{T} \vdash \mathrm{A}$ is, in some sense, *equivalent* to the proof by transitivity:

$$\frac{\mathrm{T} \vdash \mathrm{A} \quad \mathrm{A} \vdash \mathrm{B}}{\mathrm{T} \vdash \mathrm{B}}.$$

Deductive systems are also used to construct free cartesian closed categories generated by graphs, whose arrows $\mathrm{A} \to \mathrm{B}$ are equivalence classes of proofs.

We present a decision procedure for equality of arrows in the free cartesian closed category (with weak natural numbers object) generated by the empty graph; equivalently, for convertibility of expressions in the pure typed λ-calculus under consideration. This is the coherence problem for cartesian closed categories, the solution of which goes back to early work in the λ-calculus.

Finally, we study *C-monoids*, essentially monoids which may be viewed as one-object cartesian closed categories without terminal object. The category of C-monoids is shown to be equivalent (even isomorphic) to the category of untyped λ-calculi with surjective pairing. Again, this result depends on functional completeness of C-monoids.

It is shown that every C-monoid may be regarded as the monoid of endomorphisms of an object U in a cartesian closed category such that $U \times U \cong U \cong U^U$. An example of such a category with U not isomorphic to 1, due to Dana Scott, is presented.

The reader who wishes to see these results in their historical perspective is advised to look at the following comments.

Historical perspective on Part I

For the purpose of this discussion, it will suffice to define a *cartesian closed category* as a category with an object 1 and operations $(-) \times (-)$ and $(-)^{(-)}$ on objects satisfying conditions which assure that

(i) $\mathrm{Hom}(A, 1) \cong \{*\}$,
(ii) $\mathrm{Hom}(C, A \times B) \cong \mathrm{Hom}(C, A) \times \mathrm{Hom}(C, B)$,
(iii) $\mathrm{Hom}(A, C^B) \cong \mathrm{Hom}(A \times B, C)$.

Here $\{*\}$ is supposed a typical one-element set, chosen once and for all.

It will be instructive to reverse the historical process and see how combinatory logic could have been discovered by rigorous application of Occam's razor.

Condition (i) says that, for each object A, there is only one arrow $A \to 1$, hence we might as well forget about the object 1 and the arrow leading to it. However, the arrows $1 \to A$ must be preserved, let us call them *entities of type A*.

Condition (ii) says that the arrows $C \to A \times B$ are in one-to-one correspondence with pairs of arrows $C \to A$ and $C \to B$, hence we might as well forget about the arrows going into $A \times B$.

Condition (iii) says that the arrows $A \times B \to C$ are in one-to-one correspondence with the arrows $A \to C^B$, hence we might as well forget about the arrows coming out of $A \times B$ too. Consequently, we might as well forget about $A \times B$ altogether.

We end up with a category with a binary operation 'exponentiation' on objects. Of course, this will have to satisfy some conditions, but these may be a little difficult to state. It is interesting to note that Eilenberg and Kelly went on a similar *tour de force* and ended up with a category with exponentiation in which some monstrous diagrams had to commute.

We may go a little further and forget about the category structure as well, since arrows $A \to B$ are in one-to-one correspondence with entities of type B^A, which we shall write $B \Leftarrow A$ for typographical reasons. Composition of arrows is then represented by a single entity of type $((C \Leftarrow A) \Leftarrow (C \Leftarrow B)) \Leftarrow (B \Leftarrow A)$. However, we do need a binary operation on entities called 'application': given entities f of type B^A and a of type A, there is an entity $f\,{}^{\int}a$ (read 'f of a') of type B.

We have now arrived at typed combinatory logic. But even this came rather late in the thinking of logicians, although type theory had already been introduced by Russell and Whitehead. Let us continue on our journey backwards in time and apply Occam's razor still further.

An arrow $A \to B$ in a category has a source A and a target B. But what if there is only one object? Such a category is called a monoid and, indeed, the original presentation of combinatory logic by Curry does describe a monoid with additional structure. (The binary operation of multiplication is defined in terms of the primitive operation of application.) Underlying untyped combinatory logic there is a tacit ontological assumption, namely that all entities are functions and that each function can be applied to any entity.

To present the work of Schönfinkel and Curry in the modern language of universal algebra, one should think of an algebra $A = (|A|, {}^{\int}, I, K, S)$, where $|A|$ is a set, ${}^{\int}$ is a binary operation and I, K and S are elements of $|A|$ or nullary operations. According to Schönfinkel, these had to satisfy the following identities:

$$I\,{}^{\int}a = a,$$

$$(K\,{}^{\int}a)\,{}^{\int}b = a,$$

$$((S\,{}^{\int}f)\,{}^{\int}g)\,{}^{\int}c = (f\,{}^{\int}c)\,{}^{\int}(g\,{}^{\int}c),$$

for all elements a, b, c, f and g of $|A|$. (Actually, he defined I in terms of K and S, but this is beside the point here.) The reader may think of I as the identity function and of K as the function which assigns to every entity a the function with constant value a. It is a bit more difficult to put S into words and we shall refrain from doing so.

Schönfinkel (1924) discovered a remarkable result, usually called 'functional completeness'. In modern terms this may be expressed as follows: every polynomial $\varphi(x)$ in an indeterminate x over a Schönfinkel algebra A can be written in the form $f^{\varsigma}x$, where $f \in |A|$.

From now on in our exposition, the arrow of time will point in its customary direction.

Curry (1930) rediscovered Schönfinkel's results, but went further in his thinking. He discovered that a finite set of additional identities would assure that the element f representing the polynomial $\varphi(x)$ was uniquely determined. We shall not reproduce these identities here, but reserve the name 'Curry algebra' for a Schönfinkel algebra which satisfies them.

Using the terminology of Church (1941), one writes f as $\lambda_x \varphi(x)$, which must then satisfy two equations:

$$(\beta) \qquad (\lambda_x \varphi(x))^{\varsigma} a = \varphi(a),$$

$$(\eta) \qquad \lambda_x(f^{\varsigma}x) = f.$$

(Many mathematicians write $x \mapsto \varphi(x)$ in place of $\lambda_x \varphi(x)$.) A λ-calculus is a formal language built up from variables $x, y, z, \ldots$ by means of term forming operations $(-)^{\varsigma}(-)$ and $\lambda_x(-)$, the latter being assumed to bind all free occurrences of the variable x occurring in $(-)$, such that the two given identities hold. The basic entities I, K and S may then be defined formally by

$$I \equiv \lambda_x x,$$
$$K \equiv \lambda_x \lambda_y x,$$
$$S \equiv \lambda_u \lambda_v \lambda_z((u^{\varsigma} z)^{\varsigma}(v^{\varsigma} z)).$$

(Actually, Church would have called such a language a λK-calculus and Curry might have called it a $\lambda\beta\eta$-calculus, but never mind.)

Both Curry and Church realized the importance of introducing types into combinatory logic or λ-calculus. To do this one just has to observe that, if f has type $B \Leftarrow A$ and a has type A, then $f^{\varsigma}a$ has type B, as already pointed out. In particular, the basic entities I, K and S, suitably equipped with subscripts, should have prescribed types. Thus I_A, $K_{A,B}$ and $S_{A,B,C}$ have types $A \Leftarrow A$, $(A \Leftarrow B) \Leftarrow A$ and $((A \Leftarrow C) \Leftarrow (B \Leftarrow C)) \Leftarrow ((A \Leftarrow B) \Leftarrow C)$ respectively.

As pointed out in the book by Curry and Feys, these three types are precisely the axioms of intuitionistic implicational logic. Moreover, the rule which computes the type of $f^{\varsigma}a$ from those of f and a corresponds to modus ponens: from $B \Leftarrow A$ and A one may infer B. In fact, Schönfinkel's definition of I in terms of K and S is exactly the same as the known proof that $A \Leftarrow A$ may be derived from the other two axioms.

Incidentally, several early texts on propositional logic used only implication and negation as primitive connectives, having eliminated conjunction and other connectives by suitable definitions, again inspired by Occam's razor. The observation that it is more natural to retain conjunction and other connectives as primitive is probably due to Gentzen and was made again by Lawvere in a categorical context.

Curry and Feys also realized that the proof of Schönfinkel's version of functional completeness was really the same as the proof of the usual deduction theorem: if one can prove B on the assumption A then one can prove B⇐A without any assumption. In fact, it asserts that the proof of B on the assumption A is 'equivalent' to the proof by modus ponens:

$$\frac{\mathrm{B} \Leftarrow \mathrm{A} \quad \mathrm{A}}{\mathrm{B}}.$$

From our viewpoint, Curry's version of functional completeness, which insists on the uniqueness of f such that $\varphi(x)$ equals $f{}^{\ulcorner}x$, then presupposes that entities are not proofs but equivalence classes of proofs.

In connection with cartesian closed categories, the analogy with propositional logic requires that $1, A \times B$ and B^A be written as $T, A \wedge B$ and $B \Leftarrow A$ respectively. (For other structured categories, the senior author had pointed out and exploited a similar analogy with certain deductive systems, beginning with the so-called 'syntactic calculus' (see Lambek 1961b, Appendix II), which traces the idea back to joint work with George D. Findlay in 1956.) The relation between λ-calculi with product types and cartesian closed categories then suggests the observation: types = formulas, terms = proofs, or rather equivalence classes of proofs. Independently, W. Howard in 1969 privately circulated an influential manuscript on the equivalence of typed λ-terms (there called 'constructions') and derivations in various calculi, which finally appeared in the 1980 Curry Festschrift (see also Stenlund 1972).

Up to this point we have avoided discussing natural numbers. In an untyped λ-calculus natural numbers are easily defined (Church 1941). Writing

$$f \circ g \equiv \lambda_x(f{}^{\ulcorner}(g{}^{\ulcorner}x)),$$

one regards 2 as the process which assigns to every function f its iterate $f \circ f$, so $2{}^{\ulcorner}f \equiv f \circ f$. Formally, one defines

$$0 \equiv \lambda_x I, \quad 1 \equiv \lambda_x x = I, \quad 2 \equiv \lambda_x(x \circ x), \ldots .$$

The successor function and the usual operations on natural numbers are defined by

$$\begin{aligned} S^{\int} n &\equiv \lambda_y(y \circ (n^{\int} y)), \\ m + n &\equiv \lambda_y((m^{\int} y) \circ (n^{\int} y)), \\ mn &\equiv m \circ n, \\ m^n &\equiv n^{\int} m. \end{aligned}$$

Unfortunately, there are difficulties with this as soon as one introduces types. For, if a has type A, then f and g in $(f \circ g)^{\int} a$ both have types $A^A = B$ say. For $n^{\int} f$ to make sense, n will have to be of type B^B, and for $n^{\int} m$ to make sense, m will have to be of type B. If m and n are to have the same type, we are thus led to require that $B^B = B$, which is certainly not true in general, although Dana Scott (1972) showed that one may have $B^B \cong B$.

One way to get around this difficulty is to postulate a type N of natural numbers, a term 0 of type N and term forming operations $S(-)$ (successor) and $I(-,-,-)$ (iterator) such that $S(n)$ has type n and $I(a, h, n)$ has type A for all n of type N, a of type A and h of type A^A. These must satisfy suitable equations to assure that $I(a, h, n)$ means $h^{n \int} a$.

The analogous concept for cartesian closed categories is a *weak natural numbers object*: an object N with arrows $0: 1 \to N$ and $S: N \to N$ and a process which assigns to all arrows $a: 1 \to A$ and $h: A \to A$ an arrow $g: N \to A$ such that the following diagram commutes:

$$\begin{array}{ccccc} 1 & \xrightarrow{0} & N & \xrightarrow{S} & N \\ \big\downarrow & & \big\downarrow g & & \big\downarrow g \\ 1 & \xrightarrow[a]{} & A & \xrightarrow[h]{} & A \end{array}$$

Lawvere had defined a (strong) natural numbers object to be such that the arrow $g: N \to A$ with the above property is unique.

For us, a typed λ-calculus contains by definition the structure given by N, 0, S and I. In stating Theorem 11.3 on the equivalence between typed λ-calculi and cartesian closed categories, we stipulate that the latter be equipped with a weak natural numbers object. Such categories were first studied formally by Marie-France Thibault (1977, 1982), who called them 'prerecursive categories', although they are implicit in the work of logicians, e.g. in Gödel's functionals of finite type (1958).

We would have preferred to state Theorem 11.3 for strong natural numbers objects in Lawvere's sense. Unfortunately, we do not yet know how to handle the corresponding notion in typed λ-calculus equationally.

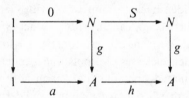

As far as we can see, the iterators appearing in the literature (e.g. Troelstra 1973) mostly correspond to weak natural numbers objects. See however Sanchis (1967).

For further historical comments the reader is referred to the end of Part I.

1 Propositional calculus as a deductive system

We recall (Part 0; Definition 1.2) that, for categorists, a *graph* consists of two classes and two mappings between them:

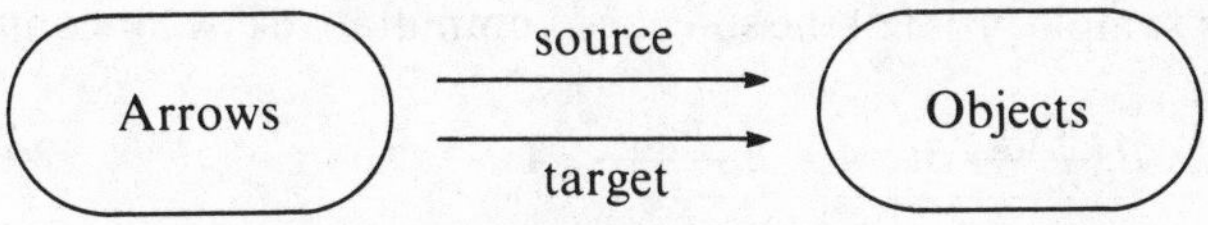

In graph theory the arrows are usually called 'oriented edges' and the objects 'nodes' or 'vertices', but in various branches of mathematics other words may be used. Instead of writing

$$\text{source}(f) = A, \quad \text{target}(f) = B,$$

one often writes $f: A \to B$ or $A \xrightarrow{f} B$. We shall look at graphs with additional structure which are of interest in logic.

A *deductive system* is a graph with a specified arrow

R1a. $A \xrightarrow{1_A} A,$

and a binary operation on arrows (*composition*)

R1b. $$\frac{A \xrightarrow{f} B \quad B \xrightarrow{g} C}{A \xrightarrow{gf} C}.$$

Logicians will think of the objects of a deductive system as *formulas*, of the arrows as *proofs* (or *deductions*) and of an operation on arrows as a *rule of inference*.

Logicians should note that a deductive system is concerned not just with unlabelled entailments or sequents $A \to B$ (as in Gentzen's proof theory), but with deductions or proofs of such entailments. In writing $f: A \to B$ we think of f as the 'reason' why A entails B.

A *conjunction calculus* is a deductive system dealing with truth and conjunction. Thus we assume that there is given a formula T (= true) and a binary operation $\wedge$ (= and) for forming the conjunction $A \wedge B$ of two given formulas A and B. Moreover, we specify the following additional arrows and rules of inference:

R2. $A \xrightarrow{O_A} \mathrm{T}$;

R3a. $A \wedge B \xrightarrow{\pi_{A,B}} A$,

R3b. $A \wedge B \xrightarrow{\pi'_{A,B}} B$,

R3c. $$\frac{C \xrightarrow{f} A \quad C \xrightarrow{g} B}{C \xrightarrow{\langle f, g \rangle} A \wedge B}.$$

Here is a sample proof of the so-called commutative law for conjunction:

$$\frac{A \wedge B \xrightarrow{\pi'_{A,B}} B \quad A \wedge B \xrightarrow{\pi_{A,B}} A}{A \wedge B \xrightarrow{\langle \pi'_{A,B}, \pi_{A,B} \rangle} B \wedge A}.$$

The presentation of this proof in tree-form, while instructive, is superfluous. It suffices to denote it by $\langle \pi'_{A,B}, \pi_{A,B} \rangle$ or even by $\langle \pi', \pi \rangle$ when the subscripts are understood.

Another example is the proof of the associative law $\alpha_{A,B,C}: (A \wedge B) \wedge C \to A \wedge (B \wedge C)$. It is given by

$$\alpha_{A,B,C} \equiv \langle \pi_{A,B} \pi_{A \wedge B,C}, \langle \pi'_{A,B} \pi_{A \wedge B,C}, \pi'_{A \wedge B,C} \rangle \rangle \tag{1.1}$$

or just by $\alpha \equiv \langle \pi\pi, \langle \pi'\pi, \pi' \rangle \rangle$.

If we compose operations on proofs, we obtain 'derived' rules of inference. For example, consider the derived rule:

$$\frac{\dfrac{A \wedge C \xrightarrow{\pi_{A,C}} A \quad A \xrightarrow{f} B}{A \wedge C \to B} \qquad \dfrac{A \wedge C \xrightarrow{\pi'_{A,C}} C \quad C \xrightarrow{g} D}{A \wedge C \to D}}{A \wedge C \xrightarrow{f \wedge g} B \wedge D}.$$

It asserts that from proofs f and g one can construct the proof

$$f \wedge g = \langle f\pi_{A,C}, g\pi'_{A,C} \rangle.$$

Thus we may write simply

$$\frac{A \xrightarrow{f} B \quad C \xrightarrow{g} D}{A \wedge C \xrightarrow{f \wedge g} B \wedge D}.$$

A *positive intuitionistic propositional calculus* is a conjunction calculus with an additional binary operation $\Leftarrow (=$ if). Thus, if A and B are formulas, so are T, $A \wedge B$ and $A \Leftarrow B$. (Yes, most people write $B \Rightarrow A$ instead.) We also specify the following new arrow and rule of inference.

R4a. $(A \Leftarrow B) \wedge B \xrightarrow{\varepsilon_{A,B}} A$,

R4b. $$\frac{C \wedge B \xrightarrow{h} A}{C \xrightarrow{h^*} A \Leftarrow B}.$$

Actually we should have written $h^* = \Lambda^C_{A,B}(h)$, but the subscripts are usually understood from the context.

We note that from R4b, with the help of R4a, one may derive

R′4b. $C \xrightarrow{\eta_{C,B}} (C \wedge B) \Leftarrow B$,

R′4c. $$\frac{D \xrightarrow{g} A}{(D \Leftarrow B) \xrightarrow{g \Leftarrow 1_B} (A \Leftarrow B)}.$$

To derive these, we put

$$\eta_{C,B} \equiv 1^*_{C \wedge B}, \quad g \Leftarrow 1_B \equiv (g\varepsilon_{D,B})^*.$$

Conversely, one may derive R4b from R′4b and R′4c by putting

$$h^* \equiv (h \Leftarrow 1_B)\eta_{C,B}.$$

For future reference, we also note the following two derived rules of inference:

$$\frac{A \xrightarrow{f} B}{\mathrm{T} \xrightarrow{\ulcorner f \urcorner} B \Leftarrow A}, \qquad \frac{\mathrm{T} \xrightarrow{g} B \Leftarrow A}{A \xrightarrow{g^\int} B},$$

where

$$\ulcorner f \urcorner \equiv (f\pi'_{1,A})^*, \quad g^\int \equiv \varepsilon_{B,A}\langle g\bigcirc_A, 1_A\rangle.$$

An intuitionistic (propositional) calculus is more than a positive one; it requires also falsehood and disjunction, that is, a formula $\perp$ ($=$ false) and an operation $\vee$ ($=$ or) on formulas, together with the following additional arrows:

R5. $\perp \xrightarrow{\Box_A} A$;

R6a. $A \xrightarrow{\kappa_{A,B}} A \vee B$,

R6b. $B \xrightarrow{\kappa'_{A,B}} A \vee B$,

R6c. $(C \Leftarrow A) \wedge (C \Leftarrow B) \xrightarrow{\zeta^C_{A,B}} C \Leftarrow (A \vee B)$.

The last mentioned arrow gives rise to and may be derived from the rule

R′6c. $$\frac{A \xrightarrow{f} C \quad B \xrightarrow{g} C}{A \vee B \xrightarrow{[f,g]} C}.$$

Indeed, we may put

$$[f,g] \equiv (\zeta^C_{A,B} \langle \ulcorner f \urcorner, \ulcorner g \urcorner \rangle)^{\int}.$$

If we want *classical* propositional logic, we must also require

R7. $\bot \Leftarrow (\bot \Leftarrow A) \to A$.

Exercises

1. For the appropriate deductive systems, obtain proofs of the following and their converses:

$A \wedge \top \to A,\ A \Leftarrow \top \to A,\ \top \Leftarrow A \to \top$;
$(A \wedge B) \Leftarrow C \to (A \Leftarrow C) \wedge (B \Leftarrow C)$;
$A \Leftarrow (B \wedge C) \to (A \Leftarrow C) \Leftarrow B$;
$A \wedge \bot \to 1,\ A \Leftarrow \bot \to \top,\ A \vee \bot \to A$;
$(A \wedge C) \vee (B \wedge C) \to (A \vee B) \wedge C$.

2. For the appropriate deductive systems, deduce the following derived rules of inference:

$$\frac{A \xrightarrow{f} B \quad C \xrightarrow{g} D}{A \Leftarrow D \xrightarrow{f \Leftarrow g} B \Leftarrow C}; \qquad \frac{A \xrightarrow{f} B \quad C \xrightarrow{g} D}{A \vee C \xrightarrow{f \vee g} B \vee D}.$$

3. Show how $\zeta^C_{A,B}$ may be defined in terms of the rule R′6c.

4. Show that, in the presence of R1 to R6, the classical axiom R7 may be replaced by

$\top \to A \vee (\bot \Leftarrow A)$.

2 The deduction theorem

The usual deduction theorem asserts:

if $A \wedge B \vdash C$ then $A \vdash C \Leftarrow B$.

This result is here incorporated into R4, with the deduction symbol $\vdash$ replaced by actual arrows in the appropriate deductive system $\mathscr{L}$:

$$\frac{h: A \wedge B \to C}{h^*: A \to C \Leftarrow B}.$$

However, at a higher level, the horizontal bar functions as a deduction symbol, and we obtain a new form of the deduction theorem. It deals with proofs from an *assumption* $x: T \to A$. In other words, we form a new deductive system $\mathscr{L}(x)$ by adjoining a new arrow $x: T \to A$ and talk about proofs $\varphi(x): B \to C$ in this new system. More precisely, $\mathscr{L}(x)$ has the same formulas (= objects) as $\mathscr{L}$ and its proofs (= arrows) $\varphi(x)$ are freely generated from those of $\mathscr{L}$ and the new arrow x by the appropriate rules of inference (= operations). Clearly, if $\mathscr{L}$ is a conjunction calculus (positive calculus, intuitionistic calculus, classical calculus), so is the new deductive system $\mathscr{L}(x)$.

Proposition 2.1. (Deduction theorem). In a conjunction, positive, intuitionistic or classical calculus, with every proof $\varphi(x): B \to C$ from the assumption $x: T \to A$ there is associated a proof $f: A \wedge B \to C$ in $\mathscr{L}$ not depending on x.

We write $f = \kappa_{x \in A} \varphi(x)$, where the subscript '$x \in A$' indicates that x is of type A.

Proof. We shall give the proof for a positive calculus. The same proof is valid for a conjunction calculus, if * is ignored. The proof goes through for an intuitionistic or classical calculus, as the additional structure is presented in the form of arrows rather than rules of inference.

We note that every proof $\varphi(x): B \to C$ from the assumption $x: T \to A$ must have one of the five forms:

(i) $k: B \to C$, a proof in $\mathscr{L}$;
(ii) $x: \mathrm{T} \to A$, with $B = \mathrm{T}$ and $C = A$;
(iii) $\langle \psi(x), \chi(x) \rangle$, where $\psi(x): B \to C'$, $\chi(x): B \to C''$, $C = C' \wedge C''$;
(iv) $\chi(x)\psi(x)$, where $\psi(x): B \to D$, $\chi(x): D \to C$;
(v) $\psi(x)^*$, where $\psi(x): B \wedge C' \to C''$, $C = C'' \Leftarrow C'$.

In all cases, $\psi(x)$ and $\chi(x)$ are 'shorter' proofs than $\varphi(x)$, and we define inductively:

(i) $\kappa_{x \in A} k = k\pi'_{A,B}$;
(ii) $\kappa_{x \in A} x = \pi_{A,\mathrm{T}}$,
(iii) $\kappa_{x \in A} \langle \psi(x), \chi(x) \rangle = \langle \kappa_{x \in A} \psi(x), \kappa_{x \in A} \chi(x) \rangle$;
(iv) $\kappa_{x \in A}(\chi(x)\psi(x)) = \kappa_{x \in A} \chi(x) \langle \pi_{A,B}, \kappa_{x \in A} \psi(x) \rangle$;
(v) $\kappa_{x \in A}(\psi(x)^*) = (\kappa_{x \in A} \psi(x) \alpha_{A,B,C'})^*$;

where $\alpha_{A,B,C'}: (A \wedge B) \wedge C' \to A \wedge (B \wedge C')$ is the proof of associativity discussed in Section 1.

The above argument was by induction on the *length* of the proof $\varphi(x)$. Formally, this may be defined as 0 in cases (i) and (ii), as the sum of the lengths of $\chi(x)$ and $\psi(x)$ plus 1 in cases (iii) and (iv) and as the length of $\psi(x)$ plus 1 in case (v).

Remark 2.1. Logicians don't usually talk of an assumption $x: \mathrm{T} \to A$ if there is a known proof $a: \mathrm{T} \to A$ or another assumption $y: \mathrm{T} \to A$; but from our algebraic viewpoint, this does not matter.

The reader is warned that we do not distinguish notationally between composition of proofs gf in $\mathscr{L}$ and in $\mathscr{L}(x)$. In $\mathscr{L}$, $\kappa_{x \in A} gf = gf\pi'_{A,B}$ and in $\mathscr{L}(x)$ it is $g\pi'_{A,B}\langle \pi_{A,B}, f\pi'_{A,B}\rangle$.

Exercise

Prove the following general form of the deduction theorem for the positive intuitionistic propositional calculus: with every proof $\varphi(x): B \to C$ from the assumption $x: D \to A$ there is associated a proof $f: (A \Leftarrow D) \wedge B \to C$. Hint: writing $f = \rho_x\varphi(x)$, put

(i) $\rho_x k = k\pi'_{A \Leftarrow D, B}$, (ii) $\rho_x x = \varepsilon_{A,B}$,

(iii) $\rho_x\langle \psi(x), \chi(x)\rangle = \langle \rho_x\psi(x), \rho_x\chi(x)\rangle$,

(iv) $\rho_x(\chi(x)\psi(x)) = \rho_x\chi(x)\langle \pi_{A \Leftarrow D, B}, \rho_x\psi(x)\rangle$,

(v) $\rho_x(\psi(x)^*) = (\rho_x\psi(x)\alpha_{A \Leftarrow D, B', B''})^*$, where $\psi(x): B' \wedge B'' \to C$.

3 Cartesian closed categories equationally presented

A *category* is a deductive system in which the following equations hold between proofs:

E1. $f1_A = f$, $1_B f = f$, $(hg)f = h(gf)$,
for all $f: A \to B$, $g: B \to C$, $h: C \to D$.

Thus, from any deductive system one may obtain a category by imposing a suitable equivalence relation between proofs.

A *cartesian category* is both a category and a conjunction calculus satisfying the additional equations:

E2. $f = \bigcirc_A$, for all $f: A \to \mathrm{T}$;

E3a. $\pi_{A,B}\langle f, g\rangle = f$,

E3b. $\pi'_{A,B}\langle f, g\rangle = g$,

E3c. $\langle \pi_{A,B}h, \pi'_{A,B}h\rangle = h$,
for all $f: C \to A$, $g: C \to B$, $h: C \to A \wedge B$.

E2 asserts T is a terminal object. One usually writes $T \equiv 1$, and we shall do so from now on. An equivalent formulation of E2 is

E′2. $1_1 = \bigcirc_1, \bigcirc_B f = \bigcirc_A$ for all $f: A \to B$.

E3 asserts that $A \wedge B$ is a product of A and B with projections $\pi_{A,B}$ and $\pi'_{A,B}$. We shall adopt the usual notation $A \wedge B \equiv A \times B$.

As a consequence of E3, let us record the *distributive law*:

$$\langle f, g \rangle h = \langle fh, gh \rangle \tag{3.1}$$

for all $f: C \to A, \quad g: C \to B, \quad h: D \to C$.

Proof. We show this as follows, omitting subscripts:

$$\begin{aligned}\langle f, g \rangle h &= \langle \pi(\langle f, g \rangle h), \pi'(\langle f, g \rangle h) \rangle \\ &= \langle (\pi \langle f, g \rangle) h, (\pi' \langle f, g \rangle) h \rangle \\ &= \langle fh, gh \rangle.\end{aligned}$$

We shall also write

$$f \times g \equiv f \wedge g = \langle f\pi_{A,C}, g\pi'_{A,C} \rangle,$$

whenever $f: A \to B$ and $g: C \to D$, and note that $\times : \mathscr{A} \times \mathscr{A} \to \mathscr{A}$ is a functor (see Part 0, Definition 1.3). Indeed, we have

$$\begin{aligned}1_A \times 1_C &= \langle 1_A \pi_{A,C}, 1_C \pi'_{A,C} \rangle \\ &= \langle \pi_{A,C}, \pi'_{A,C} \rangle \\ &= \langle \pi_{A,C} 1_{A \times C}, \pi'_{A,C} 1_{A \times C} \rangle \\ &= 1_{A \times C}\end{aligned}$$

and, omitting subscripts, by the distributive law,

$$\begin{aligned}(f \times g)(f' \times g') &= \langle f\pi, g\pi' \rangle \langle f'\pi, g'\pi' \rangle \\ &= \langle f\pi \langle f'\pi, g'\pi' \rangle, g\pi' \langle f'\pi, g'\pi' \rangle \rangle \\ &= \langle ff'\pi, gg'\pi' \rangle \\ &= ff' \times gg'.\end{aligned}$$

A *cartesian closed category* is a cartesian category $\mathscr{A}$ with additional structure R4 satisfying the additional equations

E4a. $\varepsilon_{A,B} \langle h^* \pi_{C,B}, \pi'_{C,B} \rangle = h$,

E4b. $(\varepsilon_{A,B} \langle k\pi_{C,B}, \pi'_{C,B} \rangle)^* = k$,

for all $h: C \wedge B \to A$ and $k: C \to A \Leftarrow B$.

Thus, a cartesian closed category is a positive intuitionistic propositional calculus satisfying the equations E1 to E4. This illustrates the general principle that one may obtain interesting categories from deductive systems by imposing an appropriate equivalence relation on proofs.

Inasmuch as we have decided to write $C \wedge B \equiv C \times B$, we shall also write $A \Leftarrow B \equiv A^B$. The equations E4 assure that the mapping

$$\mathrm{Hom}(C \times B, A) \xrightarrow{*} \mathrm{Hom}(C, A^B)$$

is a one-to-one correspondence. In fact, one has the following universal property of the arrow $\varepsilon_{A,B}: A^B \times B \to A$:

> given any arrow $h: C \times B \to A$, there is a unique arrow $h^*: C \to A^B$ such that
>
> $$\varepsilon_{A,B}(h^* \times 1_B) = h.$$

The reader who recalls the notion of adjoint functor will recognize that therefore $U_B = (-)^B$ is right adjoint to the functor $F_B = (-) \times B: \mathscr{A} \to \mathscr{A}$ with coadjunction $\varepsilon_B: F_B U_B \to 1_{\mathscr{A}}$ defined by $\varepsilon_B(A) = \varepsilon_{A,B}$. Thus, an equivalent description of cartesian closed categories makes use of the adjunction $\eta_B: 1_{\mathscr{A}} \to U_B F_B$ in place of *, where $\eta_B(C) = \eta_{C,B}: C \to (C \times B)^B$, and stipulates equations expressing the functoriality of U_B and the naturality of ε_B and η_B as well as the two adjunction equations. Here

$$U_B(f) = f^B \equiv f \Leftarrow 1_B = (f \varepsilon_{A,B})^*,$$

for all $f: A \to A'$. (For η_B see R′4b in Section 1.)

We shall state another useful equation, which may also be regarded as a kind of distributive law.

$$h^* k = (h \langle k\pi_{D,B}, \pi'_{D,B} \rangle)^*, \tag{3.2}$$

where $h: A \times B \to C$ and $k: D \to A$.

Proof. We show this as follows, omitting subscripts:

$$\begin{aligned} h^* k &= (\varepsilon \langle h^* k\pi, \pi' \rangle)^* \\ &= (\varepsilon \langle h^* \pi, \pi' \rangle \langle k\pi, \pi' \rangle)^* \\ &= (h \langle k\pi, \pi' \rangle)^*. \end{aligned}$$

Quite important is the following bijection, which holds in any cartesian closed category.

$$\mathrm{Hom}(A, B) \cong \mathrm{Hom}(1, B^A). \tag{3.3}$$

Proof. As in Section 1, with any $f: A \to B$ we associate $\ulcorner f \urcorner: 1 \to B^A$, called the *name* of f by Lawvere, given by

$$\ulcorner f \urcorner \equiv (f \pi'_{1,A})^*,$$

and with any $g: 1 \to B^A$ we associate $g^{\int}: A \to B$, read 'g *of*', given by

$$g^{\int} \equiv \varepsilon_{B,A} \langle g \bigcirc_A, 1_A \rangle.$$

We then calculate

$$\ulcorner f \urcorner^{\int} = f, \quad \ulcorner g^{\int} \urcorner = g.$$

Exercises

1. Show that in any cartesian category
$A \times 1 \cong A, \quad A \times B \cong B \times A, \quad (A \times B) \times C \cong A \times (B \times C).$

2. Show that in any cartesian closed category
$A^1 \cong A, \quad 1^A \cong 1, \quad (A \times B)^C \cong A^C \times B^C, \quad A^{B \times C} \cong (A^C)^B.$

3. Write down the equivalent definition of a cartesian closed category in terms of U_B, F_B, η_B and ε_B.

4. Prove the last two equations of Section 3.

4 Free cartesian closed categories generated by graphs

Given a graph $\mathscr{X}$, we may construct the positive intuitionistic calculus $\mathscr{D}(\mathscr{X})$ and the cartesian closed category $\mathscr{F}(\mathscr{X})$ *freely generated* by $\mathscr{X}$.

Informally speaking, $\mathscr{D}(\mathscr{X})$ is the smallest positive intuitionistic calculus whose formulas include the vertices of $\mathscr{X}$ and whose proofs include the arrows of $\mathscr{X}$. (Logicians may think of the latter as 'postulates', although there may be more than one way of postulating $X \to Y$, as there may be more than one arrow $X \to Y$ in $\mathscr{X}$.) More precisely, the formulas and proofs of $\mathscr{D}(\mathscr{X})$ are defined inductively as follows: all vertices of $\mathscr{X}$ are formulas, $\mathrm{T}(\equiv 1)$ is a formula, if A and B are formulas so are $A \wedge B(\equiv A \times B)$ and $B \Leftarrow A(\equiv B^A)$; the arrows of $\mathscr{X}$ and the arrows 1_A, $\bigcirc_A$, $\pi_{A,B}$, $\pi'_{A,B}$ and $\varepsilon_{A,B}$ are proofs, for all formulas A and B, and proofs are closed under the rules of inference–composition, $\langle -,- \rangle$ and $(-)^*$.

We construct $\mathscr{F}(\mathscr{X})$ from $\mathscr{D}(\mathscr{X})$ by imposing all equations between proofs which have to hold in any cartesian closed category. Another way of saying this is that we pick the smallest equivalence relation between proofs satisfying the appropriate substitution laws and respecting the equations of a cartesian closed category. The equivalence classes of proofs are then the arrows of $\mathscr{F}(\mathscr{X})$; but, as usual, we will not distinguish notationally between proofs and their equivalence classes.

Let **Grph** be the category of graphs, whose objects are graphs and whose morphisms $F: \mathscr{X} \to \mathscr{Y}$ are pairs of mappings F: Objects($\mathscr{X}$) $\to$ Objects($\mathscr{Y}$) and F: Arrows($\mathscr{X}$) $\to$ Arrows($\mathscr{Y}$) such that $f: X \to X'$ implies $F(f): F(X) \to F(X')$.

Let **Cart** be the category of cartesian closed categories, whose objects are cartesian closed categories and whose arrows are functors $F: \mathscr{A} \to \mathscr{B}$ which preserve the cartesian closed structure on the nose, that is,

$$F(1) = 1, \quad F(A \times B) = F(A) \times F(B), \quad F(A^B) = F(A)^{F(B)};$$
$$F(\bigcirc_A) = \bigcirc_{F(A)}, \quad F(\pi_{A,B}) = \pi_{F(A),F(B)}, \text{ etc.};$$
$$F(\langle f, g \rangle) = \langle F(f), F(g) \rangle \text{ etc.}$$

Let $\mathscr{U}$ be the obvious *forgetful* functor **Cart** → **Grph**. With any graph $\mathscr{X}$ we associate a morphism of graphs $H_{\mathscr{X}}: \mathscr{X} \to \mathscr{U}\mathscr{F}(\mathscr{X})$ as follows: $H_{\mathscr{X}}(X) = X$ and, if $f: X \to Y$ in $\mathscr{X}$, then $H_{\mathscr{X}}(f) = f$ (the equivalence class of f regarded as a proof in $\mathscr{D}(\mathscr{X})$). We then have the following universal property:

Proposition 4.1. Given any cartesian closed category $\mathscr{A}$ and any morphism $F: \mathscr{X} \to \mathscr{U}(\mathscr{A})$ of graphs, there is a unique arrow $F': \mathscr{F}(\mathscr{X}) \to \mathscr{A}$ in **Cart** such that $\mathscr{U}(F')H_{\mathscr{X}} = F$.

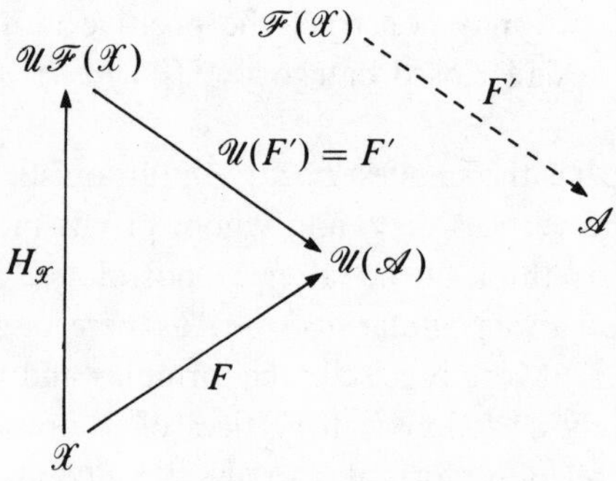

Proof. Indeed, the construction of F' is forced upon us:

$$F'(X) = F(X), \quad F'(\mathrm{T}) = 1, \quad F'(A \wedge B) = F'(A) \times F'(B), \text{ etc.};$$
$$F'(f) = F(f) \quad \text{for all } f: X \to Y, \quad F'(\bigcirc_A) = \bigcirc_{F'(A)}, \text{ etc.};$$
$$F'(\langle f, g \rangle) = \langle F'(f), F'(g) \rangle, \text{ etc.}$$

We must check that F' is well defined, that is, for all $f, g: A \to B$ in $\mathscr{F}(\mathscr{X})$, $f = g$ implies $F'(f) = F'(g)$. This easily follows because no equations hold in $\mathscr{F}(\mathscr{X})$ except those that have to hold.

The above universal property means that $\mathscr{F}$ is a functor **Grph** → **Cart** which is left adjoint to $\mathscr{U}$ with adjunction $H_{\mathscr{X}}: \mathrm{Id} \to \mathscr{U}\mathscr{F}$.

The reader will have noticed that the objects of the category **Grph** and **Cart** introduced here are classes. These may have to be regarded as sets in an appropriate universe.

Exercise

Show that the deductive system $\mathscr{L}(x)$ in Section 2 is $\mathscr{D}(\mathscr{L}_x)$, where $\mathscr{L}_x$ is the graph obtained from $\mathscr{L}$ by adjoining a new edge x between the old vertices T and A.

5 Polynomial categories

Given objects A_0 and A of a (cartesian, cartesian closed) category $\mathscr{A}$, how does one adjoin an indeterminate arrow $x: A_0 \to A$ to $\mathscr{A}$? One method is to adjoin an arrow $x: A_0 \to A$ to the underlying graph of $\mathscr{A}$ and then to form the (cartesian, cartesian closed) category freely generated by the new graph, as was done in Section 4 for cartesian closed categories. Equivalently, one could first form the deductive system (conjunction calculus, positive intuitionistic calculus $\mathscr{A}[x]$ based on the 'assumption' x, as was done in Section 2 in the special case $A_0 = \mathrm{T}$. The formulas of $\mathscr{A}[x]$ are the objects of $\mathscr{A}$ and the proofs of $\mathscr{A}[x]$ are formed from the arrows of $\mathscr{A}$ and the new arrow $x: A_0 \to A$ by the appropriate rules of inference.

To assure that $\mathscr{A}[x]$ becomes a category and that the inclusion of $\mathscr{A}$ into $\mathscr{A}[x]$ becomes a functor, one then imposes the appropriate equations between proofs. If equality of proofs is denoted by $\underset{x}{=}$, we may also regard $\underset{x}{=}$ as the smallest equivalence relation $\equiv$ between proofs such that

$$gf = h \text{ in } \mathscr{A} \text{ implies } gf \equiv h,$$

$$\psi(x) \equiv \psi'(x) \text{ and } \chi(x) \equiv \chi'(x) \text{ implies } \chi(x)\psi(x) \equiv \chi'(x)\psi'(x),$$

$$\varphi(x)1_B \equiv \varphi(x) \equiv 1_C\varphi(x),$$

$$(\chi(x)\psi(x))\varphi(x) \equiv \chi(x)(\psi(x)\varphi(x)),$$

for all $\varphi(x): B \to C$, $\psi(x), \psi'(x): C \to D$, $\chi(x), \chi'(x): D \to E$.

Note that, in view of the reflexive law, $\equiv$ and $\underset{x}{=}$ extend equality in $\mathscr{A}$. Arrows in the category $\mathscr{A}[x]$ are proofs on the assumption x modulo $\underset{x}{=}$, they may be regarded as *polynomials* in x.

The same construction works for cartesian categories or cartesian closed categories, only then $\underset{x}{=}$ must be such that $\mathscr{A}[x]$ becomes a cartesian or cartesian closed category and that the functor $\mathscr{A} \to \mathscr{A}[x]$ preserves the cartesian (closed) structure. That is, the equivalence relations $\equiv$ between proofs considered above must also satisfy:

$$\text{if } \langle f, g\rangle = h \text{ in } \mathscr{A} \text{ then } \langle f, g\rangle \equiv h, \text{ etc.;}$$

$$\text{if } \psi(x) \equiv \psi'(x) \text{ and } \chi(x) \equiv \chi'(x) \text{ then } \langle \psi(x), \chi(x)\rangle \equiv \langle \psi'(x), \chi'(x)\rangle,$$

$$\pi_{B,C}\langle \psi(x), \chi(x)\rangle \equiv \psi(x), \text{ etc.,}$$

for all $\psi(x)$, $\psi'(x): D \to B$ and $\chi(x)$, $\chi'(x): D \to C$.

By a *cartesian* (*closed*) *functor* we mean a functor which preserves the cartesian (closed) structure on the nose. Let $H_x: \mathscr{A} \to \mathscr{A}[x]$ be the (cartesian, cartesian closed) functor which sends $f: B \to C$ onto the 'constant' polynomial with the same name. This possesses the following universal property.

Proposition 5.1. Given a (cartesian, cartesian closed) category $\mathscr{A}$, an indeterminate $x: A_0 \to A$ over $\mathscr{A}$, any (cartesian, cartesian closed) functor $F: \mathscr{A} \to \mathscr{B}$ and any arrow $b: F(A_0) \to F(A)$ in $\mathscr{B}$, there is a unique (cartesian, cartesian closed) functor $F': \mathscr{A}[x] \to \mathscr{B}$ such that $F'(x) = b$ and $F'H_x = F$.

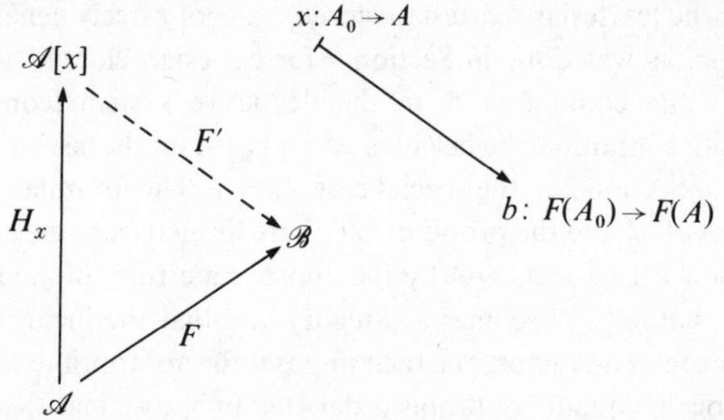

Proof. Every proof $\varphi(x)$ on the assumption x has one of the following forms:

$$k, \quad x, \quad \chi(x)\psi(x), \quad \langle \psi(x), \chi(x) \rangle, \quad \psi(x)^*.$$

where k is an arrow in $\mathscr{A}$, that is, a constant polynomial. The crucial step in the argument is to define $F'(\varphi(x))$. We define inductively:

$$F'(k) = F(k),$$
$$F'(x) = b,$$
$$F'(\chi(x)\psi(x)) = F'(\chi(x))F'(\psi(x)),$$
$$F'(\langle \psi(x), \chi(x) \rangle) = \langle F'(\psi(x)), F'(\chi(x)) \rangle,$$
$$F'(\psi(x)^*) = (F'(\psi(x)))^*.$$

It remains to show that F' is defined on polynomials, not just on proofs, that is, that $\varphi(x) \underset{x}{=} \varphi'(x)$ implies $F'(\varphi(x)) = F'(\varphi'(x))$. Write $\varphi(x) \equiv \varphi'(x)$ for the latter, then it suffices to check that $\equiv$ has the substitution property and respects all the equations of a cartesian closed category.

For example, to check that $\langle \pi_{C,D}\chi(x), \pi'_{C,D}\chi(x) \rangle \equiv \chi(x)$, we calculate

$$F'(\langle \pi_{C,D}\chi(x), \pi'_{C,D}\chi(x) \rangle) = \langle \pi_{F(C),F(D)}F'(\chi(x)), \pi'_{F(C),F(D)}F'(\chi(x)) \rangle$$
$$= F'(\chi(x)),$$

Corollary 5.2. Given a (cartesian, cartesian closed) category $\mathscr{A}$, an indeterminate $x: A_0 \to A$ over $\mathscr{A}$ and an arrow $a: A_0 \to A$ in $\mathscr{A}$, there is a unique (cartesian, cartesian closed) functor $S_x^a: \mathscr{A}[x] \to \mathscr{A}$ such that $S_x^a(x) = a$ and $S_x^a H_x = 1_{\mathscr{A}}$.

Proof. Take $F = 1_{\mathscr{A}}$ in Proposition 5.1.

S_x^a may be regarded as the process of *substituting a* for x. One usually writes

$$S_x^a(\varphi(x)) = \varphi(a).$$

The corollary shows that, in presence of an arrow $a: A_0 \to A$, the functor $H_x: \mathscr{A} \to \mathscr{A}[x]$ is faithful.

Exercise

Adjoining an indeterminate arrow $x: 1 \to \varnothing$ to the cartesian closed category $\mathscr{S}$ of sets one obtains the degenerate category $\mathscr{S}[x]$ in which $1 \cong \varnothing$.

6 Functional completeness of cartesian closed categories

If A is a commutative ring with unity, then any polynomial in an indeterminate x over A has the unique normal form $a_0 + a_1 x + \cdots + a_n x^n$. For cartesian or cartesian closed categories one has an even simpler normal form (see Corollary 6.2 below). The following result, called *functional completeness*, refines the deduction theorem of Section 2.

Proposition 6.1. (Functional completeness). For every polynomial $\varphi(x)$: $B \to C$ in an indeterminate $x: 1 \to A$ over a cartesian or cartesian closed category $\mathscr{A}$ there is a unique arrow $f: A \times B \to C$ in $\mathscr{A}$ such that $f\langle x\bigcirc_B, 1_B\rangle \underset{x}{=} \varphi(x)$.

Proof. Let $\kappa_{x\in A}\varphi(x)$ be defined as in the proof of the deduction theorem (Proposition 2.1). We check by induction on the length of the proof $\varphi(x)$ that

$$\kappa_{x\in A}\varphi(x)\langle x\bigcirc_B, 1_B\rangle \underset{x}{=} \varphi(x).$$

Indeed,

$$k\pi'_{A,B}\langle x\bigcirc_B, 1_B\rangle \underset{x}{=} k1_B \underset{x}{=} k,$$
$$\pi_{A,1}\langle x\bigcirc_1, 1_1\rangle \underset{x}{=} x\bigcirc_1 \underset{x}{=} x1_1 \underset{x}{=} x,$$
$$\langle \kappa_{x\in A}\psi(x), \kappa_{x\in A}\chi(x)\rangle\langle x\bigcirc_B, 1_B\rangle$$
$$\underset{x}{=} \langle \kappa_{x\in A}\psi(x)\langle x\bigcirc_B, 1_B\rangle, \kappa_{x\in A}\chi(x)\langle x\bigcirc_B, 1_B\rangle\rangle \underset{x}{=} \langle \psi(x), \chi(x)\rangle,$$

$$\kappa_{x\in A}\chi(x)\langle \pi_{A,B}, \kappa_{x\in A}\psi(x)\rangle\langle x\bigcirc_B, 1_B\rangle \underset{x}{=} \kappa_{x\in A}\chi(x)\langle x\bigcirc_B, \psi(x)\rangle$$
$$\underset{x}{=} \kappa_{x\in A}\chi(x)\langle x\bigcirc_D, 1_D\rangle\psi(x) \underset{x}{=} \chi(x)\psi(x),$$
$$(\kappa_{x\in A}\psi(x)\alpha_{A,B,C'})^*\langle x\bigcirc_B, 1_B\rangle$$
$$\underset{x}{=} (\kappa_{x\in A}\psi(x)\alpha\langle\langle x\bigcirc, 1\rangle\pi, \pi'\rangle)^*$$
$$\underset{x}{=} (\kappa_{x\in A}\psi(x)\alpha\langle\langle x\bigcirc, \pi\rangle, \pi'\rangle)^*$$
$$\underset{x}{=} (\kappa_{x\in A}\psi(x)\langle x\bigcirc, \langle \pi, \pi'\rangle\rangle)^*$$
$$\underset{x}{=} (\kappa_{x\in A}\psi(x)\langle x\bigcirc, 1\rangle)^* \underset{x}{=} (\psi(x))^*,$$

using (3.2) for h^*k and $\alpha\langle\langle f, g\rangle, h\rangle = \langle f, \langle g, h\rangle\rangle$.

We next verify that $\kappa_{x\in A}\varphi(x)$ depends only on the polynomial $\varphi(x)$, that is, on the equivalence class of the proof $\varphi(x)$ modulo $\underset{x}{=}$. Let us write $\varphi(x) \equiv \psi(x)$ for $\kappa_{x\in A}\varphi(x) = \kappa_{x\in A}\psi(x)$. Then it is easily checked that $\equiv$ has the substitution property and satisfies all the conditions which equality in $\mathscr{A}[x]$ should satisfy. (See the sample calculation below.) Since $\underset{x}{=}$ was defined as the smallest such equivalence relation, it follows that $\underset{x}{=}$ is contained in $\equiv$, that is, that

(*) $\qquad \varphi(x) \underset{x}{=} \psi(x)$ implies $\kappa_{x\in A}\varphi(x) = \kappa_{x\in A}\psi(x)$,

as claimed.

For example, let us check that, if $\langle f, g\rangle = h$ in $\mathscr{A}$, then $\langle f, g\rangle \equiv h$. Indeed,

$$\kappa_{x\in A}\langle f, g\rangle = \langle \kappa_{x\in A}f, \kappa_{x\in A}g\rangle = \langle f\pi', g\pi'\rangle$$
$$= \langle f, g\rangle\pi' = h\pi' = \kappa_{x\in A}h.$$

As another example, let us check that

$$\varepsilon\langle \chi(x)^*\pi, \pi'\rangle \equiv \chi(x),$$

to take the worst case. Writing $\kappa_{x\in A}\chi(x) = h$, we have

$$\kappa_{x\in A}(\text{LHS}) = \varepsilon\kappa_{x\in A}\langle \chi(x)^*\pi, \pi'\rangle$$
$$= \varepsilon\langle \kappa_{x\in A}(\chi(x)^*\pi), \kappa_{x\in A}\pi'\rangle$$
$$= \varepsilon\langle \kappa_{x\in A}\chi(x)^*\langle \pi, \kappa_{x\in A}\pi\rangle, \kappa_{x\in A}\pi'\rangle$$
$$= \varepsilon\langle (h\alpha)^*\langle \pi, \pi\pi'\rangle, \pi'\pi'\rangle$$
$$= \varepsilon\langle (h\alpha)^*\pi, \pi'\rangle\langle\langle \pi, \pi\pi'\rangle, \pi'\pi'\rangle$$
$$= h\alpha\langle\langle \pi, \pi\pi'\rangle, \pi'\pi'\rangle$$
$$= h\langle \pi, \langle \pi\pi', \pi'\pi'\rangle\rangle$$
$$= h\langle \pi, \pi'\rangle$$
$$= h1 = h.$$

Finally, to prove the uniqueness of $f: A \times B \to C$, assume that

$f\langle x\bigcirc_B, 1_B\rangle \underset{x}{=} \varphi(x)$. We claim that then $f = \kappa_{x\in A}\varphi(x)$. Indeed,

$$\begin{aligned}\kappa_{x\in A}\varphi(x) &= \kappa_{x\in A}(f\langle x\bigcirc_B, 1_B\rangle)\\ &= f\kappa_{x\in A}\langle x\bigcirc_B, 1_B\rangle\\ &= f\langle \kappa_{x\in A}(x\bigcirc_B), \kappa_{x\in A}1_B\rangle\\ &= f\langle \kappa_{x\in A}x\langle \pi_{A,B}, \kappa_{x\in A}\bigcirc_B\rangle, \kappa_{x\in A}1_B\rangle\\ &= f\langle \pi_{A,1}\langle \pi_{A,B}, \bigcirc_B\pi'_{A,B}\rangle, \pi'_{A,B}\rangle\\ &= f\langle \pi_{A,B}, \pi'_{A,B}\rangle\\ &= f1_{A\times B} = f.\end{aligned}$$

Corollary 6.2. For every polynomial $\varphi(x)\colon 1\to C$ in an indeterminate x: $1\to A$ over a cartesian or cartesian closed category $\mathscr{A}$, there is a unique arrow $g\colon A\to C$ in $\mathscr{A}$ such that $gx \underset{x}{=} \varphi(x)$. Over a cartesian closed category $\mathscr{A}$, there is a unique arrow $h\colon 1\to C^A$ such that $\varepsilon_{C,A}\langle h, x\rangle \underset{x}{=} \varphi(x)$.

Proof. To derive this from Proposition 6.1, merely put

$$g = \kappa_{x\in A}\varphi(x)\langle 1_A, \bigcirc_A\rangle, \quad h = \ulcorner g\urcorner$$

and check that $\kappa_{x\in A}(gx)\langle 1_A, \bigcirc_A\rangle = g$ and $h^{\S} = g$ (see the proof of (3.3)). Later we shall write $\lambda_{x\in A}\varphi(x)$ for the h such that $h^{\S}x \underset{x}{=} \varepsilon_{C,A}\langle h, x\rangle \underset{x}{=} \varphi(x)$.

Actually, over a cartesian closed category, the corollary is no weaker than the theorem, since the polynomials $B\to C$ are in one-to-one correspondence with the polynomials $1\to C^B$. Usually, it is this corollary, which is referred to as 'functional completeness'.

Exercises

1. Prove the following general form of functional completeness for cartesian closed categories: for every polynomial $\varphi(x)\colon B\to C$ in an indeterminate arrow $x\colon D\to A$ over a cartesian closed category $\mathscr{A}$, there is a unique arrow $f\colon A^D\times B\to C$ such that

   $$f\langle (x\pi'_{B,D})^*, 1_B\rangle \underset{x}{=} \varphi(x).$$

   Hint: see the exercise in Section 2, which establishes a general form of the deduction theorem.

2. If $\mathscr{A}$ is cartesian closed and $\mathscr{A}[x]$ is the cartesian category formed from $\mathscr{A}$ by adjoining an indeterminate $x\colon 1\to A$, show that $\mathscr{A}[x]$ is also cartesian closed.

3. Instead of adjoining a single indeterminate $x\colon 1\to A$, one can also adjoin a set of indeterminates $X = \{x_1, \ldots, x_n\}$, where $x_i\colon 1\to A_i$. Show how to do this when $n = 2$ and also that

   $$\mathscr{A}[x_1, x_2] \cong \mathscr{A}[x_1][x_2] \cong \mathscr{A}[z],$$

where $z: 1 \to A_1 \times A_2$. Prove directly that $\varphi(x_1, x_2)$ can be uniquely written in the form $g\langle x_1, x_2\rangle$ or $h^{\varsigma}\langle x_1, x_2\rangle$.

4. Fill in the details in the proof of Proposition 6.1.

7 Polynomials and Kleisli categories

In this section we take another look at the polynomial cartesian or cartesian closed category $\mathscr{A}[x]$, where x is an indeterminate arrow $1 \to A$, to show that its construction could have been carried out with tools familiar to categorists.

A *cotriple* (S, ε, δ) on a category $\mathscr{A}$ consists of a functor $S: \mathscr{A} \to \mathscr{A}$ and two natural transformations $\varepsilon: S \to 1_{\mathscr{A}}$ and $\delta: S \to S^2$ such that, for any object B of $\mathscr{A}$,

$$\varepsilon S(B)\delta(B) = 1_{S(B)} = S\varepsilon(B)\delta(B), \quad \delta S(B)\delta(B) = S\delta(B)\delta(B).$$

Of course, a cotriple on $\mathscr{A}$ is just a triple on $\mathscr{A}^{\mathrm{op}}$ (see Part 0, Section 6). Accordingly, the *Kleisli category* $\mathscr{A}_S$ of the cotriple has the same objects as $\mathscr{A}$, but arrows $f: B \to C$ in $\mathscr{A}_S$ are arrows $f: S(B) \to C$ in $\mathscr{A}$. In particular, the identity arrow 1_B in $\mathscr{A}_S$ is $\varepsilon(B)$ in $\mathscr{A}$. Moreover, if $g: C \to D$ in $\mathscr{A}_S$, the composition $g * f$ in $\mathscr{A}_S$ is defined by $g * f = gS(f)\delta(B)$. One easily verifies that $\mathscr{A}_S$ is a category.

With any object A of a cartesian or cartesian closed category $\mathscr{A}$ one may associate a cotriple $(S_A, \varepsilon_A, \delta_A)$ as follows:

$$S_A = A \times (-), \quad \varepsilon_A(B) = \pi'_{A,B}, \quad \delta_A(B) = \langle \pi_{A,B}, 1_{A \times B}\rangle.$$

Thus $S_A(B) = A \times B$ and, for $f: B \to C, S_A(f) = \langle \pi_{A,B}, f\pi'_{A,B}\rangle$. The routine calculation that this is indeed a cotriple is left to the reader.

The functional completeness of cartesian or cartesian closed categories may now be interpreted as follows:

Proposition 7.1. The category $\mathscr{A}[x]$ of all polynomials in the indeterminate $x: 1 \to A$ over the cartesian or cartesian closed category $\mathscr{A}$ is isomorphic to the Kleisli category $\mathscr{A}_A = \mathscr{A}_{S_A}$ of the cotriple $(S_A, \varepsilon_A, \delta_A)$.

First proof. Consider the functor $P: \mathscr{A}_A \to \mathscr{A}[x]$ defined for objects B of $\mathscr{A}_A$, that is of $\mathscr{A}$, and arrows $f: C \to B$ in $\mathscr{A}_A$, that is, $f: A \times C \to B$ in $\mathscr{A}$, by

$$P(B) = B, \quad P(f) = f\langle x\bigcirc_C, 1_C\rangle.$$

To check that this is a functor, we apply it to 1_B in $\mathscr{A}_A$, that is, $\varepsilon_A(B) = \pi'_{A,B}$

in $\mathscr{A}$, as well as to the composition

$$g*f = gS(f)\delta_A(B)$$
$$= g\langle \pi_{A,A\times B}, f\pi'_{A,A\times B}\rangle\langle \pi_{A,B}, 1_{A\times B}\rangle = g\langle \pi_{A,B}, f\rangle.$$

Indeed

$$P(\varepsilon_A(B)) = \pi'_{A,B}\langle x\bigcirc_B, 1_B\rangle = 1_B,$$
$$P(g*f) = g\langle \pi, f\rangle\langle x\bigcirc, 1\rangle = g\langle x\bigcirc, f\langle x\bigcirc, 1\rangle\rangle$$
$$= g\langle x\bigcirc, 1\rangle f\langle x\bigcirc, 1\rangle = P(g)P(f).$$

Finally, Proposition 6.1, tells us that P has an inverse K, where $K(B) = B$ and $K(f(x)) = \kappa_{x\in A} f(x)$.

It may be of interest to point out that the curious definition of $\kappa_{x\in A}(\chi(x)\psi(x))$ given in Proposition 2.1 is related to the curious definition of composition in a Kleisli category.

While logicians may favour the proof just given, confirmed categorists will undoubtedly prefer another proof which establishes directly that the Kleisli category $\mathscr{A}_A$ has the universal property of Proposition 5.1, and which therefore allows them to bypass the constructions in Sections 5 and 6 altogether.

Second proof. We shall confine attention to the case when $\mathscr{A}$ is cartesian, leaving the cartesian closed case as an exercise to the reader.

We first show that $\mathscr{A}_A$ is a cartesian category by defining $\bigcirc_C^A : C \to 1$, $\pi_{B,C}^A : B \times C \to B$ and $\pi'^A_{B,C} : B \times C \to C$ as follows:

$$\bigcirc_C^A = \bigcirc_{A\times C}, \quad \pi_{B,C}^A = \pi_{B,C}\pi'_{A,B\times C}, \quad \pi'^A_{B,C} = \pi'_{B,C}\pi'_{A,B\times C}$$

and by taking $\langle -,-\rangle$ as in $\mathscr{A}$. The equations of a cartesian category are easily checked, for example:

$$\pi^A * \langle f, g\rangle = \pi\pi'\langle \pi, \langle f, g\rangle\rangle = \pi\langle f, g\rangle = f,$$
$$\langle \pi^A * h, \pi'^A * h\rangle = \langle \pi\pi'\langle \pi, h\rangle, \pi'\pi'\langle \pi, h\rangle\rangle = \langle \pi h, \pi' h\rangle = h.$$

Now define the functor $H_A : \mathscr{A} \to \mathscr{A}_A$ for objects B and arrows $f : C \to B$ of $\mathscr{A}$ by

$$H_A(B) = B, \quad H_A(f) = f\pi'_{A,C}.$$

It is easily checked that H_A preserves the cartesian structure exactly. We claim that H_A has the same universal property as H_x in Proposition 5.1, with $\pi_{A,1}$ serving as the indeterminate.

Let $F : \mathscr{A} \to \mathscr{B}$ be a given cartesian functor and $b : 1 \to F(A)$ a given arrow in $\mathscr{B}$. We want to show the existence of a unique cartesian functor F': $\mathscr{A}_A \to \mathscr{B}$ such that $F'H_A = F$ and $F'(\pi_{A,1}) = b$. Define F' on objects B and arrows $f: B \to C$ in $\mathscr{A}_A$ by

$$F'(B) = B, \quad F'(f) = F(f)\langle b\bigcirc_{F(B)}, 1_{F(B)}\rangle.$$

We check that F' is a cartesian functor:

$$F'(\pi') = F(\pi')\langle b\bigcirc, 1\rangle = \pi'\langle b\bigcirc, 1\rangle = 1,$$

$$\begin{aligned} F'(g * f) &= F(g\langle \pi, f\rangle)\langle b\bigcirc, 1\rangle = F(g)\langle \pi, F(f)\rangle\langle b\bigcirc, 1\rangle \\ &= F(g)\langle b\bigcirc, F(f)\langle b\bigcirc, 1\rangle\rangle = F(g)\langle b\bigcirc, 1\rangle F(f)\langle b\bigcirc, 1\rangle \\ &= F'(g)F'(f), \end{aligned}$$

$$F'(\pi^A) = F'(\pi\pi') = F(\pi\pi')\langle b\bigcirc, 1\rangle = \pi\pi'\langle b\bigcirc, 1\rangle = \pi,$$

$$F'(\pi'^A) = \pi' \text{ similarly},$$

$$\begin{aligned} F'(\langle f, g\rangle) &= F(\langle f, g\rangle)\langle b\bigcirc, 1\rangle = \langle F(f), F(g)\rangle\langle b\bigcirc, 1\rangle \\ &= \langle F(f)\langle b\bigcirc, 1\rangle, F(g)\langle b\bigcirc, 1\rangle\rangle = \langle F'(f), F'(g)\rangle. \end{aligned}$$

Moreover, F' has the desired properties:

$$F'(H_A(B)) = F'(B) = F(B),$$

$$\begin{aligned} F'(H_A(f)) &= F'(f\pi') = F(f\pi)\langle b\bigcirc, 1\rangle = F(f)\pi\langle b\bigcirc, 1\rangle \\ &= F(f)1 = F(f), \end{aligned}$$

$$\begin{aligned} F'(\pi_{A,1}) &= F(\pi_{A,1})\langle b\bigcirc_{F(1)}, 1_{F(1)}\rangle = \pi_{A,1}\langle b\bigcirc_1, 1_1\rangle \\ &= b\bigcirc_1 = b1_1 = b. \end{aligned}$$

Finally, to show uniqueness of F', assume that F' has these properties, then

$$\begin{aligned} F'(f) &= F'(f\pi'\langle \pi, 1\rangle) = F'((f\pi') * 1) \\ &= F'(f\pi')F'(1) = F'H_A(f)F'(\langle \pi, \pi'\rangle) \\ &= F(f)F'(\langle \pi\langle \pi, \bigcirc\rangle, \pi'\rangle) = F(f)F'(\langle \pi * \bigcirc, \pi'\rangle) \\ &= F(f)\langle F'(\pi)F'(\bigcirc), F'(\pi')\rangle = F(f)\langle b\bigcirc, \pi'\rangle. \end{aligned}$$

This completes the proof.

Exercises

1. Given an indeterminate arrow $x: 1 \to A$ over a cartesian closed category $\mathscr{A}$, show that $\mathscr{A}[x]$ is isomorphic to the Kleisli category of the triple (T_A, η_A, μ_A), where $T_A = (-)^A, \eta_A(B) = \pi^*_{B,A}, \mu_A(B) = (\varepsilon\langle \varepsilon, \pi'\rangle)^*$.

2. Show that the Eilenberg–Moore category of the cotriple $(S_A, \varepsilon_A, \delta_A)$ is isomorphic to the *slice* category $\mathscr{A}/A$ whose objects are narrow $B \to A$ in $\mathscr{A}$ and whose arrows are appropriate commutative triangles.

3. Complete the second proof of Proposition 7.1 in case $\mathscr{A}$ is cartesian closed. (This will give another proof of Exercise 2 of Section 6).

8 Cartesian closed categories with coproducts

Cartesian closed categories were defined as positive intuitionistic propositional calculi satisfying certain equations between proofs. To complete the picture we define a *bicartesian closed category* as a full intuitionistic propositional calculus satisfying the following additional equations:

E5. $f = \Box_A$ for all $f: \bot \to A$;
E6a. $[f, g]\kappa_{A,B} = f$;
E6b. $[f, g]\kappa'_{A,B} = g$;
E6c. $[h\kappa_{A,B}, h\kappa'_{A,B}] = h$;

for all $f: A \to C, g: B \to C$ and $h: A \vee B \to C$.

We recall that the operation $[-,-]$ was defined in terms of the arrow $\zeta^C_{A,B}: C^A \times C^B \to C^{A \vee B}$. Thus

$$[f, g] \equiv (\zeta^C_{A,B}\langle \ulcorner f \urcorner, \ulcorner g \urcorner \rangle)^{\int}.$$

It is customary to write 0 for $\bot$ and $A + B$ for $A \vee B$. The equations assert that 0 is an initial object and $A + B$ is a coproduct of A and B with injections $\kappa_{A,B}$ and $\kappa'_{A,B}$. Thus Hom$(0, A)$ has exactly one element and

$$\mathrm{Hom}(A, C) \times \mathrm{Hom}(B, C) \cong \mathrm{Hom}(A + B, C).$$

Functional completeness holds for bicartesian closed categories. More precisely, we have:

Proposition 8.1. If $\mathscr{A}$ is a bicartesian closed category and $\mathscr{A}[x]$ is the cartesian closed category of polynomials in the indeterminate arrow $x: 1 \to A$ over $\mathscr{A}$, then $\mathscr{A}[x]$ is also a bicartesian closed category.

Proof. We refer to the one-to-one correspondence between arrows $B \to C$ in $\mathscr{A}[x]$ and arrows $A \times B \to C$ in $\mathscr{A}$. Thus 0 is an initial object in $\mathscr{A}[x]$ because

$$\mathrm{Hom}(A \times 0, B) \cong \mathrm{Hom}(0 \times A, B) \cong \mathrm{Hom}(0, B^A),$$

which has exactly one element because 0 is an initial object in $\mathscr{A}$. Again, $B + C$ is the coproduct of B and C in $\mathscr{A}[x]$ because

$$\begin{aligned}\mathrm{Hom}(A \times B, D) \times \mathrm{Hom}(A \times C, D) &\cong \mathrm{Hom}((A \times B) + (A \times C), D)\\ &\cong \mathrm{Hom}(A \times (B + C), D).\end{aligned}$$

This uses the fact that $(A \times B) + (A \times C)$ is a coproduct in $\mathscr{A}$ and also the distributive law

$$(A \times B) + (A \times C) \cong A \times (B + C),$$

see the exercises below. A slightly longer argument avoids the distributive law:

$$\text{LHS} \cong \text{Hom}(B, D^A) \times \text{Hom}(C, D^A) \cong \text{Hom}(B + C, D^A) \cong \text{RHS}.$$

It is an interesting consequence of Proposition 8.1 that the identities E5 and E6 can be stated as equations between constants, that is, without quantifying over arrows f, g and h. For example, in E6a, we may replace f by $(\pi w)^\int$ and g by $(\pi' w)^\int$, where $w: 1 \to C^A \times C^B$. Now, regarding w as an indeterminate arrow, we have

$$[(\pi w)^\int, (\pi' w)^\int]\kappa_{A,B} \underset{w}{=} (\pi w)^\int.$$

By functional completeness, w may be eliminated from this equation.

Writing $2 \equiv 1 + 1$, we may think of arrows $p: 1 \to 2$ as propositions or truth-values. In particular, we put

$$\kappa_{1,1} \equiv \top, \quad \kappa'_{1,1} \equiv \bot.$$

We shall also introduce the classical propositional connectives

$$\neg: 2 \to 2; \quad \wedge, \vee, \Rightarrow, \Leftrightarrow: 2 \times 2 \to 2.$$

(The arrows $\bot$ and $p \wedge q$ should not be confused with the objects $\bot$ and $A \wedge B = A \times B$.) For example, we shall exploit functional completeness to obtain the definition of $p \wedge q$.

We want

$$p \wedge \top = p, \quad p \wedge \bot = \bot.$$

By functional completeness, we have

$$p \wedge x \underset{x}{=} fx$$

for a certain arrow $f: 2 \to 2$, where $x: 1 \to 2$ is an indeterminate proposition. Then

$$f = [f\kappa, f\kappa'] = [p \wedge \top, p \wedge \bot] = [p, \bot],$$

hence

$$p \wedge q = fq = [p, \bot]q.$$

In this fashion, we arrive at the following:

Definition 8.2. If p and q are arrows $1 \to 2$ in a bicartesian closed category and $\top = \kappa_{1,1}$, $\bot = \kappa'_{1,1}$,

$$\neg = [\bot, \top],$$
$$p \wedge q = [p, \bot]q,$$

$$p \vee q = [\top, p]q,$$
$$p \Rightarrow q = [\top, \neg p]q,$$
$$p \Leftrightarrow q = [p, \neg p]q.$$

Somewhat surprising at first sight is the following observation about bicartesian closed categories.

Proposition 8.3. In a bicartesian closed category there is at most one arrow $A \to 0$. That is, in an intuitionistic propositional calculus there is at most one proof $A \to \bot$, up to equivalence of proofs.

Proof. In a bicartesian closed category $\mathrm{Hom}(A \times 0, C) \cong \mathrm{Hom}(0 \times A, C) \cong \mathrm{Hom}(0, C^A)$ is a one-element set. In particular, the composite arrow

$$A \times 0 \xrightarrow{\pi'_{A,0}} 0 \xrightarrow{\Box_{A \times 0}} A \times 0$$

must be the identity $1_{A \times 0}$. Now suppose there is an arrow $f: A \to 0$. Then the top arrow of

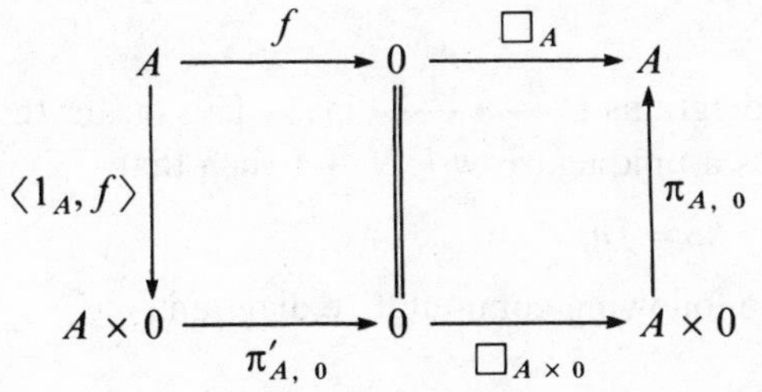

is the same as

$$A \xrightarrow{\langle 1_A, f \rangle} A \times 0 \xrightarrow{\pi_{A,0}} A,$$

namely 1_A. Since also

$$0 \xrightarrow{\Box_A} A \xrightarrow{f} 0$$

is 1_0, it follows that $A \cong 0$.

We have shown that either $\mathrm{Hom}(A, 0) = \varnothing$ or else $A \cong 0$, in which case $\mathrm{Hom}(A, 0) \cong \mathrm{Hom}(0, 0)$ consists of a single element.

Proposition 8.3 tells us in particular that it is futile to try and define 'Boolean categories', that is, bicartesian closed categories in which the obvious arrow $A \to (\bot \Leftarrow (\bot \Leftarrow A))$ is an isomorphism. Up to equivalence of categories, there are no Boolean categories other than Boolean algebras.

Exercises

1. Establish the following isomorphisms in any bicartesian closed category:

$$A+0 \cong A, \quad A \times 0 \cong 0, \quad A^0 \cong 1;$$

$$A+B \cong B+A, \quad (A+B)+C \cong A+(B+C);$$
$$(A+B)\times C \cong (A\times C)+(B\times C), \quad A^{B+C} \cong A^B \times A^C.$$

2. Write down explicit equations between arrows to replace E5 and E6, that is, eliminate f, g and h.

3. Give a detailed justification for Definition 8.2, as was done for $p \wedge q$ in the text.

4. Show that in a bicartesian closed category $0^A \cong 0$ if and only if $A \not\cong 0$.

9 Natural numbers objects in cartesian closed categories

A *natural numbers object* in a cartesian closed category $\mathscr{A}$, according to Lawvere, consists of an initial object

$$1 \xrightarrow{0} N \xrightarrow{S} N$$

in the category of all diagrams $1 \xrightarrow{a} A \xrightarrow{f} A$ in $\mathscr{A}$. This means that, for all such diagrams there is a unique arrow $h\colon N \to A$ such that

$$h0 = a \quad \text{and} \quad hS = fh,$$

as is illustrated by the following commutative diagram:

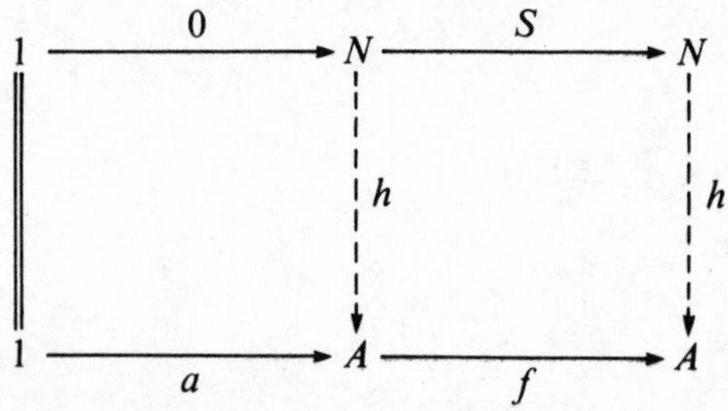

Sometimes we merely wish to assert the existence of h, never mind its uniqueness. Then we shall speak of a *weak natural numbers object*. Cartesian closed categories with a weak natural numbers object have been called 'prerecursive categories' by Marie-France Thibault. (See Exercise 2 below.)

For example, in **Sets**, the set $\mathbb{N} \equiv \{0, 1, 2, \ldots\}$ of natural numbers together with the successor function $S(x) \equiv x + 1$ forms a natural numbers object.

More generally, all toposes considered in Part II have natural numbers objects.

If $1 \xrightarrow{0} N \xrightarrow{S} N$ is a weak natural numbers object, we shall write $h \equiv J_A(a, f)$. Thus

$$J_A\colon \mathrm{Hom}(1, A) \times \mathrm{Hom}(A, A) \to \mathrm{Hom}(N, A)$$

satisfies the equations

$$J_A(a, f)0 = a, \quad J_A(a, f)S = f J_A(a, f).$$

Proposition 9.1. If the cartesian closed category $\mathscr{A}$ has a natural numbers object (weak natural numbers object) and if $x\colon 1 \to A$ is an indeterminate arrow over $\mathscr{A}$, then the cartesian closed category $\mathscr{A}[x]$ has the same natural numbers object (weak natural numbers object).

Proof. A short conceptual argument goes as follows. A (weak) natural numbers object in $\mathscr{A}$ gives rise to one in the slice category $\mathscr{A}/A$, hence in $\mathscr{A}[x]$, which comes with a full and faithful functor $K_x\colon \mathscr{A}[x] \to \mathscr{A}/A$. For the more meticulous reader, we shall now give a detailed computational proof.

First, assume the existence of a weak natural numbers object in $\mathscr{A}$. Let $\beta(x)\colon 1 \to B$ and $\varphi(x)\colon B \to B$ be given polynomials in $\mathscr{A}[x]$. We seek a polynomial $\chi(x)\colon N \to B$ such that

$$\chi(x)0 \underset{x}{=} \beta(x), \quad \chi(x)S \underset{x}{=} \varphi(x)\chi(x).$$

In view of functional completeness, these equations involving x are equivalent to the following equations not involving x:

$$\kappa_{x \in A}(\chi(x)0) = \kappa_{x \in A}\beta(x), \quad \kappa_{x \in A}(\chi(x)S) = \kappa_{x \in A}(\varphi(x)\chi(x)).$$

Writing

$$\kappa_{x \in A}\beta(x) = b\colon A \times 1 \to B, \quad \kappa_x \varphi(x) = f\colon A \times B \to B,$$

we seek

$$\kappa_{x \in A}\chi(x) = h\colon A \times N \to B$$

such that

$$h\langle \pi, 0\pi' \rangle = b, \quad h\langle \pi, S\pi' \rangle = f\langle \pi, h \rangle.$$

With b and f we may associate $b'\colon 1 \to B^A$ and $f'\colon B^A \to B^A$ given by

$$b' = (b\langle \pi', \pi \rangle)^*, \quad f' = (f\langle \pi', \varepsilon \rangle)^*.$$

Then we may find $h'\colon N \to B^A$ such that

$$h'0 = b', \quad h'S = f'h',$$

as is illustrated by the following diagram:

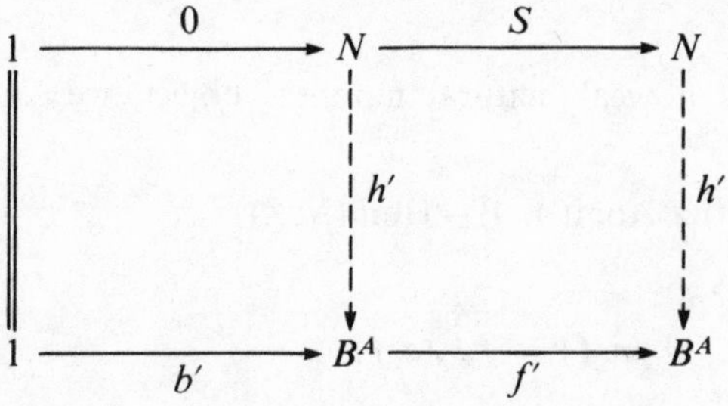

Now put $h = \varepsilon\langle h'\pi', \pi\rangle$, then routine calculations show that

$$h\langle \pi, 0\pi'\rangle = b, \quad h\langle \pi, S\pi'\rangle = f\langle \pi, h\rangle,$$

as required.

If we have a natural numbers object in $\mathscr{A}$, not just a weak one, then the arrow $h': N \to B^A$ is uniquely determined by the equations $h'0 = b'$ and $h'S = f'h'$. From this it easily follows that h is also uniquely determined by the equations it satisfies. For we may calculate h' in terms of h as follows:

$$\begin{aligned}(h\langle \pi', \pi\rangle)^* &= (\varepsilon\langle h'\pi', \pi\rangle\langle \pi', \pi\rangle)^* \\ &= (\varepsilon\langle h'\pi, \pi'\rangle)^* \\ &= h',\end{aligned}$$

and then transform the equations satisfied by h into the equations satisfied by h'.

In what follows we shall write, for indeterminate arrows $y: 1 \to B$, $v: 1 \to B^B$ and $z: 1 \to N$,

$$J_B(y, v^\int)z \equiv I_B\langle y, v, z\rangle,$$

where $\langle y, v, z\rangle$ is short for $\langle\langle y, v\rangle, z\rangle$.

Corollary 9.2 A weak natural numbers object in a cartesian closed category is given by an object N and arrows $1 \xrightarrow{0} N \xrightarrow{S} N$ and I_B: $(B \times B^B) \times N \to B$, for each object B, satisfying the identities

$$I_B\langle y, v, 0\rangle = y, \quad I_B\langle y, v, Sz\rangle = v^\int I_B\langle y, v, z\rangle,$$

where the subscripts y, v, z on the equality symbol have been omitted.

Proof. We use Proposition 9.1 with $A = B \times B^B$. Adjoining a single indeterminate $x: 1 \to A$ is equivalent to adjoining two indeterminates $y: 1 \to B$, and $v: 1 \to B^B$.

Corollary 9.2 may also be stated in terms of an arrow $N \to (B^B)^{(B^B)}$ in place of I_B.

Exercises

1. If a cartesian closed category has a natural numbers object, then this is unique up to isomorphism.

2. Determine all weak natural numbers objects in the category of sets.

3. Carry out the routine calculations mentioned in the text to show that $h\langle \pi, 0\pi' \rangle = b, \quad h\langle \pi, S\pi' \rangle = f\langle \pi, h \rangle$.

4. Show that a natural numbers object in a cartesian closed category is equivalent to the following condition: for each $g\colon A \to B$ and $f\colon B \to B$ there is a unique $k\colon N \times A \to B$ such that the following diagram commutes:

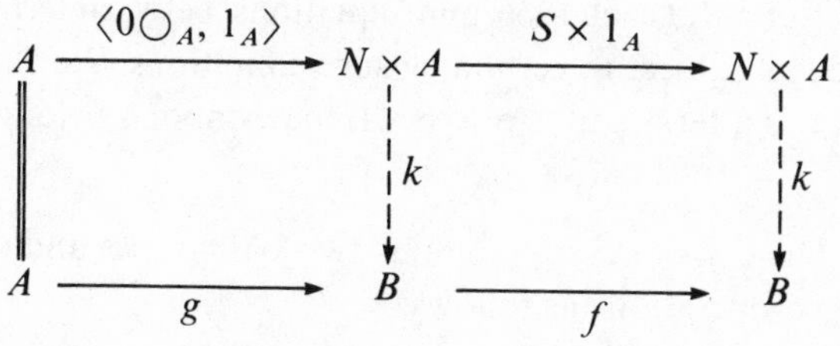

The same assertion without uniqueness holds for a weak natural numbers object. This suggests how to define (weak) natural numbers objects in cartesian categories which are not cartesian closed.

5. Show that, if a cartesian closed category $\mathscr{A}$ has a (weak) natural numbers object, then so does $\mathscr{A}/A$ for each object A of $\mathscr{A}$.

6. Verify the remark in the proof of Proposition 9.1 that, if A is an object of a cartesian closed category $\mathscr{A}$ and $x\colon 1 \to A$ an indeterminate arrow, there is a full and faithful functor $K_x\colon \mathscr{A}[x] \to \mathscr{A}/A$.

7. (Lawvere) Given a category $\mathscr{A}$, let $\mathscr{A}^{\text{loop}}$ be the category whose objects are 'endomaps' $f\colon A \to A$ and whose arrows are commutative squares. There is an obvious forgetful functor $U\colon \mathscr{A}^{\text{loop}} \to \mathscr{A}$.
   (a) Show that $\mathscr{A}^{\text{loop}}$ is equivalent to $\mathscr{A}^{\mathbb{N}}$, where $\mathbb{N}$ is regarded as the free monoid on one generator.
   (b) Assuming that $\mathscr{A}$ is cartesian closed, show that U has a left adjoint F if and only if $\mathscr{A}$ has a natural numbers object. (Hint: In one direction define $F(A) \equiv S \times 1_A\colon N \times A \to N \times A$ and use Exercise 4 above. In the other direction consider $F(1)$.)

8. Consider the cartesian closed category of limit spaces (Part 0, Section 7, Example 7.8). Show that the natural numbers object is the set $\mathbb{N}$ of natural numbers with the 'discrete' convergence structure: a sequence $\{x_n | n \in \mathbb{N}\}$ converges to $x \in \mathbb{N}$ if it is eventually constant with value x.

10 Typed λ-calculi

The purpose of this section is to associate a language with a cartesian closed category with weak natural numbers object, which will be called its 'internal language'. The kind of language we have in mind will be called a 'typed λ-calculus' for short, although it might be known from the literature more fully as a 'typed λη-calculus with product types (surjective pairing) and iterator'. This association will turn out to be an equivalence between appropriate categories.

A *typed λ-calculus* is a formal theory defined as follows. It consists of classes of 'types', 'terms' of each type, and 'equations' between terms which are said to 'hold', all subject to certain closure conditions. We shall write $a \in A$ to say that a is a term and is of type A; the symbol $\in$ belongs to the metalanguage.

(a) *Types*: The class of types contains two basic types and is closed under two operations as follows:

(a1) 1 and N are types (these are the 'basic' types).

(a2) If A and B are types so are $A \times B$ and B^A.
There may be other types not indicated by (a1) and (a2) and there may be un-expected identifications between types.

(b) *Terms*: The class of terms is freely generated from variables and certain basic constants by certain term forming operations as follows:*

(b1) For each type A there are countably many variables of type A, say $x_i^A \in A \quad (i = 0, 1, 2, \ldots)$. We shall hardly have occasion to refer to a specific variable, instead we shall frequently use the phrase 'let x be a variable of type A', abbreviated as '$x \in A$'.

(b2) $* \in 1$.

(b3) If $a \in A, b \in B$ and $c \in A \times B$, then $\langle a, b \rangle \in A \times B$, $\pi_{A,B}(c) \in A$ and $\pi'_{A,B}(c) \in B$.

(b4) If $f \in B^A$ and $a \in A$, then $\varepsilon_{B,A}(f, a) \in B$.

(b5) If $x \in A$ and $\varphi(x) \in B$, then $\lambda_{x \in A}\, \varphi(x) \in B^A$.

(b6) $0 \in N$; if $n \in N$, then $S(n) \in N$.

(b7) If $a \in A, h \in A^A$ and $n \in N$, then $I_A(a, h, n) \in A$.

* There may be other constants and term forming operations than those specified.

We shall abbreviate $\varepsilon_{B,A}(f,a)$ as $f^{\varsigma}a$ (read: 'f of a') when the type subscripts are clear from the context. There may be other terms not indicated by (b1) to (b7). Intuitively, $\varepsilon_{B,A}$ means evaluation, $\langle -,- \rangle$ means pairing and $\lambda_{x\in A}\varphi(x)$ denotes the function $x \mapsto \varphi(x)$. $\lambda_{x\in A}$ acts like a quantifier, so the variable x in $\lambda_{x\in A}(\varphi(x)$ is 'bound' as in $\forall_{x\in A}\varphi(x)$ or $\int_b^a f(x)\mathrm{d}x$. We have the usual conventions for free and bound variables and when it is permitted to substitute a term for a variable. The term a is *substitutable* for x in $\varphi(x)$ if no free occurrence of a variable in a becomes bound in $\varphi(a)$. A term is 'closed' if it contains no free variables. We usually omit subscripts in $\pi_{A,B}(-)$, $I_A(-,-,-)$ etc.

(c) *Equations*:

(c1) Equations have the form $a \underset{X}{=} a'$, where X is a finite set of variables, a and a' have the same type A, and all variables occurring freely in a or a' are elements of X.

(c2) The binary relation between terms a, a' which says that $a \underset{X}{=} a'$ holds is reflexive, symmetric and transitive and it satisfies the rule: when $X \subseteq Y$ then if $a \underset{X}{=} b$ holds one may infer that $a \underset{Y}{=} b$ holds, which will be abbreviated:

$$\frac{a \underset{X}{=} b}{a \underset{Y}{=} b}.$$

It also satisfies the usual substitution rules for all term forming operations, in particular the following:

$$\frac{a \underset{X}{=} b}{f^{\varsigma}a \underset{X}{=} f^{\varsigma}b}, \quad \frac{\varphi(x) \underset{X\cup\{x\}}{=} \varphi'(x)}{\lambda_{x\in A}\varphi(x) \underset{X}{=} \lambda_{x\in A}\varphi'(x)},$$

from which the other substitution rules follow.

All these are 'obvious' substitution rules, except perhaps the rule involving λ, which decreases the number of free variables.

(c3) The following specific equations hold:

$a \underset{X}{=} *$ for all $a \in 1$;

$\pi(\langle a,b \rangle) \underset{X}{=} a$ for all $a \in A, b \in B$,

$\pi'(\langle a,b \rangle) \underset{X}{=} b$ for all $a \in A, b \in B$,

$\langle \pi(c), \pi'(c) \rangle \underset{X}{=} c$ for all $c \in A \times B$;

$\lambda_{x\in A}\varphi(x)^{\varsigma}a \underset{X}{=} \varphi(a)$ for all $a \in A$ which are substitutable for x,

$\lambda_{x\in A}(f^{\varsigma}x) \underset{X}{=} f$ for all $f \in B^A$, provided x is not in X (hence does not occur freely in f);

$$I(a,h,0) \underset{X}{=} a, \quad \text{for all } a \in A, h \in A^A$$

$$I(a,h,S(x)) \underset{X\cup\{x\}}{=} h{}^{\varsigma} I(a,h,x); \text{ provided } x \text{ is not in } X \text{ (hence does not occur freely in } a \text{ or } h);$$

$$\lambda_{x\in A}\varphi(x) \underset{X}{=} \lambda_{x'\in A}\varphi(x'), \quad \text{if } x' \text{ is substitutable for } x.$$

There may be other equations not indicated by (c1) to (c3).

The last equation listed under (c3) may be omitted if we are willing to identify terms which differ only in the choice of bound variables.

One of the rules listed under (c2) allows us to pass from $a \underset{X}{=} b$ to $a \underset{X\cup\{x\}}{=} b$, even when x is not in X. Under certain conditions one can go in the opposite direction, as we shall see. Of course, if this were always the case, there would have been no point in putting the subscript X on the equality sign.

Proposition 10.1. In any typed λ-calculus, one may infer from $\varphi(x) \underset{X\cup\{x\}}{=} \psi(x)$ that $\varphi(a) \underset{X}{=} \psi(a)$ for any $a \in A$, provided x is not in X and all variables occurring freely in a are elements of X.

Proof. From $\varphi(x) \underset{X\cup\{x\}}{=} \psi(x)$ we infer $\lambda_{x\in A}\varphi(x) \underset{X}{=} \lambda_{x\in A}\psi(x)$, hence $\lambda_{x\in A}\varphi(x){}^{\varsigma}a \underset{X}{=} \lambda_{x\in A}\psi(x){}^{\varsigma}a$, using (c2). In view of (c3), we then obtain $\varphi(a) \underset{X}{=} \psi(a)$.

Corollary 10.2. If f and g do not contain free occurrences of the variable x of type A, then from $f \underset{X\cup\{x\}}{=} g$ we infer $f \underset{X}{=} g$, provided there exists a term a of type A such that all variables occurring freely in a are elements of X.

Proof. If x is not already in X, this follows from Proposition 10.1.

Unfortunately, it may happen that A is 'empty' that is, no closed term of type A exists (see examples 10.5 and 10.6, below). On the other hand, if there are closed terms of each type, the proviso of Corollary 10.2 is always satisfied. This is the case, for example, in the 'pure typed λ-calculus with weak natural numbers object' to be discussed presently. In such a situation the subscript X on $\underset{X}{=}$ is redundant and one may replace $\underset{X}{=}$ by just $=$.

Sometimes one may argue differently, but with the same result. Suppose $f \underset{X\cup\{x\}}{=} g$, then $f{}^{\varsigma}x \underset{X\cup\{x\}}{=} g{}^{\varsigma}x$, hence $\lambda_{x\in A}(f{}^{\varsigma}x) \underset{X}{=} \lambda_{x\in A}(g{}^{\varsigma}x)$. In view of (c3), it follows that $f \underset{X}{=} g$. The assumption here is that f and g have type B^A. We shall sum this up:

Proposition 10.3. If f and g are of type B^A and if $x \in A$ does not occur freely in f or g, then from $f \underset{X\cup\{x\}}{=} g$ one may infer $f \underset{X}{=} g$.

We shall consider three examples of typed λ-calculi with weak natural numbers object.

Example 10.4. Suppose there are no types, terms and equations other than those indicated by the closure rules (and also no nontrivial identifications between types), then we obtain the *pure typed λ-calculus with weak natural numbers object* called $\mathscr{L}_0$.

Example 10.5. Given a graph $\mathscr{G}$, the λ-calculus $\Lambda(\mathscr{G})$ *generated* by $\mathscr{G}$ is defined as follows. Its types are generated inductively by the type forming operations $(-)\times(-)$ and $(-)^{(-)}$ from the basic types 1, N and the vertices of $\mathscr{G}$ (which now count as basic types). Its terms are generated inductively from the basic terms x_i^A, 0 and $*$ by the old term forming operations $\langle -,-\rangle$, $\pi(-)$, $\pi'(-)$, $\varepsilon(-,-)$, $\lambda_{x\in A}(-)$, $S(-)$ and $I(-,-,-)$ together with the new term forming operations: if $a\in A$ then $fa\in B$, for each arrow $f: A\to B$ in $\mathscr{G}$. Finally, its equations are precisely those which follow from (c1) to (c3) and no others. Note that there are plenty of empty types, for instance, all the vertices of $\mathscr{G}$. Clearly, Example 10.5 includes 10.4 if $\mathscr{G}$ is the empty graph.

We now come to the principal example of this section.

Example 10.6. The *internal language* $\mathbf{L}(\mathscr{A})$ of a cartesian closed category $\mathscr{A}$ with weak natural numbers object is defined as follows. Its types are the objects of $\mathscr{A}$, with 1, N, $A\times B$ and B^A having the obvious meanings. Terms of type A are those polynomial expressions $\varphi(x_1,\ldots,x_n): 1\to A$ in the indeterminates $x_i: 1\to A_i$ which are obtained from variables, namely indeterminates, and basic constants, namely arrows $1\to A$ in $\mathscr{A}$, by the term forming operations:

$$\frac{a:1\to A \quad b:1\to B}{\langle a,b\rangle: 1\to A\times B}, \frac{a:1\to A}{fa: 1\to B}, \frac{\varphi(x): 1\to B}{\lambda_{x\in A}\varphi(x): 1\to B^A},$$

where $f: A\to B$ and $\lambda_{x\in A}\varphi(x)\equiv \ulcorner\kappa_{x\in A}\varphi(x)\langle 1_A, \bigcirc_A\rangle\urcorner$ as in the proof of Corollary 6.2. Moreover, we write $*$ for $\bigcirc_1$, $\pi_{A,B}(c)$ for $\pi_{A,B}c$, $\varepsilon_{B,A}(f,a)$ for $\varepsilon_{B,A}\langle f,a\rangle$, etc. Finally, if a and b are polynomial expressions whose free variables are in X, $a =_X b$ is said to hold in $\mathbf{L}(\mathscr{A})$ if $a =_X b$ as polynomials in $\mathscr{A}[X]$.

We shall now introduce morphisms $\Phi: \mathscr{L}\to\mathscr{L}'$ of typed λ-calculi, to be called *translations*.

(d1) Φ sends types of $\mathscr{L}$ to types of $\mathscr{L}'$ and terms of $\mathscr{L}$ to terms of $\mathscr{L}'$ so that if $a\in A$, then $\Phi(a)\in\Phi(A)$; but we insist that if a is closed, so is $\Phi(a)$ and that Φ sends the ith variable of type A to the ith variable of type $\Phi(A)$.

(d2) Φ preserves the specified type operations on the nose, for example:

$$\Phi(1) = 1, \quad \Phi(A \times B) = \Phi(A) \times \Phi(B), \ldots;$$

and the specified term forming operations up to 'equality holding', e.g. the following equations hold in $\mathscr{L}'$:

$$\Phi(\pi_{A,B}(c)) = \pi_{\Phi(A),\Phi(B)}(\Phi(c)); \quad \Phi(\lambda_{x \in A}\varphi(x)) = \lambda_{\Phi(x) \in \Phi(A)}\Phi(\varphi(x)).$$

(d3) Moreover, Φ preserves equations:

if $a \underset{X}{=}$ holds in $\mathscr{L}$ then $\Phi(a) \underset{\Phi(X)}{=} \Phi(b)$ holds in $\mathscr{L}'$.

In view of (d3), Φ really acts on equivalence classes of terms (modulo the equivalence relation described in (c2)). We shall say that two translations are *equal* if they have the same effect on such equivalence classes. Thus $\Phi = \Psi$ provided $\Phi(a) \underset{\Phi(X)}{=} \Psi(a')$ holds whenever $a \underset{X}{=} a'$ holds.

We thus obtain a category **λ-Calc** whose objects are typed λ-calculi and whose arrows are translations.

Let $\mathbf{Cart}_\mathrm{N}$ be the category of cartesian closed categories with weak natural numbers object and cartesian closed functors preserving weak natural numbers objects on the nose. The proof of the following is left to the reader.

Proposition 10.7. Let $\mathbf{L}(\mathscr{A})$ be the internal language of $\mathscr{A}$ and, for any morphism $F: \mathscr{A} \to \mathscr{A}'$, let $\mathbf{L}(F)$ be defined by $\mathbf{L}(F)(A) = F(A)$, $\mathbf{L}(F)(x_i) = x_i'$, $\mathbf{L}(F)(\varphi(X)) = F_X(\varphi(X))$, where x_i' is the ith variable of type $F(A)$ and F_X is the unique arrow in $\mathbf{Cart}_\mathrm{N}$ such that the following diagram commutes:

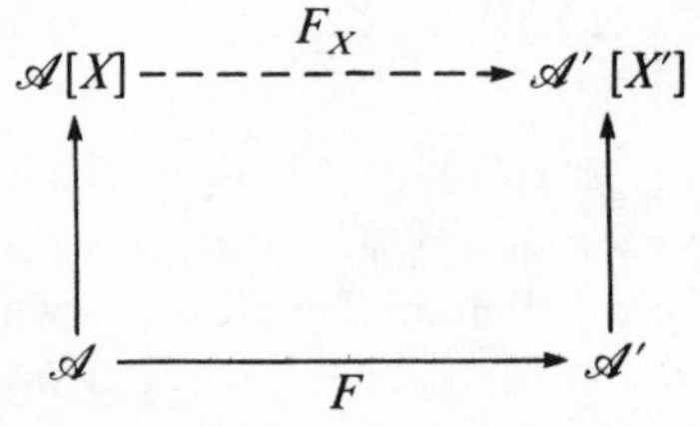

Then $\mathbf{L}$ is a functor from $\mathbf{Cart}_\mathrm{N}$ to **λ-Calc.**

$\mathscr{L}_0$ is an initial object in **λ-Calc**, that is, for any typed λ-calculus $\mathscr{L}$ there is a unique translation $\mathscr{L}_0 \to \mathscr{L}$. In particular, for any $\mathscr{A}$ in $\mathbf{Cart}_\mathrm{N}$ there is a unique morphism $\mathscr{L}_0 \to \mathbf{L}(\mathscr{A})$. This may be called the *interpretation* of $\mathscr{L}_0$ in $\mathscr{A}$.

The reader may have noticed that the languages discussed in this section may be proper classes in the sense of Gödel–Bernays. If necessary, one may work in a set theory with universes, in which 'classes' are replaced by 'sets in a sufficiently large universe'.

Exercises

1. Verify that Λ: **Grph** → λ-**Calc** (see Example 10.5) is a functor left adjoint to the obvious 'forgetful' functor V: λ-**Calc** → **Grph.** (The underlying graph of a λ-calculus has as vertices the types and as arrows $A \to B$ suitable equivalence classes of pairs $(x, \varphi(x))$, where $\varphi(x)$ is a term of type B with no free variables other than x, which is of type A.)
2. By a *classification* we mean two classes and a mapping between them:

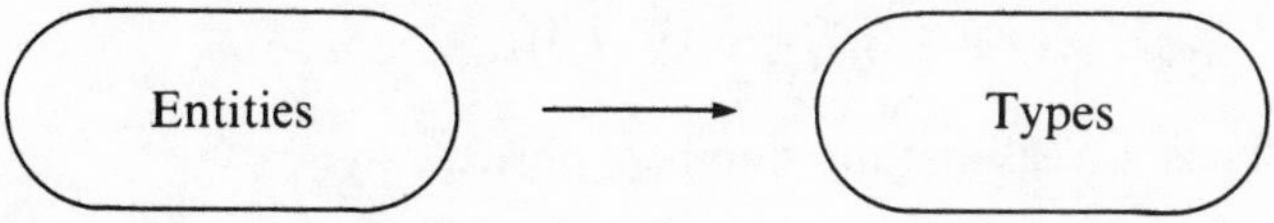

The mapping assigns to each entity its type, and we write '$a \in A$' for 'the type of a is A'. Morphisms Φ between classifications are defined in the obvious way: $a \in A$ should imply $\Phi(a) \in \Phi(A)$. The category of small classifications is thus equivalent to $\mathbf{Sets}^{\mathbf{2}}$, where **2** is the category consisting of two objects and one arrow between them. Show that the obvious forgetful functor from λ-**Calc** to the category of classifications has a left adjoint.
3. If $\mathscr{C}$ is a Heyting algebra considered as a cartesian closed category, show that there may be unexpected identifications between types in $\mathbf{L}(\mathscr{C})$.
4. Verify that $\mathbf{L}(\mathscr{A})$ in Example 10.6 is a typed λ-calculus.

11 The cartesian closed category generated by a typed λ-calculus

To show that the functor **L** in Section 10 is an equivalence of categories we shall obtain a functor **C** in the opposite direction.

Given a typed λ-calculus $\mathscr{L}$, we construct a cartesian closed category $\mathbf{C}(\mathscr{L})$ with weak natural numbers object as follows:

The objects of $\mathbf{C}(\mathscr{L})$ are the types of $\mathscr{L}$.

The arrows $A \to B$ of $\mathbf{C}(\mathscr{L})$ are (equivalence classes of) pairs $(x \in A, \varphi(x))$, with x a variable of type A and $\varphi(x)$ a term of type B with no free variables other than x. (Think of the function $x \mapsto \varphi(x)$.)

Equality of arrows is defined by: $(x \in A, \varphi(x)) = (x' \in A, \psi(x'))$ if and only if $\varphi(x) \underset{x}{=} \psi(x)$ holds, where $\underset{x}{=}$ abbreviates $\underset{\{x\}}{=}$.

The identity arrow $A \to A$ is the pair $(x \in A, x)$.

The composition of $(x \in A, \varphi(x)): A \to B$ and $(y \in B, \psi(y)): B \to C$ is given by $(x \in A, \psi(\varphi(x))): A \to C$, $\varphi(x)$ having been substituted for y in $\psi(y)$.

The cartesian closed structure of $\mathbf{C}(\mathscr{L})$ is obtained as follows:

$$\bigcirc_A \equiv (x \in A, *),$$
$$\pi_{A,B} \equiv (z \in A \times B, \pi(z)),$$
$$\pi'_{A,B} \equiv (z \in A \times B, \pi'(z)),$$
$$\langle (z \in C, \varphi(z)), (z \in C, \psi(z)) \rangle \equiv (z \in C, \langle \varphi(z), \psi(z) \rangle),$$
$$(z \in A \times B, \chi(z))^* \equiv (x \in A, \lambda_{y \in B} \chi(\langle x, y \rangle)),$$
$$\varepsilon_{C,A} \equiv (y \in C^A \times A, \varepsilon_{C,A}(\pi(y), \pi'(y))).$$

$\mathbf{C}(\mathscr{L})$ has a weak natural numbers object:

$$0 \equiv (x \in 1, 0),$$
$$S \equiv (x \in N, S(x)),$$
$$I_B \equiv (w \in (B \times B^B) \times N, I(\pi(\pi(w)), \pi'(\pi(w)), \pi'(w))).$$

It is easy to make $\mathbf{C}$ into a functor λ-**Calc** $\to$ $\mathbf{Cart}_{\mathrm{N}}$. Indeed, suppose $\Phi: \mathscr{L} \to \mathscr{L}'$ is a translation, define $\mathbf{C}(\Phi): \mathbf{C}(\mathscr{L}) \to \mathbf{C}(\mathscr{L}')$ as follows:

If A is an object of $\mathbf{C}(\mathscr{L})$, that is, a type of $\mathscr{L}$, $\mathbf{C}(\Phi)\ (A) = \Phi(A)$ is the corresponding type of $\mathscr{L}'$, hence an object of $\mathbf{C}(\mathscr{L}')$.

If $f = (x \in A, \varphi(x))$ is an arrow $A \to B$ in $\mathbf{C}(\mathscr{L})$, that is, $\varphi(x)$ is a term of type B in $\mathscr{L}, \mathbf{C}(\Phi)\ (f) = \Phi(x) \in \Phi(A)$, $\Phi(\varphi(x))$ is the corresponding arrow $\Phi(A) \to \Phi(B)$ in $\mathbf{C}(\mathscr{L}')$.

To sum up:

Proposition 11.1. $\mathbf{C}$ is a functor from λ-**Calc** to $\mathbf{Cart}_{\mathrm{N}}$.

Instead of adjoining an indeterminate arrow $x: 1 \to A$ to the cartesian closed category $\mathbf{C}(\mathscr{L})$, one may adjoin a 'parameter' x of type A to the language $\mathscr{L}$. To be precise, if $\mathscr{L}$ is a typed λ-calculus and x is a variable of type A, one may form another language $\mathscr{L}(x)$ by *adjoining the parameter* x as follows:

$\mathscr{L}(x)$ has exactly the same types as $\mathscr{L}$ and also the same terms, except that x is no longer counted as a variable. In other words, the closed terms of $\mathscr{L}(x)$ are terms $\varphi(x)$ in $\mathscr{L}$ which contain no free variables other than x. In the same spirit, $\underset{X}{=}$ in $\mathscr{L}(x)$ means $\underset{X \cup \{x\}}{=}$ in $\mathscr{L}$: just make sure that x is not in X.

Some dictionaries define a 'parameter' as a 'variable constant'. For us it is a variable kept constant.

Proposition 11.2. $\mathbf{C}(\mathscr{L})[x] \cong \mathbf{C}(\mathscr{L}(x))$.

Proof. We show that $\mathbf{C}(\mathscr{L}(x))$ has the universal property of $\mathbf{C}(\mathscr{L})[x]$ (see Section 5):

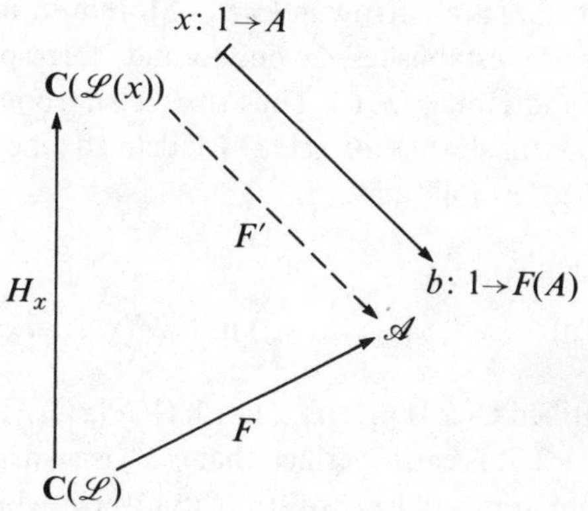

The indeterminate $x: 1 \to A$ is defined by $(y \in 1, \quad x)$. H_x is $\mathbf{C}$ of the inclusion of $\mathscr{L}$ into $\mathscr{L}(x)$, which may necessitate some relabelling of variables. Suppose $F: \mathbf{C}(\mathscr{L}) \to \mathscr{A}$ is any cartesian closed functor preserving the weak natural numbers object, and given any arrow $b: 1 \to F(A)$ in $\mathscr{A}$, we claim that there is a unique such functor $F': \mathbf{C}(\mathscr{L}(x)) \to \mathscr{A}$ such that $F'H_x = F$ and $F'(x) = b$.

Indeed, put $F'(B) = F(B)$ for each object B of $\mathbf{C}(\mathscr{L})$, that is type in $\mathscr{L}$. Suppose $f = (y \in B, \varphi(x, y))$ is any arrow $B \to C$ in $\mathbf{C}(\mathscr{L}(x))$, that is, $\varphi(x, y)$ is any term of type C in $\mathscr{L}$ with free variables $x \in A$ and $y \in B$. Define $F'(f)$: $F(B) \to F(C)$ in $\mathscr{A}$ as follows:

First note that $\varphi(x, y) \underset{\{x,y\}}{=} \psi\lambda(y)^{\int} x$ holds, where $\psi(y)$ is $\lambda_{x \in A}\varphi(x, y)$. Thus $f = \varepsilon_{C,A}\langle g, x \bigcirc_B \rangle$, where $g = (y \in B, \psi(y)): B \to C^A$ in $\mathbf{C}(\mathscr{L})$. Now define

$$F'(f) = \varepsilon_{F(C),F(A)}\langle F(g), b \bigcirc_{F(B)} \rangle.$$

That this definition has the right property is easily seen. Moreover, it is clearly forced upon us, since $F'(g) = F(g)$ and $F'(x) = b$. We now establish the main result of this section.

Theorem 11.3. The categories λ-**Calc** and $\mathbf{Cart}_\mathrm{N}$ are equivalent, in fact $\mathbf{CL} \cong \mathrm{id}$ and $\mathbf{LC} \cong \mathrm{id}$.

Proof. (i) Consider the natural transformation $\varepsilon: \mathbf{CL} \to \mathrm{id}$ defined for each $\mathscr{A}$ in $\mathbf{Cart}_{\mathrm{N}}$ by $\varepsilon(\mathscr{A}): \mathbf{CL}(\mathscr{A}) \to \mathscr{A}$ as follows:

An object of $\mathbf{CL}(\mathscr{A})$ is a type of $\mathbf{L}(\mathscr{A})$, that is, an object of $\mathscr{A}$. Put $\varepsilon(\mathscr{A})(A) = A$.

An arrow $B \to C$ in $\mathbf{CL}(\mathscr{A})$ has the form $f = (y \in B, \varphi(y))$, where $\varphi(y) \in C$ in $\mathbf{L}(\mathscr{A})$. Put $\varepsilon(\mathscr{A})\ (f) =$ the unique arrow $g: B \to C$ such that $gy \underset{y}{=} \varphi(y)$, using functional completeness.

It is easily verified that $\varepsilon(\mathscr{A})$ is an arrow in $\mathbf{Cart}_{\mathrm{N}}$. Moreover, in view of functional completeness, it establishes a one-to-one correspondence between $\mathrm{Hom}_{\mathbf{CL}(\mathscr{A})}\ (B, C)$ and $\mathrm{Hom}_{\mathscr{A}}\ (B, C)$. Thus $\varepsilon(\mathscr{A})$ is an isomorphism.

(ii) Consider the natural transformation $\eta: \mathrm{id} \to \mathbf{LC}$ defined for each $\mathscr{L}$ in λ-**Calc** by $\eta(\mathscr{L}): \mathscr{L} \to \mathbf{LC}(\mathscr{L})$ as follows:

$$\eta(\mathscr{L})(A) \equiv A;$$

$$\eta(\mathscr{L})(\varphi(x_1, \ldots, x_n)) \equiv (z \in 1, \varphi(x_1, \ldots, x_n)) \text{ in } \mathbf{C}(\mathscr{L}(x_1, \ldots, x_n)).$$

Note that we have identified $\mathbf{C}(\mathscr{L})[x_1, \ldots, x_n]$ with $\mathbf{C}(\mathscr{L}(x_1, \ldots, x_n))$ as is justified by Proposition 11.2. It is easily verified that $\eta(\mathscr{L})$ is an arrow in λ-**Calc.** To see that $\eta(\mathscr{L})$ is an isomorphism, construct its inverse, which sends $(z \in 1, \varphi(z))$ onto $\varphi(*)$.

Corollary 11.4. $\mathbf{C}(\mathscr{L}_0)$, the free cartesian closed category with weak natural numbers object generated by the pure typed λ-calculus, is an initial object in $\mathbf{Cart}_{\mathrm{N}}$.

The initial object of $\mathbf{Cart}_{\mathrm{N}}$ may also be obtained by the methods of Section 4.

We end this section with a remark concerning the problem of how to interpret languages in categories. In the present context this is explained quite easily: an *interpretation* of a typed λ-calculus $\mathscr{L}$ in a cartesian closed category $\mathscr{A}$ with weak natural numbers object is just a translation $\mathscr{L} \to \mathbf{L}(\mathscr{A})$. By Theorem 11.3 (or just by adjointness, see Exercise 3 below), this is essentially the same as a cartesian closed functor $\mathbf{C}(\mathscr{L}) \to \mathscr{A}$. As already observed after Proposition 10.7, $\mathscr{L}_0$ has a unique interpretation in any cartesian closed category with weak natural numbers object.

Exercises

1. Show how to obtain the free cartesian closed category with a weak natural numbers object generated by any classification. (See Exercise 2 of Section 10.)
2. In the spirit of this section, find a new method for constructing the free

cartesian closed category with a weak natural numbers object generated by a graph.

3. Show that $\mathbf{C}$ is left adjoint to $\mathbf{L}$ with adjunction η and ε.

4. Prove that $I_B\langle\langle y, v\rangle, x\rangle = (t\in 1, \quad I(y, v, x))$ in $\mathbf{C}(\mathscr{L}(y, v, x))$.

12 The decision problem for equality

Let us look at the cartesian closed category with weak natural numbers object freely generated by the empty graph, as in Section 4, but with weak natural numbers object, or as in Exercise 2 of Section 11. Since both are initial objects in $\mathbf{Cart}_{\mathrm{N}}$ (see Corollary 11.4), they are isomorphic. We shall write $\mathscr{C}_0$ for this initial object. $\mathscr{C}_0$ is of interest to logicians, as it gives a version of Gödel's primitive recursive functionals of finite type, and to categorists, as it is related to the so-called 'coherence problem' for $\mathbf{Cart}_{\mathrm{N}}$. This problem asks when diagrams in a category commute or, equivalently, when two arrows between two given objects are equal. Indeed if one wants to compute $\mathrm{Hom}(A, B)$ in $\mathscr{C}_0$, two problems arise:

(I) Find an algorithm for obtaining all arrows $A \to B$ in $\mathscr{C}_0$ (that is, all proofs $A \to B$ in the corresponding deductive system).

(II) Find an algorithm for deciding when two arrows $A \to B$ are equal (better: when two proofs describe the same arrow).

We shall here address ourselves to the second problem. Looking at the proof of the distributive law

$$\langle f, g\rangle h = \langle fh, gh\rangle$$

for cartesian closed categories given in Section 3, we note that both sides must be expanded to be shown equal. It seems easier to consider $\mathscr{C}_0$ as given by $\mathbf{C}(\mathscr{L}_0)$ rather than as constructed by the method of Section 4.

Two arrows $f, g\colon A \to B$ in $\mathscr{C}_0$ are thus given by two terms $\varphi(x)$ and $\psi(x)$ of type B in $\mathscr{L}_0$ with a free variable x of type A. We want to decide whether $\varphi(x) \underset{x}{=} \psi(x)$ holds, or equivalently, $\lambda_{x\in A}\phi(x) = \lambda_{x\in A}\psi(x)$ holds in $\mathscr{L}_0$. Let us call two terms a and b of $\mathscr{L}_0$ whose free variables are contained in X *convertible* if the equation $a \underset{X}{=} b$ holds in $\mathscr{L}_0$. Terms of $\mathscr{L}_0$ are, of course, defined inductively, as the reader will recall. Thus Problem (II) has been reduced to deciding when two closed terms of type B in $\mathscr{L}_0(x)$ or $\mathscr{L}_0$ are convertible. In view of the fact that there are closed terms of each type in $\mathscr{L}_0$, we need not distinguish between $\underset{x}{=}$ and $=$, as was pointed out in Section 10.

Actually, we shall solve the decision problem for convertibility not in $\mathscr{L}_0$ but in $\mathscr{L}'_0$, which is like $\mathscr{L}_0$ but without type 1. In other words, $\mathscr{L}'_0$ is a variant of pure typed λ-calculus in which the only basic type is N. This may be done without loss in generality for the following reason: a closed term of type B in $\mathscr{L}_0$ or $\mathscr{L}_0(x)$ corresponds to an arrow $1 \to B$ in $\mathscr{C}_0$ or $\mathscr{C}_0[x]$, where the object B is canonically isomorphic to either 1 (which case may be dismissed as uninteresting) or an object whose inductive construction does not contain 1 at all. This is so in view of the canonical isomorphisms $C \times 1 \cong C \cong 1 \times C$, $C^1 \cong C$ and $1^C \cong 1$. The last mentioned isomorphism presupposes that $\mathrm{Hom}(1, C)$ is not empty, which is the case in $\mathscr{C}_0$, as there are closed terms of each type in $\mathscr{L}_0$.

To solve the decision problem for convertibility of terms in $\mathscr{L}'_0$, we shall replace convertibility by a finer relation, that of reducibility. However, it becomes tedious to distinguish between terms which differ only in the choice of bound variables. We shall call two such terms a and a' *congruent* and write $a \equiv a'$.

First we shall define a relation $a > a'$ between terms of type A in $\mathscr{L}_0$ or $\mathscr{L}'_0$ (actually, congruence classes) and say 'a basically reduces to a''. There are eight basic reductions; in each of the basic equations of typed λ-calculus the left hand side *basically reduces* to the right hand side:

B1.	$a > *$	$(a \in 1, a \not\equiv *)$; (not used in $\mathscr{L}'_0$)
B2.	$\pi(\langle a, b\rangle) > a$	$(a \in A, b \in B)$;
B3.	$\pi'(\langle a, b\rangle) > b$	$(a \in A, b \in B)$;
B4.	$\langle \pi(c), \pi'(c)\rangle > c$	$(c \in A \times B)$;
B5.	$\lambda_{x \in A}\varphi(x)^{\varsigma} a > \varphi(a)$	$(a \in A)$;
B6.	$\lambda_{x \in A}(f^{\varsigma} x) > f$	$(f \in B^A, x$ not free in $f)$;
B7.	$I(a, h, 0) > a$	$(a \in A, h \in A^A)$;
B8.	$I(a, h, S(n)) > h^{\varsigma} I(a, h, n)$	$(a \in A, h \in A^A, n \in N)$.

We shall say that b *reduces to* b' *in one step* and write $b \underset{1}{\geq} b'$ provided b' is obtained from b by replacing a single occurrence of a subterm a in b by a', where $a > a'$. For example,

$$\lambda_{x \in A}\langle \pi(\langle x, y\rangle), y\rangle \underset{1}{\geq} \lambda_{x \in A}\langle x, y\rangle$$

because $\pi(\langle x, y\rangle) > x$.

We shall say that b *reduces to* b' *in* n *steps* and write $b \underset{n}{\geq} b'$ provided

$$b \equiv b_0 \underset{1}{\geq} b_1 \underset{1}{\geq} \cdots \underset{1}{\geq} b_n \equiv b'.$$

In particular, $b \underset{0}{\geq} b'$ means that $b \equiv b'$. We shall also say b *reduces to* b' and write $b \geqslant b'$ provided there is a natural number n such that $b \underset{n}{\geq} b'$.

The convertibility relation between terms is of course the smallest equivalence relation containing $\geqslant$, that is, the equivalence relation generated by $\geqslant$. This takes a particularly simple form in view of the following:

Proposition 12.1. (Church–Rosser Theorem). In $\mathscr{L}'_0$, if $b \geqslant c$ and $b \geqslant d$ then there exists a term e such that $c \geqslant e$ and $d \geqslant e$.

We postpone the proof of this until later and only note its consequence:

Corollary 12.2. If b and b' are terms of type B, then $b = b'$ holds in $\mathscr{L}_0$ if and only if there is a term $d \in B$ such that $b \geqslant d$ and $b' \geqslant d$.

Proof. It suffices to check that the relation between b and b' which holds whenever there exists d such that $b \geqslant d$ and $b' \geqslant d$ is an equivalence relation.

We shall call a term b *irreducible*, or *in normal form*, if there does not exist a term b' such that $b \underset{1}{>} b'$, that is, if for no subterm a of b there exists a' with $a > a'$. Another way of saying this is that $b \geqslant b'$ implies $b \equiv b'$ for all terms b'.

Remark 12.3. In $\mathscr{L}_0$ there are irreducible closed terms $\kappa(A)$ of each type A, defined inductively as follows:

$$\kappa(1) \equiv *, \kappa(N) \equiv 0, \kappa(A \times B) \equiv \langle \kappa(A), \kappa(B) \rangle, \kappa(B^A) \equiv \lambda_{x \in A} \kappa(B).$$

We shall leave the easy verification of this to the reader and pass on to some further obvious consequences of the Church–Rosser Theorem.

Clearly, a sufficient condition for $b = b'$ to hold in $\mathscr{L}_0$ or $\mathscr{L}'_0$ is that b and b' reduce to congruent irreducible terms (or have congruent normal forms). Call b *normalizable* if there exists an irreducible b^* such that $b \geqslant b^*$.

Corollary 12.4. In $\mathscr{L}'_0$, if b is normalizable, then its normal form is unique up to congruence. Two normalizable terms are convertible if and only if they have congruent normal forms.

One might think that this gives a decision procedure for convertibility of normalizable terms: reduce each to normal form and see whether these irreducible terms are congruent. Unfortunately, there is still a problem of how to reduce a given term to normal form. While one sequence of one-step reductions may end up with an irreducible term after a finite number of steps, it is conceivable that another sequence of one-step reductions will never terminate, and we may have no way of telling beforehand whether we are on the right track.

We shall call a term *bounded* (some authors say 'strongly normalizable') if there is a number n so that no sequence of one-step reductions has more than n steps. The *bound* of t, written $\mathrm{bd}(t)$, is the smallest such n. For example, the bound of an irreducible term is 0. Clearly, if a term is bounded, every sequence of one-step reductions will terminate after a finite number of steps. (The converse of this statement is also true, in view of König's Lemma; but we shall not need this.) In particular, every bounded term is normalizable. Note that if $t \underset{1}{>} t'$ then $\mathrm{bd}(t) > \mathrm{bd}(t')$.

We thus have an algorithm for deciding convertibility of two bounded terms in $\mathscr{L}'_0$: just reduce both of them at random until irreducible terms are reached and then compare these to see whether they are congruent.

We shall prove in Section 13 that the Church–Rosser Theorem holds for bounded terms and in Section 14 that all terms are bounded. We shall thus obtain an algorithm for deciding convertibility of terms in $\mathscr{L}'_0$ and therefore for deciding equality of arrows in $\mathscr{C}_0 = \mathbf{C}(\mathscr{L}_0)$.

For the moment, let us just make an observation that will be useful later.

Lemma 12.5. Suppose $\varphi(x)$ is a term in $\mathscr{L}_0$ with no free variables other than x of type A and a is a closed term of type A such that $\varphi(a)$ is bounded. Then $\varphi(x)$ is bounded.

Proof. If $\varphi(x) \underset{1}{>} \psi(x)$ by virtue of B2 to B8, then surely also $\varphi(a) \underset{1}{>} \psi(a)$. However, when the basic reduction $x \underset{1}{>} *$ is used, then $\varphi(*) \equiv \psi(*)$ are the same terms. This unfortunate exception complicates the proof somewhat. Still, $\varphi(x) \geqslant \psi(x)$ implies $\varphi(a) \geqslant \psi(a)$. Consider the set Γ of all terms $\psi(x)$ such that $\varphi(x) \geqslant \psi(x)$. For any $\psi(x)$ in Γ it thus follows that $\varphi(a) \geqslant \psi(a)$. Since $\varphi(a)$ is bounded, the set Δ of all terms b such that $\varphi(a) \geqslant b$ is finite. (Remember that we do not distinguish between congruent terms, that is, terms that differ only in the choice of bound variables.) Moreover, for each b in Δ, the set Γ_b of all $\psi(x)$ such that $\psi(a) \equiv b$ is finite. Hence $\Gamma \subseteq \bigcup_{b \in \Delta} \Gamma_b$ is also finite, and therefore $\psi(x)$ is bounded.

Exercises

1. Show that all irreducible closed terms of $\mathscr{L}_0$ have the form $*$, $S^k(0)$, $\langle a, b \rangle$ (where a and b are necessarily irreducible) or $\lambda_{x \in A}\varphi(x)$ (where $\varphi(x)$ is necessarily irreducible). Thus closed terms of the form $\pi(c)$, $\pi'(c)$, $f^{\varsigma}a$ or $I(a, h, n)$ are never irreducible.
2. Show that $\langle a, b \rangle$ is bounded if and only if a and b are bounded.

13 The Church–Rosser theorem for bounded terms

In this section we shall prove the Church–Rosser Theorem (Proposition 12.1) for the special case when b is a bounded term of $\mathscr{L}'_0$ or, more generally, of $\mathscr{L}_0$ but without any subterm of type 1 other than $*$. As we shall prove in Section 14 that every term is bounded, this will establish Proposition 12.1.

Proposition 13.1. If b is bounded and $b \underset{m}{>} c$ and $b \underset{n}{>} d$, then there is a term e such that $c \geqslant e$ and $d \geqslant e$.

Proof. We argue by induction on the bound of b and reduce the problem to the case $m \leqslant 1, n \leqslant 1$. The case $m = 1$, $n = 1$, is handled by Lemma 13.2 below. If $m = 0$, or $n = 0$, there is nothing to prove.

So suppose $m > 1$ or $n > 1$. We then have

$$b \underset{1}{>} c_1 >_{m-1} c, \quad b \underset{1}{>} d_1 >_{n-1} d,$$

where $m - 1 > 0$ or $n - 1 > 0$. By Lemma 13.2 below we find e_1 so that $c_1 \geqslant e_1$ and $d_1 \geqslant e_1$. Now c_1 and d_1 have smaller bound than b; so, by inductional assumption, we can find c_2 and d_2 so that $c \geqslant c_2, e_1 \geqslant c_2$, $e_1 \geqslant d_2$ and $d \geqslant d_2$. Again, e_1 has smaller bound than b; so we can find e such that $c_2 \geqslant e$ and $d_2 \geqslant e$. By transitivity, $c \geqslant e$ and $d \geqslant e$.

This proof is illustrated by the following diagram:

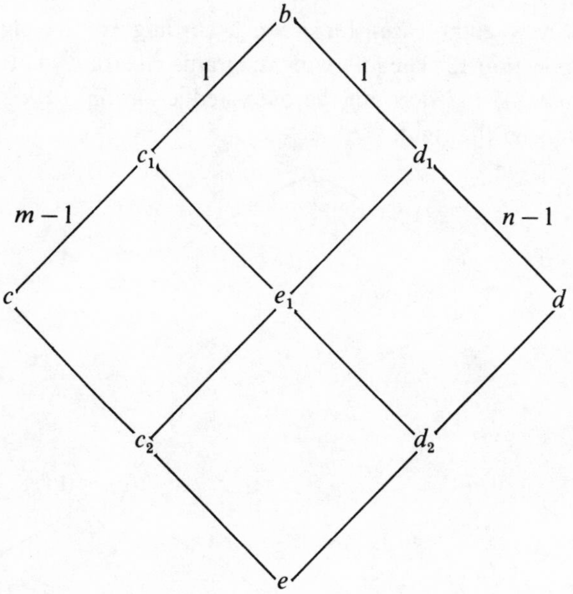

It remains to prove the lemma.

Lemma 13.2. If $b \underset{1}{>} c$ and $b \underset{1}{>} d$ then there is a term e such that $c \geqslant e$ and $d \geqslant e$.

Proof. The reduction of b to c depends on the basic reduction of a subterm a of b to a', and the reduction of b to d depends on the basic reduction of a subterm f of b to f'. If a and f do not overlap, we have

$$b \equiv \ldots a \ldots f \ldots,$$

$$c \equiv \ldots a' \ldots f \ldots,$$

$$d \equiv \ldots a \ldots f' \ldots.$$

If we now take

$$e \equiv \ldots a' \ldots f' \ldots,$$

then clearly $c \geqslant e$ and $d \geqslant e$.

If the subterms a and f of b do overlap, one of them must be a subterm of the other, say f is a subterm of a. Without loss in generality, we may assume that $a \equiv b$. So b reduces in two ways: an 'outer' reduction on the whole term b and an 'inner' reduction on the subterm f. Thus we have achieved a reduction of the problem to the following special case:

If $b \underset{1}{>} c$ (*outer* reduction) and $b \underset{1}{>} d$ (*inner* reduction) then there is a term e such that $c \geqslant e$ and $d \geqslant e$. (Recall that $b \underset{1}{>} c$ means that there is a basic reduction of b to c.)

There are now eight cases for $b \underset{1}{>} c$ according to the eight basic reductions in Section 12. The following diagrams illustrate what we do in these eight cases. We always put the outer reduction on the left and the inner reduction on the right.

B1.

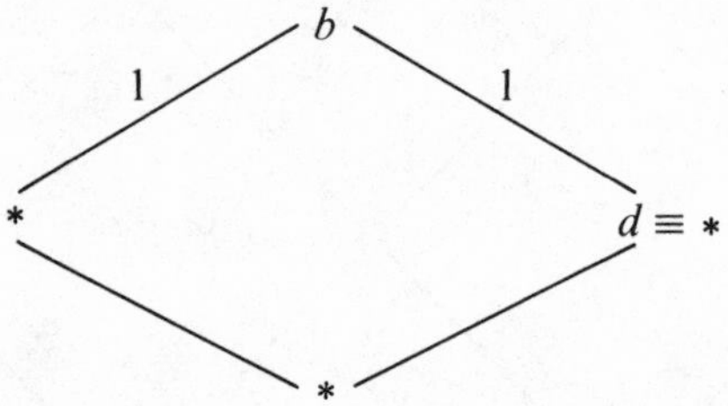

B2.

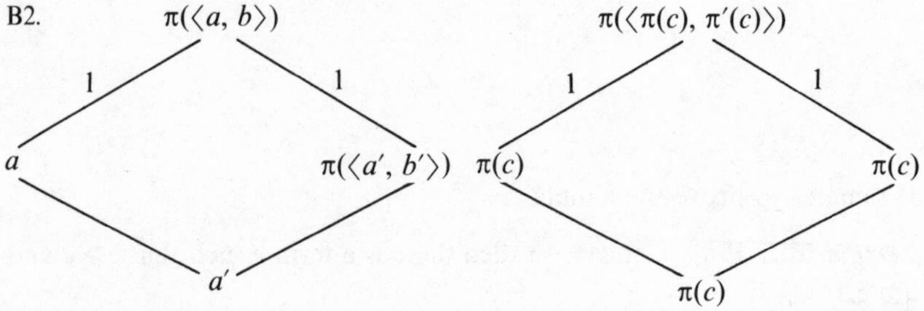

In the first subcase of B2, the inner reduction takes a to a' (whence $b' \equiv b$) or b to b' (whence $a' \equiv a$). In the second subcase of B2, the inner reduction takes $\langle a, b\rangle$ to c, provided $a \equiv \pi(c)$ and $b \equiv \pi'(c)$.

B3. Similar

B4.

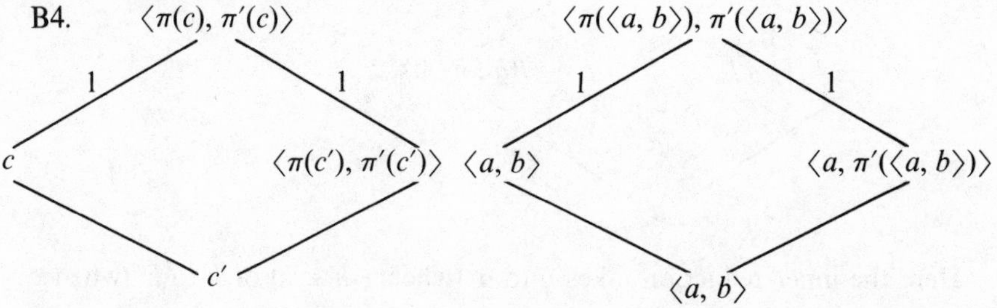

In the first subcase of B4, the inner reduction takes c to c'. In the second subcase of B4, the inner reduction takes $\pi(c)$ to a, provided $c \equiv \langle a, b\rangle$. There is another subcase of B4, not shown, in which the inner reduction takes $\pi'(c)$ to b.

B5.

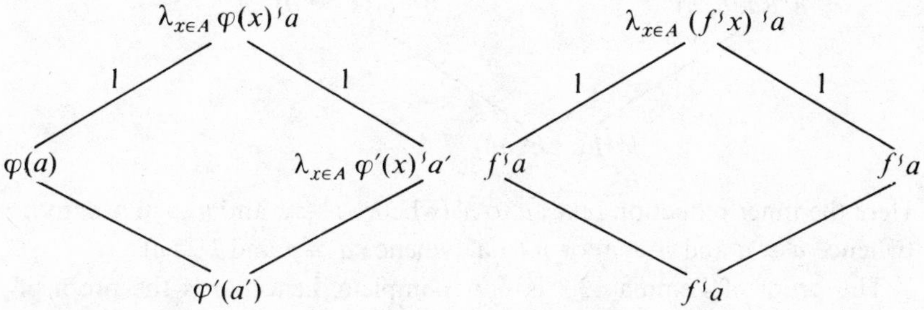

In the first subcase of B5, the inner reduction takes $\varphi(x)$ to $\varphi'(x)$ (whence $a' \equiv a$) or a to a' (whence $\varphi'(x) \equiv \varphi(x)$). Note that if $\varphi(x) \geqslant \varphi'(x)$ then $\varphi(a) \geqslant \varphi'(a)$. In the second subcase of B5, the inner reduction takes $\lambda_{x\in A}\varphi(x)$ to f, provided $\varphi(x) \equiv f\,ʻx$ and x is not free in f.

B6.

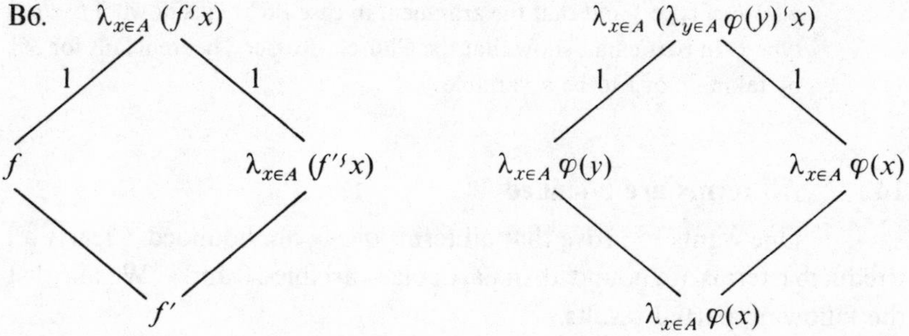

In the first subcase of B6, the inner reduction takes f to f'. In the second subcase of B6, the inner reduction takes $f^{\int} x$ to $\varphi(x)$, provided $f \equiv \lambda_{y \in A}\varphi(y)$.

B7.

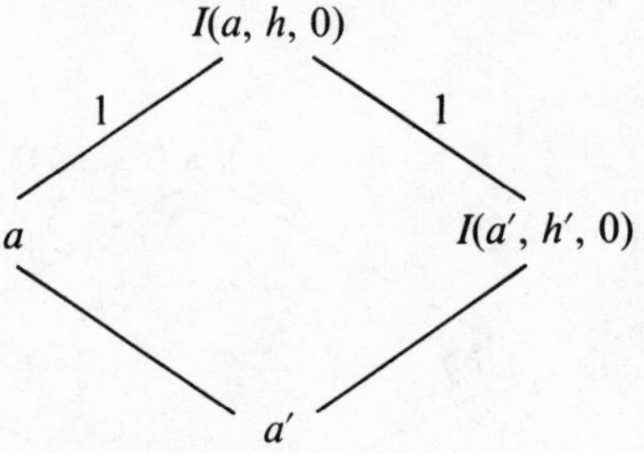

Here the inner reduction takes a to a' (whence $h' \equiv h$) or h to h' (whence $a' \equiv a$).

B8.

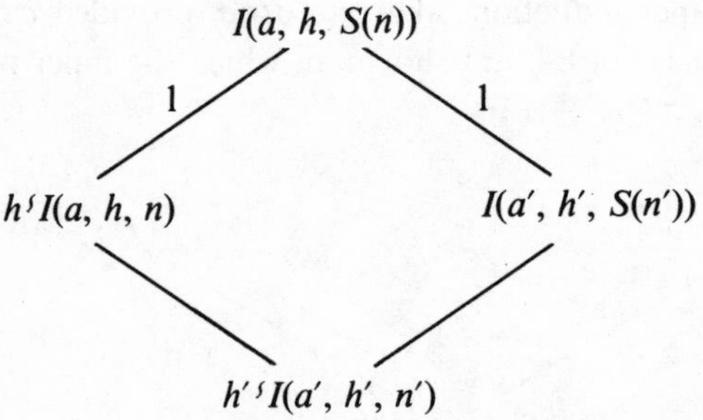

Here the inner reduction takes a to a' (whence $h' \equiv h$ and $n' \equiv n$) or h to h' (whence $a' \equiv a$ and $n' \equiv n$) or n to n' (whence $a' \equiv a$ and $h' \equiv h$).

The proof of Lemma 13.2 is now complete, hence so is the proof of Proposition 13.1.

Exercise

(Obtulowicz). Show that the argument in case B4 breaks down if $\pi(c)$ or $\pi'(c)$ is of type 1 and that the argument in case B6 breaks down if $f^{\int} x$ has type 1. In particular, show that the Church–Rosser Theorem fails for $\mathscr{L}_0$ by taking c or f to be a variable.

14 All terms are bounded

One wants to prove that all terms of $\mathscr{L}_0$ are bounded. Clearly all irreducible terms are bounded, in particular variables, 0 and $*$. We may list the following partial results.

Remark 14.1.

(1) $\langle a, b\rangle$ is bounded if a and b are.

(2) $\pi(c)$ and $\pi'(c)$ are bounded if c is.

(3) $\lambda_{x\in A}\varphi(x)$ is bounded if $\varphi(x)$ is.

(4) $S(n)$ is bounded if n is.

Proof. We shall prove (1), for example, the others being similar. We argue by induction on $\mathrm{bd}(a) + \mathrm{bd}(b)$. Clearly, it suffices to show that c is bounded whenever $\langle a, b\rangle \underset{1}{>} c$. If $c \equiv \langle a', b\rangle$ with $a \underset{1}{>} a'$ or $c \equiv \langle a, b'\rangle$ with $b \underset{1}{>} b'$, c is bounded by inductional assumption. The only other case is $a \equiv \pi(c)$, $b \equiv \pi'(c)$, but then c is bounded because it is a subterm of a.

Unfortunately, this kind of argument does not extend to terms of the form $b^{\int} a$ and $I(a, h, n)$. Note that in the basic reductions

$$\lambda_{x\in A}\varphi(x)^{\int} a > \varphi(a), \quad I(a, h, S(n)) > h^{\int} I(a, h, n)$$

the right hand side may be more complicated than the left hand side.

We shall follow Tait and replace boundedness by an apparently stronger notion, that of 'computability', which is defined by induction on types. We first confine attention to closed terms.

Definition 14.2. A closed term c is *computable* provided one of the following cases holds:

(i) $c \in 1$ or N and c is bounded;

(ii) $c \in A \times B$ and $\pi(c)$ and $\pi'(c)$ are computable;

(iii) $c \in B^A$, c is bounded and $c^{\int} a$ is computable for all computable closed $a \in A$.

Here are two immediate consequences of the definition.

Lemma 14.3. Assume c and c' closed.

(1) If c is computable, then c is bounded.

(2) If c is computable and $c \geqslant c'$, then c' is computable.

Proof. We shall prove (1) and leave (2) as an exercise, as it is never used. The proof of (1) goes by induction on types. We need only look at the case $c \in A \times B$. Since c is computable, so is $\pi(c) \in A$, by Definition 14.2. By inductional assumption, the result holds for A, hence $\pi(c)$ is bounded. Therefore c, being a subterm of $\pi(c)$, is also bounded.

Lemma 14.4. (I) A closed term c is computable if one of the following three cases holds:

(1) $c \equiv \langle a, b\rangle$ and a and b are computable;

(2) $c \equiv \lambda_{x \in A} \varphi(x)$ and $\varphi(a)$ is computable for all computable closed $a \in A$,

(3) c is neither of the above and, for all closed c', if $c \underset{1}{>} c'$ then c' is computable.

(II) For all types C, $\kappa(C)$ is computable.

Proof. For the purpose of this proof only we shall make two definitions. Call a closed term *a pre-computable* if it satisfied (1), (2) or (3) above. Call a type C *nice* if all pre-computable $c \in C$ are computable. We may thus restate (I) as:

(I′) All types are nice.

Before proving this, let us make an observation:

(III) If C and all subtypes of C are nice then $\kappa(C)$ is computable.

'Being a subtype' is of course the transitive relation generated by: A and B are subtypes of $A \times B$ and B^A.

Indeed, (III) is easily shown by induction on the type C. By Definition 14.2, $\kappa(1) \equiv *$ and $\kappa(N) \equiv 0$ are clearly computable, because they are bounded. Moreover, $\kappa(A \times B) \equiv \langle \kappa(A), \kappa(B) \rangle$ is computable by (1) if $\kappa(A)$ and $\kappa(B)$ are, and $\kappa(B^A) \equiv \lambda_{x \in A} \kappa(B)$ is computable by (2) if $\kappa(B)$ is, so that the induction hypothesis applies.

As we are planning to prove (I′), (II) will follow immediately from (III). It remains to prove (I′), which we shall do by induction on types. To this purpose, let us adopt the following assumption:

> **(A)** All proper subtypes of C are nice and the closed term $c \in C$ is pre-computable.

We wish to establish the following conclusion:

> **(C)** c is computable.

We shall look at the cases $C = 1, N, A \times B$ and B^A separately, but first let us note this preliminary conclusion:

> **(P)** c is bounded.

Indeed, by assumption (**A**), c satisfies (1), (2) or (3). In case (1), c is bounded by Remark 14.1 and Lemma 14.3, because a and b are bounded. In case (2), it will follow from Remark 14.1 that c is bounded if we show that $\varphi(x)$ is bounded. Now $\kappa(A)$ is computable by the assumption (**A**) and (III). Therefore, $\varphi(\kappa(A))$ is bounded by (2) and Lemma 14.3, hence $\varphi(x)$ is bounded by Lemma 12.5. In case (3), c is evidently bounded, because, whenever $c \underset{1}{>} c'$, then c' is bounded by Lemma 14.3.

We are now ready to prove the conclusion (**C**). When $C = 1$ or N, computable means bounded, and so we refer to the preliminary conclusion (**P**).

Suppose $C = A \times B$. According to Definition 14.2, we must show that $\pi(c) \in A$ and $\pi'(c) \in B$ are computable, for example, the former. We shall proceed by induction on bd(c). By assumption (**A**), A is nice, so we need only show that $\pi(c)$ is pre-computable. Since $\pi(c)$ is neither a pair nor a λ-term, we only have to check (3).

So suppose $\pi(c) \underset{1}{>} a'$, we must show that a' is computable. There are two cases. If $a' \equiv \pi(c')$ and $c \underset{1}{>} c'$, a' is computable by inductional assumption, since $\text{bd}(c') < \text{bd}(c)$. If $a' \equiv a$ and $c \equiv a$ and $c \equiv \langle a, b \rangle$, a' is computable by (1).

Next, suppose $C = B^A$. According to Definition 14.2, we must show that $c\,{}^{\varsigma}a \in B$ is computable for all computable closed $a \in A$, as we already know that c is bounded by (**P**). We shall proceed by induction on $\text{bd}(c) + \text{bd}(a)$. By assumption B is nice, so we need only show that $c\,{}^{\varsigma}a$ is pre-computable. Since $c\,{}^{\varsigma}a$ is neither a pair nor a λ-term, we only have to check (3).

So suppose $c\,{}^{\varsigma}a \underset{1}{>} b'$, we must show that b' is computable. There are two cases. If $b' \equiv c'\,{}^{\varsigma}a$ with $c \underset{1}{>} c'$ or $b' \equiv c\,{}^{\varsigma}a'$ with $a \underset{1}{>} a'$, b' is computable by inductional assumption, since $\text{bd}(a') < \text{bd}(a)$ and $\text{bd}(c') < \text{bd}(c)$. If $b' \equiv \varphi(a)$ and $c \equiv \lambda_{x \in A}\varphi(x)$, b' is computable by (2).

We have thus established the conclusion (**C**) and the proof of Lemma 14.4 is complete.

Lemma 14.5. If $a \in A$, $h \in A^A$ and $n \in N$ are computable closed terms, then $I(a, h, n)$ is computable.

Proof. We proceed by induction on $\text{bd}(a) + \text{bd}(h) + \text{bd}(n) + \sigma(n)$, where $\sigma(n)$ is the number of occurrences of the symbol S in the normal form of n. (Recall that n computable implies n bounded.) Since $I(a, h, n)$ is neither a pair nor a λ-term, we need only check case (3) of Lemma 14.4.

So suppose $I(a, h, n) \underset{1}{>} d$; we must show that d is computable. There are three cases. If $d \equiv I(a', h, n)$ with $a \underset{1}{>} a'$ or $d \equiv I(a, h', n)$ with $h \underset{1}{>} h'$ or $d \equiv I(a, h, n')$ with $n \underset{1}{>} n'$, d is computable by inductional assumption, since $\text{bd}(a') < \text{bd}(a)$, $\text{bd}(h') < \text{bd}(h)$ and $\text{bd}(n') < \text{bd}(n)$ but $\sigma(n') = \sigma(n)$. If $d \equiv a$ with $n \equiv 0$, d is given to be computable. Finally, if $d \equiv h\,{}^{\varsigma}I(a, h, m)$ with $n \equiv S(m)$, we have $\sigma(m) < \sigma(n)$ and $\text{bd}(m) = \text{bd}(n)$, so $I(a, h, m)$ is computable by inductional assumption. Since h is given to be computable, d is computable by Definition 14.2.

We now extend the notion of computability to open terms.

Definition 14.6. A term $t \equiv \varphi(x_1, \ldots, x_n)$, with no free variables other than $x_1, \ldots, x_n$, is *computable* provided, for all computable closed terms $a_1, \ldots, a_n$ of appropriate types, the closed term $\bar{t} \equiv \varphi(a_1, \ldots, a_n)$ is computable.

Theorem 14.7. All terms of $\mathscr{L}_0$ are computable, hence bounded.

Proof. We proceed by induction on the length of terms. For the constants $*$ and 0 and for all variables the result holds trivially. It remains to prove the following six statements.

(1) If a and b are computable, so is $\langle a, b \rangle$.

Indeed, let $\bar{a}$ and $\bar{b}$ be computable, then so is $\langle \bar{a}, \bar{b} \rangle$, by Lemma 14.4.

(2) If c is computable, then so are $\pi(c)$ and $\pi'(c)$.

Indeed, let $\bar{c}$ be computable, then so are $\pi(\bar{c})$ and $\pi'(\bar{c})$ by Definition 14.2.

(3) If $f \in B^A$ and $a \in A$ are computable, then so is $f{}^{\varsigma} a$.

Indeed, let $\bar{f}$ and $\bar{a}$ be computable, then so is $\bar{f}{}^{\varsigma}\bar{a}$, by Definition 14.2 and Lemma 14.3.

(4) If $\varphi(x, x_1, \ldots, x_n)$ is computable, so is $\lambda_{x \in A}\varphi(x, x_1, \ldots, x_n)$.

Indeed, let $\bar{\varphi}(x) \equiv \varphi(x, a_1, \ldots, a_n)$ for computable closed $a_1, \ldots, a_n$ and assume that $\bar{\varphi}(a) \in B$ is computable for all computable closed $a \in A$. Then $\lambda_{x \in A}\bar{\varphi}(x)$ is computable, by Lemma 14.4.

(5) If $n \in N$ is computable, so is $S(n)$.

Indeed, let $\bar{n}$ be computable, that is, bounded. Then so is $S(\bar{n})$.

(6) If $a \in A$, $h \in A^A$ and $n \in N$ are computable, then so is $I(a, h, n)$.

Indeed, let $\bar{a}, \bar{h}$ and $\bar{n}$ be computable. Then so is $I(\bar{a}, \bar{h}, \bar{n})$, by Lemma 14.5.

The first person to prove Theorem 14.7 in essentially the generality given here was R.C. de Vrijer. Our proof is closer to Tait's original proof, but depends crucially on an idea of de Vrijer, which is here embedded in condition (3) of Lemma 14.4 and the use that is made of it.

Exercises

1. Prove (2) to (4) of Remark 14.1.
2. Prove (2) of Lemma 14.3.

15 C-monoids

A small category with one object is a *monoid*, that is, a semigroup with a unity element. (See Part 0, Section 1, Example C2′.) If a small cartesian closed category has only one object, it is a rather uninteresting monoid. For, if 1 is the terminal object, Hom(1, 1) has only one element. However, if we delete the terminal object, we obtain an interesting class of monoids.

A *C-monoid* (C for Curry, Church, combinatory or cartesian) is a monoid $\mathscr{M}$ with extra structure $(\pi, \pi', \varepsilon, *, \langle\,\rangle)$, where π, π', and ε are elements of $\mathscr{M}$ (nullary operations), $(-)^*$ is a unary operation and $\langle -,- \rangle$ is a binary operation satisfying the following identities:

C1. $\pi\langle a, b\rangle = a$,

C2. $\pi'\langle a, b\rangle = b$,

C3. $\langle \pi c, \pi' c\rangle = c$,

C4. $\varepsilon\langle h^*\pi, \pi'\rangle = h$,

C5. $(\varepsilon\langle k\pi, \pi'\rangle)^* = k$,

for all a, b, c, h and k. These are the axioms of a cartesian closed category without terminal object, with the type subscripts erased.

We list some easy consequences of the above definition:

C3a. $\langle a, b\rangle c = \langle ac, bc\rangle$,

C3b. $\langle \pi, \pi'\rangle = 1$,

C4a. $\varepsilon\langle h^* a, b\rangle = h\langle a, b\rangle$,

C5a. $h^* k = (h\langle k\pi, \pi'\rangle)^*$,

C5b. $\varepsilon^* = 1$,

for all a, b, c, h and k.

Proof.

$$\begin{aligned}
\langle a, b\rangle c &= \langle \pi\langle a, b\rangle c, \pi'\langle a, b\rangle c\rangle && \text{by C3,}\\
&= \langle ac, bc\rangle && \text{by C3.}\\
\langle \pi, \pi'\rangle &= \langle \pi 1, \pi' 1\rangle = 1 && \text{by C3.}\\
\varepsilon\langle h^* a, b\rangle &= \varepsilon\langle h^*\pi\langle a, b\rangle, \pi'\langle a, b\rangle\rangle && \text{by C1 and C2,}\\
&= \varepsilon\langle h^*\pi, \pi'\rangle\langle a, b\rangle && \text{by C3a,}\\
&= h\langle a, b\rangle && \text{by C4.}\\
h^* k &= (\varepsilon\langle h^* k\pi, \pi'\rangle)^* && \text{by C5,}\\
&= (\varepsilon\langle h^*\pi\langle k\pi, \pi'\rangle, \pi'\langle k\pi, \pi'\rangle\rangle)^* && \text{by C1 and C2,}
\end{aligned}$$

$$= (\varepsilon\langle h^*\pi, \pi'\rangle\langle k\pi, \pi'\rangle)^* \qquad \text{by C3a,}$$
$$= (h\langle k\pi, \pi'\rangle)^* \qquad \text{by C4.}$$
$$\varepsilon^* = (\varepsilon\langle \pi, \pi'\rangle)^* \qquad \text{by C3b,}$$
$$= (\varepsilon\langle 1\pi, \pi'\rangle)^* = 1 \qquad \text{by C5.}$$

Definition 15.1. In any C-monoid we may write

$$a \times b = \langle a\pi, b\pi'\rangle, \quad g^f = (g\varepsilon\langle \pi, f\pi'\rangle)^*$$

for all elements a, b, f, g.

Of course, this definition is motivated by the corresponding equations in a cartesian closed category. The following consequences of this definition are left as an exercise to the reader.

C6. $(a \times b)(c \times d) = ac \times bd$,

C7. $g^f h = (g\varepsilon\langle h\pi, f\pi'\rangle)^*$,

C8. $g^f k^h = (gk)^{(hf)}$.

C-monoids are the objects of a category whose arrows are *C-homomorphisms*, that is, mappings which preserve the operations π, π', ε, * and $\langle\rangle$.

Given a C-monoid $\mathscr{A}$, we may form the *polynomial* C-monoid $\mathscr{A}[x]$ by the usual construction of universal algebra: the elements of $\mathscr{A}[x]$ are polynomials, that is, words built up from x and the elements of $\mathscr{A}$ using the C-monoid operations modulo the smallest congruence relation which satisfies C1 to C5 and which assures that the mapping $h: \mathscr{A} \to \mathscr{A}[x]$ which sends every element of $\mathscr{A}$ onto the corresponding constant polynomial is a C-homomorphism. In particular, if $\langle a, b\rangle = c$ in $\mathscr{A}$ then also $\langle a, b\rangle = c$ in $\mathscr{A}[x]$, etc.

The canonical C-homomorphism $h: \mathscr{A} \to \mathscr{A}[x]$ has the usual universal property: for every C-homomorphism $f: \mathscr{A} \to \mathscr{B}$ and every element $b \in \mathscr{B}$, there exists a unique C-homomorphism $f_b: \mathscr{A}[x] \to \mathscr{B}$ such that $f_b h = f$ and $f_b(x) = b$.

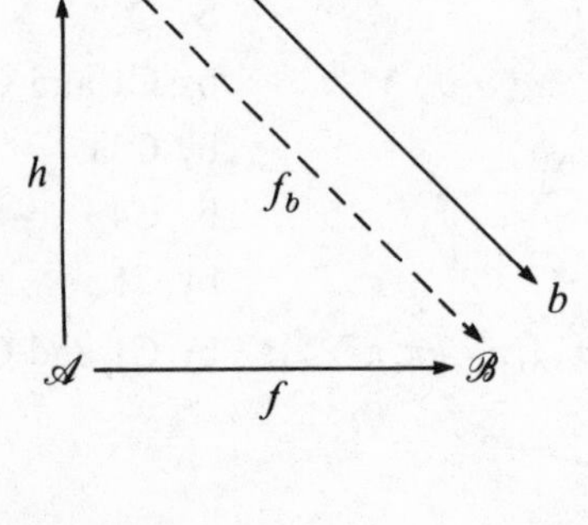

In particular, when $\mathscr{B} = \mathscr{A}$ and $f = 1_{\mathscr{A}}$, $f_b(\varphi(x)) = \varphi(b)$.

C-monoids have the important property of *functional completeness*:

Theorem 15.2. If $\varphi(x)$ is a polynomial in the indeterminate x over a C-monoid $\mathscr{A}$, there exists a unique constant f in $\mathscr{A}$ such that $f\langle(x\pi')^*, 1\rangle = \varphi(x)$ in $\mathscr{A}[x]$.

Proof. Define $\rho_x\varphi(x)$ by induction on the length of the word $\varphi(x)$:

(i) $\rho_x k \equiv k\pi'$ if $k \in \mathscr{A}$;

(ii) $\rho_x x \equiv \varepsilon$;

(iii) $\rho_x\langle\psi(x), \chi(x)\rangle \equiv \langle\rho_x\psi(x), \rho_x\chi(x)\rangle$;

(iv) $\rho_x(\chi(x)\psi(x)) \equiv \rho_x\chi(x)\langle\pi, \rho_x\psi(x)\rangle$;

(v) $\rho_x(\psi(x)^*) \equiv (\rho_x\psi(x)\alpha)^*$,

where $\alpha \equiv \langle\pi\pi, \langle\pi'\pi, \pi'\rangle\rangle$. (We have written $\equiv$ for identity of words, while $=$ is reserved for equality of polynomials, that is, equivalence classes of words. See also the exercise at the end of Section 2.)

We next show that ρ_x really applies to polynomials, that is,

(*) if $\varphi(x) = \psi(x)$ then $\rho_x\varphi(x) = \rho_x\psi(x)$.

Now $=$ is the smallest congruence relation between words in x satisfying C1 to C5 and the conditions assuring that h is a homomorphism. Therefore it suffices to show that for each of these $\rho_x\text{LHS} = \rho_x\text{RHS}$, e.g., for C4 that

$$\rho_x(\varepsilon\langle\chi(x)^*\pi, \pi'\rangle) = \rho_x\chi(x).$$

This is shown by routine calculations, which are left as an exercise to the reader. Note also that, if $\langle a, b\rangle = c$ in $\mathscr{A}$, then $\rho_x\langle a, b\rangle = \langle\rho_x a, \rho_x b\rangle = \langle a\pi', b\pi'\rangle = \langle a, b\rangle\pi' = c\pi' = \rho_x c$. Equally routine is the verification that

$$\rho_x\varphi(x)\langle(x\pi')^*, 1\rangle = \varphi(x),$$

so that the existence part of the theorem has been established.

Finally, to show the uniqueness of f in the theorem, we suppose that $f\langle(x\pi')^*, 1\rangle = \varphi(x)$ and wish to show that $\rho_x\varphi(x) = f$. Indeed, a straightforward calculation gives

$$\begin{aligned}
\rho_x(f\langle(x\pi')^*, 1\rangle) &= f\rho_x\langle(x\pi')^*, 1\rangle && \text{by (iv) and (i),}\\
&= f\langle\rho_x(x\pi')^*, \rho_x 1\rangle && \text{by (iii),}\\
&= f\langle(\rho_x(x\pi')\alpha)^*, \pi'\rangle && \text{by (v) and (ii),}\\
&= f\langle(\rho_x x\langle\pi, \rho_x\pi'\rangle\alpha)^*, \pi'\rangle && \text{by (iv),}\\
&= f\langle(\varepsilon\langle\pi, \pi'\pi'\rangle\alpha)^*, \pi'\rangle && \text{by (ii) and (i),}\\
&= f\langle(\varepsilon\langle\pi\alpha, \pi'\pi'\alpha\rangle)^*, \pi'\rangle && \text{by C3a,}\\
&= f\langle(\varepsilon\langle\pi\pi, \pi'\rangle)^*, \pi'\rangle && \text{by definition of } \alpha,
\end{aligned}$$

$$= f\langle \pi, \pi' \rangle \qquad \text{by C5,}$$
$$= f \qquad \text{by C3b.}$$

This completes the proof of the theorem.

If we now define

$$g{}^{\varsigma}a \equiv \varepsilon\langle g(a\pi')^*, 1 \rangle,$$

we have the following immediate consequence:

Corollary 15.3. If $\varphi(x)$ is a polynomial in the indeterminate x over a C-monoid $\mathcal{M}$, then there exists a unique $g \in \mathcal{M}$ such that $g{}^{\varsigma}x = \varphi(x)$.

Proof. Take $g = f^* = (\rho_x \varphi(x))^*$ and use C4a.

It is suggestive to write

$$\lambda_x \varphi(x) \equiv (\rho_x \varphi(x))^*,$$

then the corollary may be expressed by the equations:

$$\lambda_x \varphi(x){}^{\varsigma}x = \varphi(x), \quad \lambda_x(g{}^{\varsigma}x) = g.$$

Of course, the universal property of $\mathcal{M}[x]$ allows us to obtain from the first of these that, for all $a \in \mathcal{M}$,

$$\lambda_x \varphi(x){}^{\varsigma}a = \varphi(a).$$

Let us also mention the following 'fixed point theorem' for C-monoids, which is behind Russell's paradox.

Proposition 15.4. For every polynomial $\varphi(x)$ in $\mathcal{M}[x]$ there exists an element $a \in \mathcal{M}$ such that $\varphi(a) = a$.

Proof. Put $b \equiv \lambda_x \varphi(x{}^{\varsigma}x)$ and $a \equiv b{}^{\varsigma}b$, then

$$\varphi(a) \equiv \varphi(b{}^{\varsigma}b) = b{}^{\varsigma}b \equiv a.$$

This result explains why C-monoids cannot incorporate propositional logic. For, if we had an operation $\neg$ of negation, if would follow that $\neg p = p$ for some element p.

For future references we shall list some further derived identities of C-monoids:

C9 $\quad (a\pi')^*k = (a\pi')^*$;

C10. $\quad (fg){}^{\varsigma}a = f{}^{\varsigma}(g{}^{\varsigma}a)$;

C11. $\quad \lambda_x(f{}^{\varsigma}a) = f\lambda_x a$, if x is not free in $f{}^{\varsigma}a$;

C12. $\quad 1{}^{\varsigma}a = a$;

C13. $\quad (\lambda_x a)k = \lambda_x a$, if x is not free in a.

Proof.

$$\begin{aligned}
(a\pi')^*k &= (a\pi'\langle k\pi, \pi'\rangle)^* && \text{by C5a,}\\
&= (a\pi')^* && \text{by C2;}\\
(fg)^{\int} a &= \varepsilon\langle fg(a\pi')^*, 1\rangle, &&\\
&= \varepsilon\langle f(\varepsilon\langle g(a\pi')^*\pi, \pi'\rangle)^*, 1\rangle && \text{by C5,}\\
&= \varepsilon\langle f(\varepsilon\langle g(a\pi')^*\pi', \pi'\rangle)^*, 1\rangle && \text{by C9,}\\
&= \varepsilon\langle f(\varepsilon\langle g(a\pi')^*, 1\rangle\pi')^*, 1\rangle && \text{by C3a,}\\
&= \varepsilon\langle f((g^{\int} a)\pi')^*, 1\rangle &&\\
&= f^{\int}(g^{\int} a); &&\\
(\lambda_x(f^{\int} a))^{\int} x &= f^{\int} a, &&\\
&= f^{\int}((\lambda_x a)^{\int} x), &&\\
&= (f\lambda_x a)^{\int} x && \text{by C10;}\\
1^{\int} a = \varepsilon\langle (a\pi')^*, 1\rangle &= \varepsilon\langle (a\pi')^*\pi, \pi'\rangle\langle 1, 1\rangle &&\\
&= a\pi'\langle 1, 1\rangle = a; &&\\
((\lambda_x a)k)^{\int} x &= (\lambda_x a)^{\int}(k^{\int} x) = a = (\lambda_x a)^{\int} x. &&
\end{aligned}$$

Exercises

1. Prove C6, C7 and C8.
2. Define a *weak C-monoid* like a C-monoid, except that C3, C4 and C5 are replaced by C3a, C4a and C5a respectively. Check that C6 to C8 hold in a weak C-monoid.
3. Fill in the details in the proof of Theorem 15.2. Check that the existence (but not the uniqueness) of f holds in any weak C-monoid.
4. How unique is the C-structure of a monoid? If σ is any invertible element of the C-monoid $\mathscr{M}$, one can define a new pairing $\sigma\langle a, b\rangle$ with new projections $\pi\sigma^{-1}$ and $\pi'\sigma^{-1}$. Show that all possible pairings in $\mathscr{M}$ are obtained in this way.
5. Even if the operations $\langle -, -\rangle$, π and π' of a C-monoid are laid down, ε and $(-)^*$ may be changed: if τ is any invertible element, one may replace ε by $\varepsilon\langle\tau\pi, \pi'\rangle$ and h^* by $\tau^{-1}h^*$. Show that this is the only way in which ε and $(-)^*$ may be changed.
6. If $\varphi(x, y) \in \mathscr{A}[x, y] \cong \mathscr{A}[x][y]$ is a polynomial in two indeterminates over the C-monoid $\mathscr{A}$, show that there is a unique element g of $\mathscr{A}$ such that $\varphi(x, y) = g^{\int}\langle x, y\rangle$.

16 C-monoids and cartesian closed categories

A C-monoid looks like a cartesian closed category, but lacks a terminal object. Essentially, this is all it lacks, as we shall see.

An element A of a monoid is called an *idempotent* if $AA = A$. For reasons that will become clear presently, we denote idempotents by capital letters. If A and B are idempotent elements of a C-monoid, so are $A \times B$ and B^A, as follows immediately from C6 and C8 of Section 15 respectively. We also note that the elements π'^* and 1 are idempotents, the former by C9, (We shall denote them by capital letters later too).

Following Dana Scott, we shall write $f: A \to B$ for $BfA = f$ or, equivalently, $Bf = f$ and $fA = f$. One may note at this stage:

C14. If $f: A \to B$ and $g: C \to D$ then $f \times g: A \times C \to B \times D$ and $f^g: A^D \to B^C$.

But this will become evident later, once we have established the following:

Theorem 16.1. (Dana Scott). The idempotents of a C-monoid $\mathscr{M}$ are the objects of a cartesian closed category $K(\mathscr{M})$ with arrows $f: A \to B$ given by elements f of $\mathscr{M}$ such that $BfA = f$.

Proof. We shall denote an arrow $A \to B$ by a triple (A, f, B). Then it is easily seen that $K(\mathscr{M})$ is a category with identities (A, A, A) and composition $(B, g, C)\,(A, f, B) = (A, gf, C)$. Actually, $K(\mathscr{M})$ is a special case of a category $K(\mathscr{A})$ associated with any category $\mathscr{A}$, called the 'Karoubi envelope' or 'idempotent splitting' envelope of $\mathscr{A}$. Properties of this general construction will be considered in the exercises at the end of this section.

To obtain a cartesian closed structure for $K(\mathscr{M})$ we define $\bigcirc_A$: $A \to \pi'^*$, $\pi_{A,B}: A \times B \to A$, $\pi'_{A,B}: A \times B \to B$, $\varepsilon_{B,A}: B^A \times A \to A$ by

$$\bigcirc_A = (A, \pi'^*, \pi'^*),$$
$$\pi_{A,B} = (A \times B, A\pi(A \times B), A),$$
$$\pi'_{A,B} = (A \times B, B\pi'(A \times B), B),$$
$$\varepsilon_{B,A} = (B^A \times A, B\varepsilon(B^A \times A), B),$$

and verify that

$$\frac{f: C \to A \quad g: C \to B}{\langle f, g \rangle: C \to A \times B},$$

$$\frac{h: A \times B \to C}{h^*: A \to C^B}$$

The equations of a cartesian closed category (Section 3, E2 to E4b) are now easily checked.

Note that the terminal object of $K(\mathcal{M})$ is π'^*, not 1. To avoid confusion, we had better write T and U in place of π'^* and 1 respectively when they are regarded as objects.

Corollary 16.2. The full subcategory $K_0(\mathcal{M})$ of $K(\mathcal{M})$ consisting of all objects isomorphic to T or U is a cartesian closed category.

Proof. First note that any full subcategory of a cartesian closed category containing T and closed under $\times$ and exponentiation is also a cartesian closed category. In the present situation one easily verifies that

$$U \times U = U, \quad U^U = U,$$

using Definition 15.1, and one recalls that

$$A \times T \cong A \cong T \times A, \quad A^T \cong A, \quad T^A \cong T,$$

for any object A in a cartesian closed category, in particular, when $A \cong T$ or $A \cong U$.

We remark that $K_0(\mathcal{M})$ has at most two non-isomorphic objects. (When can one manufacture a cartesian closed category with exactly two objects out of $\mathcal{M}$ in such a way that $\mathcal{M}$ can be recaptured? We do not know.)

Proposition 16.3. Let $\mathcal{A}$ be any locally small cartesian closed category with an object U such that $U \times U \cong U$ and $U^U \cong U$, then End(U) is a C-monoid. In case $\mathcal{A}$ is $K(\mathcal{M})$ or $K_0(\mathcal{M})$, End$(U) \cong \mathcal{M}$.

Proof. Let $\sigma\colon U \times U \to U$ and $\tau\colon U^U \to U$ be the given isomorphisms. We shall endow End(U) with the structure of a C-monoid $(\pi, \pi', \varepsilon, \dagger, \{\})$, it being temporarily necessary to distinguish $\dagger$ from $*$ and $\{\}$ from $\langle\rangle$. We define

$$\begin{aligned}
&\pi = \pi_{U,U}\sigma^{-1},\\
&\pi' = \pi'_{U,U}\sigma^{-1},\\
&\varepsilon = \varepsilon_{U,U}(\tau^{-1} \times 1_U)\sigma^{-1},\\
&\{a, b\} = \sigma\langle a, b\rangle,\\
&h\dagger = \tau(h\sigma)^*.
\end{aligned}$$

It is now easy to check that the equations C1 to C5 of a C-monoid are satisfied. Here, for example, is a proof of C4:

$$\begin{aligned}
\varepsilon\{h\dagger\pi, \pi'\} &= \varepsilon_{U,U}(\tau^{-1} \times 1_U)\sigma^{-1}\sigma\langle\tau(h\sigma)^*\pi_{U,U}\sigma^{-1}, \pi'_{U,U}\sigma^{-1}\rangle\\
&= \varepsilon_{U,U}\langle(h\sigma)^*\pi_{U,U}, \pi'_{U,U}\rangle\sigma^{-1}\\
&= (h\sigma)\sigma^{-1} = h.
\end{aligned}$$

The argument uses the following easily proved equation in a cartesian closed category:

$$(a \times b)\langle c, d\rangle = \langle ac, bd\rangle,$$

as well as the equations (3.1) and E4a of Section 3.

Now suppose $\mathscr{A} = K(\mathscr{M})$ or $K_0(\mathscr{M})$. Then $U = 1$, $U \times U = U$ and $U^U = U$, so $\sigma = 1_U = (U, U, U)$ and $\tau = 1_U$ likewise. The isomorphism $\mathscr{M} \simeq \mathrm{End}(U)$ we seek is clearly given by $f \mapsto (U, f, U)$. For example, the element π of $\mathrm{End}(U)$ is defined as

$$\begin{aligned}\pi_{U,U}\sigma^{-1} &= (U \times U, U\pi(U \times U), U)(U, U, U)^{-1}\\ &= (U, \pi, U),\end{aligned}$$

which is the image of the element π of $\mathscr{M}$ in $\mathrm{End}(U)$.

Proposition 16.3 is useful in providing examples of C-monoids, while Theorem 16.2 tells us that all C-monoids must be of this form. So far, we have not even seen a single non-trivial C-monoid, but we shall find one in Section 18.

Exercises.

1. If $f^2 = f: A \to A$ is an idempotent arrow in a category $\mathscr{A}$, show that the following statements are equivalent:
   (i) f *splits*, that is, there exist $m_f: B_f \to A$, the *image* of f, and $e_f: A \to B_f$, the *co-image* of f, such that $m_f e_f = f$ and $e_f m_f = 1_B$.
   (ii) The pair of arrows $f, 1_A: A \rightrightarrows A$ has an equalizer $m_f: B_f \to A$.

2. Given any category $\mathscr{A}$, form the category $K(\mathscr{A})$, the *Karoubi envelope* of $\mathscr{A}$, as follows: its objects are idempotent arrows of $\mathscr{A}$ and its arrows $f \to g$, where $f^2 = f: A \to A$ and $g^2 = g: B \to B$, are triples (f, φ, g), where $\varphi: A \to B$ is such that $\varphi f = \varphi = g\varphi$ or, equivalently, $g\varphi f = \varphi$. Show that $K(\mathscr{A})$ is a category in which all idempotents split.

3. Suppose $\mathscr{M}$ is a monoid (considered as a category with one object), show the following:
   (*a*) In $K(\mathscr{M})$ every object is a retract of 1.
   (*b*) $K(\mathscr{M})$ is cartesian closed if and only if $\mathscr{M}$ is a weak C-monoid.

4. Consider the category **Split**: its objects are small categories in which to each idempotent arrow $f^2 = f: A \to A$ there is associated a given splitting $A \xrightarrow{e_f} B_f \xrightarrow{m_f} A$ such that the splitting of $1_A: A \to A$ is $A \xrightarrow{1_A} A \xrightarrow{1_A} A$; its arrows are functors which preserve splittings. Show that the forgetful functor from **Split** to **Cat** has a left adjoint.

17 C-monoids and untyped λ-calculus

The untyped λ-calculi we shall study here are an extension of the (untyped) $\lambda\eta$-calculus in the literature. The absence of types assures that application is unrestricted.

Definition 17.1. An *extended λ-calculus* is given by a set of terms and equations which are said to 'hold'. The set of terms is freely generated from countably many variables and a possibly empty set of constants by certain term forming operations including the following: if c is a term so are $\pi(c)$ and $\pi'(c)$; if a and b are terms so are $b^{\backprime}a$ and (a, b); if $\varphi(x)$ is a term, possibly containing the variable x freely, then $\lambda_x\varphi(x)$ is a term in which all occurrences of x are bound. The binary relation between terms a and b which says that the equation $a = b$ holds is an equivalence relation which satisfies the usual rules* allowing substitution of equals for equals, including

$$\frac{\varphi(x) = \psi(x)}{\lambda_x\varphi(x) = \lambda_x\psi(x)}, \frac{a = b}{f^{\backprime}a = f^{\backprime}b},$$

from which the other substitution rules follow, and the following rule allowing substitution of terms for free variables:

L0. $$\frac{\varphi(x) = \psi(x)}{\varphi(a) = \psi(a)},$$

provided no free occurrence of a variable in a becomes bound in $\varphi(a)$ or $\psi(a)$. Finally, the following specific equations are postulated, for a, b, c, f and $\varphi(x)$:

L1. $\lambda_x\varphi(x)^{\backprime}x = \varphi(x)$;

L2. $\lambda_x(f^{\backprime}x) = f$, if x is not free in f;

L3. $\pi((a, b)) = a$;

L4. $\pi'((a, b)) = b$;

L5. $(\pi(c), \pi'(c)) = c$.

Note that from L1 and L0 one obtains

L1′. $\lambda_x\varphi(x)^{\backprime}a = \varphi(a)$,

provided no free occurrence of a variable in a becomes bound in $\varphi(a)$.

A *$\lambda\eta$-calculus* is defined similarly, but without the term forming

* As usual, rules are interpreted as saying: if the hypotheses hold, so does the conclusion.

operations $\pi(-)$, $\pi'(-)$ and $(-,-)$, hence without L3, L4 and L5. Some authors furthermore drop L2, calling the resulting system a *λ-calculus.*

The *pure* extended λ-calculus (λη-calculus, λ-calculus) is such a system in which there are no terms other than those defined inductively and in which the relation which says that equality holds is the smallest equivalence relation having all the required properties.

Remark 17.2. It is known that in any λ-calculus one can define term forming operations $\pi(-)$, $\pi'(-)$ and $(-,-)$ to satisfy L3 and L4 but not L5. For example, one may put:

$$\pi(c) = c^{\int} i, \quad \text{where } i = \lambda_x \lambda_y x,$$

$$\pi'(c) = c^{\int} j, \quad \text{where } j = \lambda_x \lambda_y y,$$

$$(a, b) = \lambda_z((z^{\int} a)^{\int} b).$$

Indeed, we may then calculate

$$\pi((a,b)) = (a,b)^{\int} i = (i^{\int} a)^{\int} b = a, \quad \pi'((a,b)) = (a,b)^{\int} j = (j^{\int} a)^{\int} b = b.$$

Unfortunately, it does not follow from these definitions that $(\pi(c), \pi'(c)) = c$. In fact, no definition of such a 'surjective pairing' is possible. (See Barendregt's book, Exercise 15.4.4.)

Proposition 17.3. Every C-monoid $\mathscr{M}$ gives rise to an extended λ-calculus $L(\mathscr{M})$: the terms of $L(\mathscr{M})$ in the free variables $x_1, \ldots, x_n$ are words constructed from elements of $\mathscr{M}$ and the indeterminates $x_1, \ldots, x_n$ by the following operations:

$$\pi(c) \equiv \pi c, \qquad \pi'(c) \equiv \pi' c,$$

$$(a, b) \equiv \langle a, b \rangle,$$

$$f^{\int} a \equiv \varepsilon \langle f(a\pi')^*, 1 \rangle,$$

$$\lambda_x \varphi(x) \equiv (\rho_x \varphi(x))^*.$$

Finally, $a = b$ holds in $L(\mathscr{M})$ if $a = b$ in the polynomial C-monoid $\mathscr{M}[X]$, where X contains all the variables occurring freely in a and b.

Compare this with Example 10.6. For the last two definitions see Section 15. Alternative definitions will be discussed later. Note that the equality relation in $\mathscr{M}[x_1, \ldots, x_{n+1}]$ is a faithful extension of that in $\mathscr{M}[x_1, \ldots, x_n]$. Indeed, if we suppose that $f = g$ in $\mathscr{M}[X][y]$ and substitute 1 for y we obtain $f = g$ in $\mathscr{M}[X]$.

L1 and L2 were already established in Corollary 15.3. L3 to L5 are easily shown to hold:

$$\pi((a, b)) = \pi \langle a, b \rangle = a, \qquad (\pi(c), \pi'(c)) = \langle \pi c, \pi' c \rangle = c.$$

Proposition 17.4. Every extended λ-calculus $\mathscr{L}$ gives rise to a C-monoid $M(\mathscr{L})$, whose elements are (equivalence classes of) closed terms of $\mathscr{L}$ and where

$$\begin{aligned} 1 &\equiv \lambda_x x, \\ gf &\equiv \lambda_x(g^{\ulcorner}(f^{\ulcorner}x)), \\ \pi &\equiv \lambda_x \pi(x), \\ \pi' &\equiv \lambda_x \pi'(x), \\ \langle f, g\rangle &\equiv \lambda_x(f^{\ulcorner}x, g^{\ulcorner}x), \\ \varepsilon &\equiv \lambda_z(\pi(z)^{\ulcorner}\pi'(z)), \\ h^* &\equiv \lambda_x\lambda_y(h^{\ulcorner}(x, y)). \end{aligned}$$

Moreover, $a = b$ in $M(\mathscr{L})$ if $a = b$ holds in $\mathscr{L}$.

Proof. The equations of a C-monoid are easily checked. For example,

$$\begin{aligned} (\varepsilon\langle h^*\pi, \pi'\rangle)^{\ulcorner}x &= \varepsilon^{\ulcorner}(\langle h^*\pi, \pi'\rangle^{\ulcorner}x) \\ &= \varepsilon^{\ulcorner}((h^*\pi)^{\ulcorner}x, \pi'(x)) \\ &= ((h^*\pi)^{\ulcorner}x)^{\ulcorner}(\pi'^{\ulcorner}x) \\ &= (h^{*\ulcorner}\pi(x))^{\ulcorner}(\pi'(x)) \\ &= h^{\ulcorner}(\pi(x), \pi'(x)) \\ &= h^{\ulcorner}x, \end{aligned}$$

hence $\varepsilon\langle h^*\pi, \pi'\rangle = h$. The other equations are left as exercises.

The correspondences $\mathscr{M} \mapsto L(\mathscr{M})$ and $\mathscr{L} \mapsto M(\mathscr{L})$ discussed in Propositions 17.3 and 17.4 may be extended to functors between appropriate categories. Unfortunately, it is not obvious that these give an equivalence of categories. The reason for this is that in Proposition 17.3 the definitions of $\pi(-)$, $\pi'(-)$ and $(-,-)$ were badly chosen.

Let us change the term forming operations $\pi(-)$, $\pi'(-)$ and $(-,-)$ in Proposition 17.3 as follows:

$$\begin{aligned} \pi(c) &\equiv \pi^{\ulcorner}c, \\ \pi'(c) &\equiv \pi'^{\ulcorner}c, \\ (a, b) &\equiv \langle \lambda_x a, \lambda_x b\rangle^{\ulcorner}1, \end{aligned}$$

where it is assumed that x is not free in a or b. We now calculate:

$$\begin{aligned} \pi((a, b)) &= \pi^{\ulcorner}(\langle \lambda_x a, \lambda_x b\rangle^{\ulcorner}1), \\ &= (\pi\langle \lambda_x a, \lambda_x b\rangle)^{\ulcorner}1 && \text{by C10}, \\ &= (\lambda_x a)^{\ulcorner}1 && \text{by C1}, \\ &= a; \end{aligned}$$

$$(\pi(c), \pi'(c)) = \langle \lambda_x(\pi^\varsigma c), \lambda_x(\pi'^\varsigma c)\rangle^\varsigma 1,$$
$$= \langle \pi\lambda_x c, \pi'\lambda_x c\rangle^\varsigma 1 \quad \text{by C11},$$
$$= (\lambda_x c)^\varsigma 1 \qquad = c.$$

In the following, M is taken as in Proposition 17.4 and L as in the revised form of Proposition 17.3. Moreover, we now make the convention that terms a and b of $\mathscr{L}$ are identified if $a = b$ holds in $\mathscr{L}$.

Theorem 17.5. M and L establish a one-to-one correspondence between C-monoids $\mathscr{M}$ and extended λ-calculi $\mathscr{L}$:

$$ML(\mathscr{M}) = \mathscr{M}, \quad LM(\mathscr{L}) = \mathscr{L}.$$

Proof. (*a*) The elements of $ML(\mathscr{M})$ are (equivalence classes of) closed terms of $L(\mathscr{M})$, hence the same as the elements of $\mathscr{M}$. Let us provisionally distinguish operations in $ML(\mathscr{M})$ from those in $\mathscr{M}$ by a subscript #. Then

$$1_\# = \lambda_x x = 1, \qquad \text{by C12 and Corollary 15.3},$$
$$(fg)_\# = \lambda_x(f^\varsigma(g^\varsigma x)),$$
$$= \lambda_x((fg)^\varsigma x) \qquad \text{by C10},$$
$$= fg;$$
$$\pi_\# = \lambda_x \pi(x) = \lambda_x(\pi^\varsigma x) = \pi.$$

Similarly $\pi'_\# = \pi'$. We could also show that $\langle a, b\rangle_\# = \langle a, b\rangle$ by a direct calculation, but it is easier to argue that both are the unique c such that $\pi^\varsigma c = a$ and $\pi'^\varsigma c = b$.

Before showing that $\varepsilon_\# = \varepsilon$, let us note that in $L(\mathscr{M})$

C15. $\varepsilon^\varsigma(a, b) = a^\varsigma b.$

Indeed,

$$\varepsilon^\varsigma(a, b) = \varepsilon^\varsigma(\langle \lambda_x a, \lambda_x b\rangle^\varsigma 1),$$
$$= (\varepsilon\langle \lambda_x a, \lambda_x b\rangle)^\varsigma 1 \qquad \text{by C10},$$
$$= (\varepsilon\langle \lambda_x a, 1\rangle \lambda_x b)^\varsigma 1 \qquad \text{by C13},$$
$$= (\varepsilon\langle (a\pi')^*, 1\rangle \lambda_x b)^\varsigma 1 \qquad \text{as } (a\pi')^{*\varsigma} x = a \text{ by C4a},$$
$$= (\varepsilon\langle (a\pi')^*, 1\rangle)^\varsigma((\lambda_x b)^\varsigma 1) \qquad \text{by C10},$$
$$= (1^\varsigma a)^\varsigma b,$$
$$= a^\varsigma b \qquad \text{by C12}.$$

Therefore

$$\varepsilon_\# = \lambda_z(\pi(z)^\varsigma \pi'(z))$$
$$= \lambda_z(\varepsilon^\varsigma(\pi(z), \pi'(z)))$$
$$= \lambda_z(\varepsilon^\varsigma z)$$
$$= \varepsilon.$$

Thus $\varepsilon_\# = \varepsilon$. As for $h_\#^*$ and h^*, both are the unique k such that $\varepsilon \langle k\pi, \pi' \rangle = h$, hence $h_\#^* = h^*$.

Thus $ML(\mathcal{M})$ and $\mathcal{M}$ have not only the same elements, but also the same operations, hence they are the same C-monoid.

(b) The terms of the language $LM(\mathcal{L})$ with free variables $x_1, \ldots, x_n$ are, by definition of L, polynomials in $M(\mathcal{L})$ $[x_1, \ldots, x_n]$. Let us assert, for the moment without justification, that there is an equality

(*) $M(\mathcal{L})[x_1, \ldots, x_n] = M(\mathcal{L}(x_1, \ldots, x_n)).$

Here $\mathcal{L}(x_1, \ldots, x_n)$ is the language obtained from $\mathcal{L}$ by adjoining parameters $x_1, \ldots, x_n$. This is to say, the closed terms of $\mathcal{L}(x_1, \ldots, x_n)$ are open terms of $\mathcal{L}$ in the free variables $x_1, \ldots, x_n$. We shall justify the assertion (*) later.

Thus $LM(\mathcal{L})$ has the same terms as $\mathcal{L}$; we shall now compare their term forming operations. Temporarily we shall distinguish these by placing a subscript # on those of $LM(\mathcal{L})$.

First,

$$(f\,^\mathsf{s} a)_\# = \varepsilon \langle f(a\pi')^*, 1 \rangle$$
$$= \lambda_x(\varepsilon^\mathsf{s}(\langle f(a\pi')^*, 1 \rangle^\mathsf{s} x))$$
$$= \lambda_x(\varepsilon^\mathsf{s}((f(a\pi')^*)^\mathsf{s} x, 1^\mathsf{s} x)).$$

Now

$$(f(a\pi')^*)^\mathsf{s} x = f^\mathsf{s}((a\pi')^* {}^\mathsf{s} x)$$
$$= f^\mathsf{s}\lambda_y((a\pi')^\mathsf{s}(x, y))$$
$$= f^\mathsf{s}\lambda_y(a^\mathsf{s}(\pi^\mathsf{s}(x, y)))$$
$$= f^\mathsf{s}\lambda_y(a^\mathsf{s} y)$$
$$= f^\mathsf{s} a.$$

Hence

$$(f\,^\mathsf{s} a)_\# = \lambda_x(\varepsilon^\mathsf{s}(f^\mathsf{s} a, x)),$$
$$= \lambda_x((f^\mathsf{s} a)^\mathsf{s} x) \quad \text{by C15,}$$
$$= f^\mathsf{s} a.$$

Next,

$$(\lambda_x \varphi(x))_\# = \lambda_x \varphi(x),$$

since both are the unique f such that $f^\mathsf{s} x = \varphi(x)$. Moreover,

$$\pi_\#(c) = (\pi^\mathsf{s} c)_\# = \pi^\mathsf{s} c = \pi(c),$$

and similarly for $\pi'_\#$. Finally,

$$(a, b)_\# = \langle \lambda_x a, \lambda_x b \rangle^\mathsf{s} 1$$
$$= (\lambda_y((\lambda_x a)^\mathsf{s} y, (\lambda_x b)^\mathsf{s} y))^\mathsf{s} 1$$

$$= ((\lambda_x a)^{\ulcorner} 1, (\lambda_x b)^{\ulcorner} 1)$$
$$= (a, b).$$

It remains to justify the assertion (*). For argument's sake, take $n = 1$. It is fairly clear that we have an isomorphism

(**) $\quad M(\mathscr{L})[x] \cong M(\mathscr{L}(x)).$

For it is easily verified that $M(\mathscr{L}(x))$ has the required universal property: for any C-homomorphism $f: M(\mathscr{L}) \to \mathscr{M}$ and any element $a \in \mathscr{M}$, there is a unique C-homomorphism $f^+: M(\mathscr{L}(x)) \to \mathscr{M}$ such that its restriction to $M(\mathscr{L})$ is f and $f^+(x) = a$. Indeed, we need only take

$$f^+(\varphi(x)) = f(\lambda_x \varphi(x))^{\ulcorner} a.$$

Now why can we replace the isomorphism in (**) by equality? This depends on how indeterminates are defined. While the standard construction of $\mathscr{M}[x]$ has been described in Section 15, alternative constructions are possible. Thus, in the special case $\mathscr{M} = M(\mathscr{L})$, we may define

(***) $\quad M(\mathscr{L})[x] = M(\mathscr{L}(x)),$

in view of the universal property discussed above. This establishes (*) and therefore Theorem 17.5.

Once Theorem 17.5 has been established, we may define indeterminates in general as variables by putting

(****) $\quad \mathscr{M}[x] = M(L(\mathscr{M})(x)).$

From this the special case (***) may be recaptured thus:

$$M(\mathscr{L})[x] = M(LM(\mathscr{L})(x)) = M(\mathscr{L}(x)).$$

The category of C-monoids has as objects C-monoids and as morphisms *C-homomorphisms*, that is, monoid homomorphisms which preserve the C-structure.

The category of extended λ-calculi has as objects λ-calculi and as morphisms *translations*, that is, mappings which send variables to variables, closed terms to closed terms, and which preserve the term forming operations and equations: e.g. $t(\langle a, b \rangle) = \langle t(a), t(b) \rangle$ holds in $\mathscr{L}'$ and if $a = b$ holds in $\mathscr{L}$ then $t(a) = t(b)$ holds in $\mathscr{L}'$, for any translation $t: \mathscr{L} \to \mathscr{L}'$. Moreover, if t and t' are translations, we write $t = t'$ if $t(a) = t(a')$ holds in $\mathscr{L}'$ for all terms a of $\mathscr{L}$.

Corollary 17.6. The category of C-monoids is isomorphic to the category of extended λ-calculi.

Proof. We extend L and M to functors inverse to one another.

If $f: \mathscr{M} \to \mathscr{M}'$ is a C-homomorphism, the translation $L(f): L(\mathscr{M}) \to L(\mathscr{M}')$

is defined as follows: for any polynomial $\varphi(X)$, where $X = \{x_1, \ldots, x_n\}$, we put $L(f)\,(\varphi(X)) = f_X(\varphi(X))$, where $f_X: \mathcal{M}[X] \to \mathcal{M}'[X]$ is the unique C-homomorphism extending f such that $f_X(x_i) = x_i$ for $i = 1, \ldots, n$.

If $t: \mathcal{L} \to \mathcal{L}'$ is a translation, the C-homomorphism $M(t): M(\mathcal{L}) \to M(\mathcal{L}')$ is defined thus: for any $a \in M(\mathcal{L})$, $M(t)(a) \equiv t(a)$.

In particular we have (taking $n = 1$)

$$(ML)(f)(a) = M(L(f))(a) = L(f)(a) = f(a),$$
$$(LM)(t)(\varphi(x)) = L(M(t))(\varphi(x)) = M(t)_x(\varphi(x)) = t(\varphi(x)).$$

The very last equation requires some explanation. It is easily seen that $M(t)_x = M(t_x)$, where t_x is the restriction of t to $\mathcal{L}(x)$ (considering $\mathcal{L}(x)$ as a subset of $\mathcal{L}$), hence

$$M(t)_x(\varphi(x)) = M(t_x)(\varphi(x)) = t_x(\varphi(x)) = t(\varphi(x)).$$

Thus ML and LM are both identity functors and not just isomorphic to identity functors. This shows that the two categories are isomorphic and not just equivalent, for what it is worth.

Exercises

1. Check the remaining equations of a C-monoid in the proof of 17.4.
2. Show that every λ-calculus gives rise to a weak C-monoid (as in Proposition 17.4, using Remark 17.2) and conversely (as in Proposition 17.3). Is this a one-to-one correspondence?
3. Complete the proof of (**) above by checking that $f^+(c) = f(c)$.

18 A construction by Dana Scott

In this section we shall construct a cartesian closed category with an object U such that

(*) $\quad U \times U \cong U \cong U^U, \quad U \not\cong 1.$

As we saw in Section 16, End(U) will then be a non-trivial C-monoid. According to Section 17, it will also provide us with non-trivial models of various versions of untyped λ-calculus. The original construction is due to Dana Scott; we present it here in a categorical setting due to Plotkin and Smyth.

We shall begin by considering the category $\mathcal{P}$ of *ω-posets.* Its objects are partially ordered sets in which every countable ascending chain of elements has a supremum. Its morphisms are order preserving mappings which preserve these sups. We denote the supremum of the ascending chain $a_0 \leqslant a_1 \leqslant a_2 \leqslant \ldots$ by $\bigvee_n a_n$.

Proposition 18.1. $\mathscr{P}$ is a cartesian closed category.

Proof (sketched). The cartesian structure is inherited from **Sets.** Thus 1 is a one-element set with the trivial order and $A \times B$ is the cartesian product with order and sups defined componentwise. Thus,

$$\bigvee_n (a_n, b_n) = (\bigvee_n a_n, \bigvee_n b_n).$$

Note that the unique mapping $A \to 1$ and the usual projections $A \times B \to A$ and $A \times B \to B$ are morphisms in $\mathscr{P}$. B^A is $\mathrm{Hom}(A, B)$ in $\mathscr{P}$ with order and sups defined componentwise:

$$f \leqslant g \quad \text{if and only if } f(a) \leqslant g(a) \quad \text{for all } a \in A,$$
$$(\bigvee_n f_n)(a) = \bigvee_n f_n(a).$$

The usual evaluation mapping $\varepsilon_{B,A}: B^A \times A \to B$ given by

$$\varepsilon_{B,A}(f, a) = f(a)$$

is easily seen to be a morphism, as are the mappings obtained by the rules

$$\frac{C \xrightarrow{f} A \quad C \xrightarrow{g} B}{C \xrightarrow{\langle f, g\rangle} A \times B}, \quad \frac{A \times B \xrightarrow{h} C}{A \xrightarrow{h^*} C^B}$$

and defined by

$$\langle f, g\rangle(c) = (f(c), g(c)), \quad h^*(a)(b) = h(a, b),$$

for all $c \in C$, $a \in A$ and $b \in B$. The equations of a cartesian closed category hold as in **Sets.**

Ultimately, we wish to find an object U of $\mathscr{P}$ satisfying (*). For the time being, we have at least the following:

Lemma 18.2. There is an object V of $\mathscr{P}$ such that

(**) $\quad V \times V \cong V \ncong 1$ and V has a smallest element.

Proof. Let $A \ncong 1$ be any nontrivial object of $\mathscr{P}$ with a smallest element 0 and put $V = A^{\mathbb{N}}$ for the product of countably many copies of A, with componentwise order and sup. Consider the mappings $\alpha, \beta: V \to V$ defined by

$$\alpha(v)(n) = v(2n), \quad \beta(v)(n) = v(2n + 1),$$

for all $n \in \mathbb{N}$. It is easily verified that α and β are morphisms in $\mathscr{P}$ and that $\sigma = \langle \alpha, \beta\rangle: V \to V \times V$ is an isomorphism in $\mathscr{P}$, with inverse σ^{-1} given by

$$\sigma^{-1}(v_1, v_2)(2n) = v_1(n), \quad \sigma^{-1}(v_1, v_2)(2n + 1) = v_2(n),$$

for all $n \in \mathbb{N}$. Thus $V \times V \cong V$. Moreover, clearly $V \ncong 1$ and V has a smallest element 0 given by $0(n) = 0$ for all $n \in \mathbb{N}$.

In addition to $\mathscr{P}$, we shall make use of a related category $\mathscr{P}^\#$.

Definition 18.3. The category $\mathscr{P}^\#$ has the same objects as $\mathscr{P}$, but arrows $A \to B$ in $\mathscr{P}^\#$ are *decreasing retractions*, that is, pairs (f, g) with $f: A \to B$ and $g: B \to A$ in $\mathscr{P}$ such that

$$gf = 1_A, \quad fg \leqslant 1_B.$$

Composition is defined by

$$(f', g')(f, g) = (f'f, gg'),$$

when $(f, g): A \to B$ and $(f', g'): B \to C$. The identity arrows are $(1_A, 1_A): A \to A$.

Remark 18.4. If $(f, g): A \to B$ is a decreasing retraction, then g is uniquely determined by f.

Proof. Suppose we also have $(f, g'): A \to B$ in $\mathscr{P}^\#$. From $fg \leqslant 1_B$ and $g'f = 1_A$ we infer that $g = g'fg \leqslant g'$. But, similarly $g' \leqslant g$, and so $g = g'$.

Remark 18.5. For any object V of $\mathscr{P}$, the contravariant (!) representable functor $V^{(-)} = \mathrm{Hom}(-, V)$ gives rise to a covariant endofunctor F_V: $\mathscr{P}^\# \to \mathscr{P}^\#$ which sends $(f, g): A \to B$ in $\mathscr{P}^\#$ to $(V^g, V^f): V^A \to V^B$ in $\mathscr{P}^\#$.

Proof. We shall verify that

$$V^f V^g = 1_{V^A}, \quad V^g V^f \leqslant 1_{V^B}.$$

Indeed, let $h: A \to V$ and $k: B \to V$ in $\mathscr{P}$, then

$$(V^f V^g)(h) = V^f(hg) = hgf = h1_A = h,$$

$$(V^g V^f)(k) = V^g(kf) = kfg \leqslant k1_B = k.$$

To find the required object U of $\mathscr{P}$ satisfying (*), we shall apply the following useful lemma to $\mathscr{P}^\#$.

Lemma 18.6. Let $\mathscr{A}$ be a category in which every countable sequence

$$A_0 \to A_1 \to A_2 \to \cdots$$

has a direct limit, and let $F: \mathscr{A} \to \mathscr{A}$ be an endofunctor of $\mathscr{A}$ which preserves such direct limits. Given any object V and any arrow $h: V \to F(V)$, the direct limit U of the sequence

$$V \xrightarrow{h} F(V) \xrightarrow{F(h)} F^2(V) \to \cdots$$

is a *fixed point* of F, that is, $F(U) \cong U$.

Proof. $F(U)$ is the direct limit of the sequence

$$F(V) \xrightarrow{F(h)} F^2(V) \xrightarrow{F^2(h)} F^3(V) \rightarrow \cdots$$

Since this is the tail of the sequence in the lemma, $F(U) \cong U$.

We are now ready to obtain the main result of this section.

Theorem 18.7. There is an object U of $\mathscr{P}$ such that

(*) $\quad U \times U \cong U \cong U^U, \quad U \not\cong 1.$

Proof. Beginning with any nontrivial object A of $\mathscr{P}$ with a smallest element, we form the object $V = A^{\mathbb{N}}$ as in Lemma 18.2 and the endofunctor F_V as in Remark 18.5. We then apply Lemma 18.6 to the category $\mathscr{A} = \mathscr{P}^{\#}$ and the endofunctor $F = F_V$. To do so, we must verify the conditions of the lemma:

(i) $\mathscr{P}^{\#}$ has direct limits of countable sequences;
(ii) F_V preserves them;
(iii) there is an arrow (f, g): $V \rightarrow F_V(V) = V^V$ in $\mathscr{P}^{\#}$.

We shall defer the proofs of (i) and (ii) to the end of this section, but establish (iii) now. We define $f: V \rightarrow V^V$ and $g: V^V \rightarrow V$ as follows:

$$f(v)(w) = v, \quad g(t) = t(0),$$

for all $v, w \in V$ and $t \in V^V$, where 0 is the smallest element of V. It is easily checked that

$$gf = 1_V, \quad fg \leqslant 1_{V^V}.$$

Indeed,

$$(gf)(v) = g(f(v)) = f(v)(0) = v,$$
$$(fg)(t) = f(g(t)) = f(t(0)) \leqslant t,$$

since, for any $w \in V$,

$$f(t(0))(w) = t(0) \leqslant t(w).$$

By Lemma 18.6, F_V has a fixed point U, that is, $V^U \cong U$. Since $V \not\cong 1$, also $U \not\cong 1$. Moreover,

$$U \times U \cong V^U \times V^U \cong (V \times V)^U \cong V^U \cong U,$$
$$U^U \cong (V^U)^U \cong V^{U \times U} \cong V^U \cong U,$$

and this establishes (*).

It remains to verify (i) and (ii). Consider a sequence in $\mathscr{P}^{\#}$:

(S) $\quad A_0 \xrightarrow{(i_0, j_0)} A_1 \xrightarrow{(i_1, j_1)} A_2 \longrightarrow \cdots,$

where

(1) $j_n i_n = 1_{A_n}, \quad i_n j_n \leqslant 1_{A_{n+1}}.$

We wish to find the direct limit of (S) in $\mathscr{P}^{\#}$.

Note that (S) gives rise to the 'inverse' sequence

(S') $A_0 \xleftarrow{j_0} A_1 \xleftarrow{j_1} A_2 \longleftarrow \cdots$

in $\mathscr{P}$. We can find the inverse limit (A, p_n) of (S') in $\mathscr{P}$, where $p_n\colon A \to A_n$ has the usual universal property. This is done as in **Sets**, by taking

(2) $A = \{a \in \prod_{n \in \mathbb{N}} A_n \mid \forall_{n \in \mathbb{N}}\, j_n(a_{n+1}) = a_n\}$

and by defining $p_n\colon A \to A_n$ by

(3) $p_n(a) = a_n.$

It is easily checked that A has sups of countable ascending chains and that p_n preserves them. Therefore,

(I) (A, p_n) is the inverse limit of (S') in $\mathscr{P}$.

Consider now the arrows $f_{m,n}\colon A_m \to A_n$ in $\mathscr{P}$ defined as follows:

(4) $$f_{m,n} = \begin{cases} i_{n-1} i_{n-2} \cdots i_m & \text{if } m < n, \\ 1_{A_m} & \text{if } m = n, \\ j_n j_{n+1} \cdots j_{m-1} & \text{if } m > n. \end{cases}$$

It is easily verified that the following triangles commute:

(5) 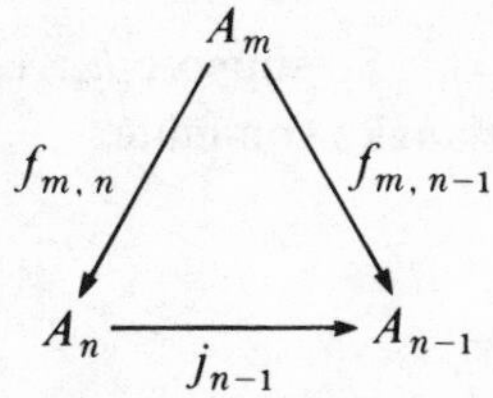

Therefore, by the universal property of the inverse limit (A, p_n), there exists a unique morphism $q_m\colon A_m \to A$ in $\mathscr{P}$ so that

(6) $p_n q_m = f_{m,n}.$

We assert that $(q_n, p_n)\colon A_n \to A$ in $\mathscr{P}^{\#}$, that is,

(7) $p_n q_n = 1_{A_n}, \quad q_n p_n \leqslant 1_A.$

Indeed, the equation follows from (6) and (4), while the inequality is contained in the following stronger assertion:

(8) $\bigvee_m q_m p_m = 1.$

This is proved by showing that, for each $a \in A$, if $a_m = a(m) = p_m(a)$,

$$\bigvee_m q_m(a_m) = a,$$

that is, for each $n \in \mathbb{N}$,

$$\bigvee_m p_n q_m(a_m) = a_n.$$

Indeed, by (6)

$$p_n q_m(a_m) = f_{m,n}(a_m)\begin{cases} \leqslant a_n & \text{if } m < n, \\ = a_n & \text{if } m \geqslant n, \end{cases}$$

as an easy consequence of (4).

There is a gap in the above argument. To calculate (8) one must first check that one has an ascending chain

$$q_0 p_0 \leqslant q_1 p_1 \leqslant \ldots .$$

As in the above argument, this translates into showing that

$$f_{0,n}(a_0) \leqslant f_{1,n}(a_1) \leqslant \ldots .$$

We leave this as an exercise.

Now (S) also gives rise to the sequence

$$(S'') \qquad A_0 \xrightarrow{i_1} A_1 \xrightarrow{i_2} A_2 \longrightarrow \cdots$$

in $\mathscr{P}$. We claim that

(II) (A, q_n) is the direct limit of (S'') in $\mathscr{P}$.

Indeed, it is easily verified that $q_{n+1} i_n = q_n$. So suppose $h_n: A_n \to B$ are given arrows in $\mathscr{P}$ so that the following triangles commute:

(9)

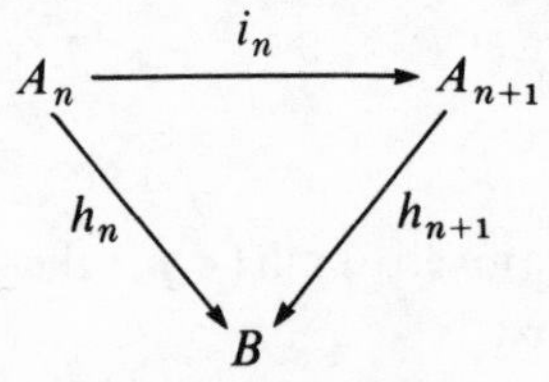

We shall prove that there is a unique arrow $\varphi: A \to B$ in $\mathscr{P}$ such that

$$(10) \qquad \varphi q_n = h_n.$$

On the one hand, if (10) holds, we may calculate φ as follows:

$$(11) \qquad \varphi = \varphi \bigvee_n q_n p_n \quad \text{by (8)},$$

$$= \bigvee_n \varphi q_n p_n,$$

$$= \bigvee_n h_n p_n \quad \text{by (10)}.$$

On the other hand, if φ is defined by (11) (first check that $h_n p_n$ increases with n), we have

$$\text{(10)}\qquad \varphi q_m = \bigvee_n h_n p_n q_m = \bigvee_n h_n f_{m,n} = h_m,$$

Since

$$\text{(12)}\qquad h_n f_{m,n} \begin{cases} = h_m & \text{if } m \leqslant n, \\ \leqslant h_m & \text{if } m > n, \end{cases}$$

as an easy consequence of (4), (9), and (1).

This completes the proof of (II).

We shall prove next that

(III) $(A,(q_n, p_n))$ is the direct limit of (S) in $\mathscr{P}^{\#}$.

Indeed, we showed in (7) that $(q_n, p_n)\colon A_n \to A$ in $\mathscr{P}^{\#}$. Moreover, the following triangles commute in $\mathscr{P}^{\#}$:

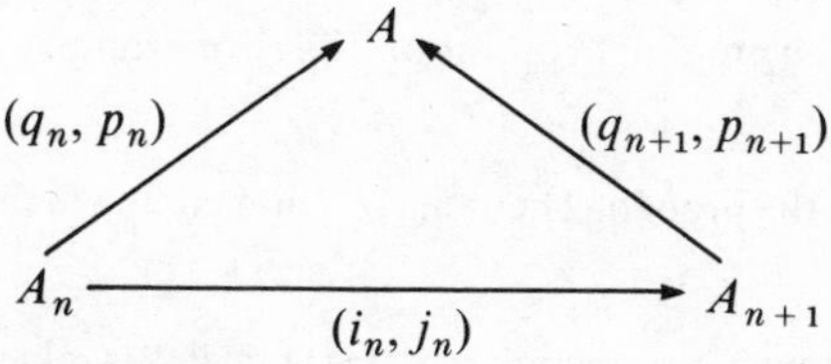

in view of (I) and (II).

To prove the universal property of $(q_n, p_n)\colon A_n \to A$, suppose that $(h_n, k_n)\colon A_n \to B$ in $\mathscr{P}^{\#}$ are such that the appropriate triangles commute. By (I), there is a unique $\psi\colon B \to A$ such that $p_n \psi = k_n$ and, by (II), there is a unique $\varphi\colon A \to B$ such that $\varphi q_n = h_n$. One easily calculates that

$$\psi\varphi = 1_A, \quad \varphi\psi \leqslant 1_B.$$

Hence $(\varphi, \psi)\colon A \to B$ is the unique morphism in $\mathscr{P}^{\#}$ such that $(\varphi, \psi)(q_n, p_n) = (h_n, k_n)$. This completes the proof of (III), hence of (i).

It remains only to prove (ii). Applying F_V to the direct limit of (S) in $\mathscr{P}^{\#}$, we obtain

$$F_V(A,(q_n, p_n)) = (V^A, (V^{p_n}, V^{q_n}))$$

by Remark 18.5. According to (III), this is the direct limit of

$$(S''')\qquad V^{A_0} \xrightarrow{(V^{j_0}, V^{i_0})} V^{A_1} \xrightarrow{(V^{j_1}, V^{i_1})} V^{A_2} \longrightarrow \cdots,$$

provided V^{p_n} and V^{q_n} play the same rôles with regard to (S''') as q_n and p_n do with regard to (S') respectively.

Indeed, the contravariant functor $V^{(-)}$ sends direct limits in $\mathscr{P}$ to inverse limits in $\mathscr{P}$. So, by (II), (V^A, V^{q_n}) is the inverse limit of

$$V^{A_0} \xleftarrow{V^{i_0}} V^{A_1} \xleftarrow{V^{i_1}} V^{A_2} \longleftarrow \cdots,$$

hence V^{q_n} plays the rôle of p_n. Now it follows from (6), if m and n are interchanged, that

$$V^{q_n} V^{p_m} = V^{f_{n,m}}.$$

Since the rôle of $f_{m,n}$ in (5) is now played by $V^{f_{n,m}}$, it follows that V^{p_m} plays the rôle of q_m.

Thus we have

$$F_V(\underset{\rightarrow}{\lim}(S)) \cong \underset{\rightarrow}{\lim}(S''') = \underset{\rightarrow}{\lim} F_V(S),$$

and this establishes (ii). The proof of the theorem is now complete.

Exercises

1. Fill in the details in the proof of Proposition 18.1. For example, show that $\varepsilon_{B,A}$ preserves sups.
2. Check (5) and (12) in the proof of Theorem 18.7 and complete the proof of (8).
3. Show that the argument of this section remains valid if $\mathscr{P}$ is taken to be the category of all partially ordered sets in which every upward directed nonempty set has a supremum, the morphisms being required to preserve these sups.
4. (R.E. Hoffman.) Show that the category $\mathscr{P}$ of Exercise 3 is the Eilenberg–Moore category of the following triple (T, η, μ) on the category of posets: for any poset A, $T(A)$ is the set of ideals (dual filters) of A; $\eta(A)\,(a)$ is the principal ideal generated by $a \in A$; $\mu(A)$ assigns to every ideal of $T(A)$ its union, an ideal of A.
5. Show that the category of sup-complete semilattices and sup-preserving mappings is not cartesian closed.

Historical comments on Part I

Section 1

Deductive systems, at least freely generated ones, probably first appeared in the proof-theoretical studies of Gentzen (see e.g. Szabo 1969, Kleene 1952 and Prawitz 1965, 1971). The view that deductive systems are graphs with additional structure or categories with missing equations has been propounded by one of us and Fred Szabo for many years. Our presentation here has been influenced by categorical considerations. In particular, our choice of axioms and rules of inference are motivated by adjoint functors, as has been suggested by Lawvere (1969*a*, 1970).

Section 2

The version of the deduction theorem in the text differs from others in the literature in various respects, but mainly in that it applies to deductive systems which are not necessarily freely generated from axioms.

Section 3

While cartesian closed categories may now be subsumed under the general notion of 'graphical algebra' (Burroni 1981), the equational presentation of cartesian closed categories and other structured categories had already been emphasized by Lambek (1968, 1969, 1972), who regarded them as certain kinds of deductive systems with an equivalence relation between proofs. Prawitz (1971) had also studied an equivalence relation between proofs in intuitionistic logic for completely different motives. It was shown by Mann (1975) and rediscovered by Seely (1977) that, for intuitionistic propositional calculus, the two equivalence relations are essentially the same. This confirms our view that category theory may serve as useful motivation for many constructions in logic.

Section 4

The construction of free algebras from words is well-known. The construction of free structured categories from formulas and proofs is in (Lambek, *op. cit.*) and was further developed by Szabo (1974*b*, 1978).

Section 5

The usual construction of polynomial algebras in an indeterminate x also finds a parallel here in the construction of a polynomial structured category in an indeterminate arrow x considered as an assumption, as in (Lambek 1972, 1974, 1980*a*, *b*). The presentation here has greatly benefited from improvements suggested by Bill Hatcher.

Section 6

The functional completeness of cartesian closed categories had been shown by the senior author in 1972 under certain conditions and in 1974 without extra conditions.

Section 7

Lambek also showed (1974) that polynomial categories may be viewed as Kleisli categories of a certain cotriple and that this gives another proof of functional completeness. That the latter categories may be used for introducing indeterminates and polynomials has also been noticed by Volger (1975). It should be emphasized that, as long as equalizers are excluded from the definition of cartesian closed categories, adjoining an indeterminate of type A is not the same as forming the slice category $\mathscr{A}/A$, but it is once equalizers are included. The latter was observed by Grothendieck and Joyal (see Part II, Section 16, Exercise 2).

Section 8

The observation that one can do propositional calculus in a bicartesian closed category appears to be due to Lawvere. The present treatment follows Lambek (1974). The fact that there is at most one arrow $A \to 0$ was noticed by Joyal (oral communication), but our proof follows Freyd (1972). For the fact that there are no Boolean categories other than Boolean algebras, see also (Szabo 1974*a*).

Section 9

The story of natural numbers objects has already been told in the Historical Perspective on Part I.

Section 10

An exhaustive discussion of the history of the λ-calculus before cartesian closed categories will be found in the book by Barendregt; see also the book by Hindley, Lercher and Seldin. Our categorical viewpoint stresses the importance of product types and surjective pairing, usually ignored by logicians. The subscript X in $\underset{X}{=}$ is to allow for the possibility of 'empty' types, which may arise in the internal language of a cartesian closed category. Incidentally, as long as such categories may be large, we must admit proper classes of types and terms. The internal language of a cartesian closed category depends crucially on functional completeness. Lawvere (1969*b*) may have anticipated this language when he called the correspondence from $\mathrm{Hom}(1, B^A)$ to $\mathrm{Hom}(A, B)$ 'λ-conversion'.

Section 11

While our internal languages may resemble what logicians call 'diagram languages' (e.g. Chang and Keisler 1973), the cartesian closed category generated by a typed λ-calculus is essentially what they call a 'term model'. The proof of the categorical equivalence between typed λ-calculi and cartesian closed categories with weak natural numbers object appears here for the first times in detail.

Section 12

The coherence problem for monoidal categories was solved by MacLane (1963), who proved that there 'all diagrams' (composed of canonical arrows) 'commute'. As this statement fails for biclosed monoidal categories and because of the necessity to define 'canonical', Lambek (1968, 1969) reformulated the coherence problem in the form (II): to find an algorithm for deciding when two arrows $A \to B$ in a free biclosed monoidal category are equal. He had first solved the preliminary problem (I): to find an algorithm for obtaining all arrows $A \to B$, by applying cut elimination *à la* Gentzen to proofs $A \to B$ in the free syntactic calculus, which had been studied (1961) because of its interest for Linguistics. These methods were extended by Szabo (1974*b*, 1978) to other structured categories, including cartesian closed categories (but without natural numbers), by Kelly and MacLane (1971), by Voreadou (1977), and, most recently by Minc (1977, 1979) and his students (e.g. Babaev 1981 and Solov'ov 1981, see also the survey article by MacLane 1982). That Gentzen's cut elimination is essentially equivalent to normalization of terms in typed λ-calculus is well-known to modern proof theorists (see e.g. Howard 1980, Prawitz 1971, Girard 1971, 1972, Minc 1975, Zucker 1974 and Pottinger 1977). A direct solution of the coherence problem for cartesian closed categories, without passing to typed λ-calculus, has quite recently been obtained by Obtułowicz.

Section 13

Although the decision problem for convertibility in the untyped λ-calculus is recursively unsolvable, it gave rise to the original Church–Rosser Theorem, which easily extends to bounded (usually called 'strongly normalizable') terms in the typed λ-calculus. The difficulties with type 1 were pointed out by Adam Obtułowicz.

Section 14

Strong normalizability for typed λ-calculus without product types was first proved by Tait (1967). Many people have extended his methods, e.g. Troelstra (1973). Different methods were employed by Howard (1970),

Pottinger (1978, 1981) and Gandy (1980*b*). De Vrijer (1982), using Troelstra's methods, was the first to show how to handle surjective pairing and iterators. Our proof is closer to Tait's original proof, for example, in defining computability for *closed* terms first; but it could not have been developed without the help of de Vrijer's manuscript, which inspired the crucial condition (3) of Lemma 14.4.

Section 15

Church (1937) made the important observation that Curry's combinatory logic (in our terminology: a free Curry algebra) may be viewed as a monoid with unsolvable word problem. The C-monoids treated here also have surjective pairing; they are essentially one-object cartesian closed categories without terminal object. The idea to consider C-monoids occurred to Dana Scott and the senior author upon comparison of their respective papers in the Curry Festschrift (1980). Their brief collaboration on this topic culminated in a seminar at Amsterdam in 1981. For further variants of Curry algebras see Meyer (1982).

Section 16

The discovery that the Karoubi envelope of a C-monoid is a cartesian closed category is due to Dana Scott (1980*b*). Incidentally, the Karoubi envelope first appears in Freyd's book on Abelian categories, though not under this name. (See also Artin *et al.* 1972 page 413.) The fact that a cartesian closed category with two non-isomorphic objects suffices was noticed by us.

Section 17

Our proof of the isomorphism between the categories of C-monoids and λ-calculi with surjective pairing was circulated in 1982. Further work, particularly on weak C-monoids, has been done by Adachi (1983), Curien (1983, 1984), Hayashi (1983), and Koymans (1984).Categorical aspects of the λ-calculus were also studied by Obtułowicz (1979) and Obtułowicz and Wiweger (1982).

Section 18

An obvious question to ask about the untyped λ-calculus, as originally defined or as extended by us, is what its models, that is, Curry algebras or C-monoids, look like. In particular, are there any models other than the trivial one with only one element? This is the old question: can one consistently posit a universe of functions which apply to all functions, including themselves, as arguments? The classical Church–Rosser Theorem implies, among other things, that the free Curry algebra generated by the empty set is not trivial. On the other hand, we know from Section 16 that all we need for a nontrivial C-monoid is a cartesian closed category with an object U, not isomorphic to 1, such that $U^U \cong U \cong U \times U$. Dana Scott has constructed a number of such examples (and others allowing U^U and $U \times U$ to be mere retracts of U, giving models of λ-calculus without Rule (η), i.e. weak C-monoids). His original model (1972) used continuous lattices, but several other people pointed out that ω-posets would suffice (e.g. Fleischer 1972 and Egli 1973). We follow Plotkin and Smyth (1978). Many other models have been considered since (see Barendregt 1981 and Koymans 1984).

II

Type theory and toposes

Introduction to Part II

We present two versions of type theory with product types and special types N for natural numbers and Ω for truth values. The first version incorporates the usual intuitionistic predicate calculus, while in the second version logical connectives and quantifiers are defined in terms of equality. These two versions are shown to be equivalent, although the second is useful for describing the internal language of a topos.

An important example of a type theory is the internal language of a topos. For us, a topos is assumed to have a natural numbers object. It follows that Peano's rules for arithmetic are provable in this internal language. We put this language to work to show that various categorical notions, such as equality of arrows, monomorphisms and injectivity, can be handled linguistically, as is usually done in the topos of sets. In particular, we show that Russell's theory of descriptions applies to toposes: if one can prove in the internal language of a topos that $\forall_{x\in A}\exists!_{y\in B}\varphi(x,y)$, then there is a unique arrow $g: A \to B$ such that $\forall_{x\in A}\varphi(x, gx)$ holds in the topos. We also show that, if a type theory satisfies the rule of choice, then the Aristotelian or Boolean axiom $\forall_{x\in\Omega}(x \vee \neg x)$ is provable. (For the internal language of a topos, this is due to Diaconescu.)

There is a way to analyse the internal language of a topos by discussing which statements hold at stage C, where C is an object of the topos. This viewpoint originates from a special case implicit in the work of Kripke and was extended to its present generality by Joyal. We discuss a number of cases in some detail: functor categories (including Kripke models and $\mathscr{M}$-sets) and sheaf categories.

Not only is there a type theory associated with a topos $\mathscr{T}$, its internal language $L(\mathscr{T})$, but there is also a topos associated with any type theory $\mathfrak{L}$, the topos $T(\mathfrak{L})$ generated by $\mathfrak{L}$. While not every type theory is the internal language of a topos, every topos is equivalent to one generated by a type theory.

We introduce categories **Lang** and **Top** whose objects are type theories and toposes respectively and extend L and T to functors L: **Top** → **Lang** and T: **Lang** → **Top**. In some loose sense, T is left adjoint to L; but to make this precise, we replace **Top** by a reflective subcategory $\mathbf{Top}_0$ whose objects are toposes 'with canonical subobjects'. This adjointness is exploited for a number of constructions, e.g. to form the so-called 'free' topos (an initial object of $\mathbf{Top}_0$), to adjoin an indeterminate arrow to a topos and to divide a topos by a filter of propositions (i.e. subobjects of 1).

For us, an *interpretation* of a type theory $\mathfrak{L}$ in a topos $\mathcal{T}$ is a morphism $\mathfrak{L} \to L(\mathcal{T})$ in **Lang** or, in view of adjointness, a morphism $T(\mathfrak{L}) \to \mathcal{T}$ in $\mathbf{Top}_0$. If the terminal object of $\mathcal{T}$ is an indecomposable projective we call the interpretation $\mathfrak{L} \to L(\mathcal{T})$, or just the topos $\mathcal{T}$, a *model* of $\mathfrak{L}$. A suitable generalization of Henkin's completeness theorem for higher order logic may be viewed as saying that every type theory has enough models or that every topos is a subdirect product of toposes whose terminal objects are indecomposable projectives.

In presence of the rule of choice, the last result is refined to show that every topos is equivalent to the topos of global sections of a sheaf of 'local' toposes. This theorem and its proof are quite similar to a well-known theorem about commutative rings.

Finally, we discuss some constructivist principles which are demanded by intuitionists on philosophical grounds, but which appear here as metatheorems about pure type theory. For example, the *disjunction property* asserts that if p ∨ q is provable then either p is provable or q is provable. Again, the *existence property* asserts that if $\exists_{x \in A} \varphi(x)$ is provable then $\varphi(a)$ is provable for some closed term a of type A. These metatheorems are here proved with the help of an ingenious method of Freyd, who translates them into asserting that the terminal object in the free topos, namely the topos generated by pure type theory, is an indecomposable projective. Similar proofs are given for other intuitionistic principles: the disjunction property with parameters, Troelstra's uniformity rule, independence of premisses and Markov's rule.

Historical perspective on Part II

Aristotle had asserted the principle of the excluded third: for every statement p, either p or not p. An equivalent formulation by the Stoics said that two negations make an affirmation. Surprisingly, this assertion went essentially unchallenged for more than 2000 years, until it turned out to be the culprit behind such nonconstructive arguments as lead from $\neg\forall_x \neg\varphi(x)$ to $\exists_x \varphi(x)$.

Brouwer, in his criticism of classical mathematics, felt that an existential statement $\exists_x \varphi(x)$ should be admitted as true only if there were an entity a about which it was known that $\varphi(a)$. While Brouwer may have been opposed to formalism originally, many of his present-day disciples are formalists; and there now exist formal languages of intuitionistic logic which differ from their classical counterparts only by lacking Aristotle's principle of the excluded third, here also referred to as the Boolean axiom. For these formal intuitionistic languages, it then becomes desirable to establish a metatheorem which asserts that, if $\exists_x \varphi(x)$ can be proved, then there must be a term a such that $\varphi(a)$ can be proved. It is a principal result of Part II that this metatheorem and others of a similar nature hold for pure intuitionistic type theory.

Types are inherent in everyday language, for example, when we distinguish between 'who' and 'what' or between 'somebody' and 'something'. In a formal language one might, in the same spirit, be led to replace $\exists_x \varphi(x)$ by $\exists_{x \in A} \varphi(x)$, where A indicates the type of entities that the variable x ranges over, be they natural numbers, truth values or sets of natural numbers, etc. Historically, this step was first undertaken by Bertrand Russell, who wanted to save Frege's formal system of Cantorian set theory (as well as his own system) from paradox.

Russell's own formulation of 'ramified' type theory was too complicated to catch on among practising mathematicians, in spite of various attempts to simplify it (see Hatcher's book for an account of the history). Instead, mathematicians have been leaning on other foundations (Gödel–Bernays or Zermelo–Fraenkel), even though elegant type theories were proposed and studied by Church (1940) and Henkin (1950).

Category theory, invented by Eilenberg and MacLane (1945), is somewhat like type theory in its very nature, since every arrow comes equipped with two 'types', its source and its target. It was Lawvere (1964) who first realized that category theory can also be used as a foundation for mathematics and he presented an axiomatic description of the category of sets. It soon became apparent that one could delete some of the less natural of these axioms and still describe categories of functors and, more generally, categories of (generalized) sheaves, known as Grothendieck toposes, that were being studied by the Paris school of algebraic geometry. This observation led to the axiomatic description of elementary toposes by Lawvere and Tierney, see Lawvere (1971) and Tierney (1972).

Nothing could have been further from the minds of the founders of topos theory than the philosophy of intuitionism. Yet, it was soon realized that with each topos there was associated a type theory, its internal language, which did not necessarily satisfy Aristotle's principle of the excluded third. Consequently, even people not embracing the philosophy of intuitionism became motivated to make mathematical arguments conform to intuitionistic restrictions. For, any theorem about sets provable in this manner would hold not just in the category of sets but in any topos whatsoever. Such a theorem could then be specialized to yield a new classical result, for example, about sheaves in place of sets (see Mulvey 1974).

According to Lawvere (1975*a*), sheaves might be perceived as continuously variable sets, perhaps changing in time, in a manner reminiscent of Heraclitus and implicit in Brouwer's original view and, more formally, in Kripke's interpretation of intuitionistic logic. More generally, one might think of arbitrary toposes as alternative mathematical universes. However, let us confine attention to those toposes $\mathscr{M}$, called *models*, which resemble the usual category of sets in having the following properties, the first of which merely serves to exclude the trivial topos with only one object:

(*a*) no contradiction holds in $\mathscr{M}$;
(*b*) if $\mathrm{p} \vee \mathrm{q}$ holds in $\mathscr{M}$, then either p holds in $\mathscr{M}$ or q does;
(*c*) if $\exists_{x \in A} \varphi(x)$ holds in $\mathscr{M}$, then there is an entity a of type A in $\mathscr{M}$ (actually an arrow $a: 1 \to A$) such that $\varphi(a)$ holds in $\mathscr{M}$.

In the presence of the Boolean axiom, these models are essentially the same as Henkin's 'nonstandard' models; and his completeness theorem (1950) may be generalized to say that every type theory has 'enough' such models in the sense that a statement is provable if and only if it holds in every model.

By *a* type theory we here mean any (applied) higher order logic which includes a type N for the natural numbers and in which Peano's axioms for arithmetic hold. (While one can weaken Peano's axioms considerably and replace them by a suitable axiom of infinity, there seems to be no particular advantage in doing so.) Among type theories there is a distinguished one, *pure type theory*, which contains no types, terms and assumptions other than those which it has to contain by virtue of being a type theory.

Every type theory gives rise to a topos, the topos generated by it, essentially what logicians might call its term model. It is constructed linguistically; for example, its objects are equivalence classes of terms and arrows between them are terms denoting functional relations. In particular, the topos generated by pure type theory is the so-called *free topos*, which admits a unique 'logical morphism' into any topos.

One of the principal results of Part II can be summarized as saying that the free topos is a model, that is, that it satisfies conditions (*a*), (*b*) and (*c*) above. Thus we have metatheorems about pure intuitionistic type theory which codify basic intuitionistic principles (as was discussed for (*c*) earlier). While these metatheorems could be proved using standard proof theoretic methods, namely Kleene–Friedmann realizability or Gentzen–Girard cut elimination, they are here proved with the help of a categorical technique due to Peter Freyd (which however turns out to be equivalent to realizability).

What are the philosophical implications of these last results? While extreme intuitionists might reject type theory altogether, we may think of models as possible worlds acceptable to moderate intuitionists (at least those models in which all numerals are standard). Among these models the free topos stands out as a kind of ideal world in the Platonic sense. But, as the construction of the free topos was linguistic, this is also a justification of the formalist point of view. We have thus reconciled three apparently competing traditional philosophies of mathematics:

(i) intuitionism, according to which only knowable statements are true,
(ii) Platonism (or realism), which asserts that mathematical expressions refer to entities whose existence is independent of the knowledge we have of them,
(iii) formalism, whose principal concern is with expressions in the formal language of mathematics.

For the sake of completeness, we mention that there is yet another basic philosophy of mathematics:

(iv) logicism, which says that all of mathematics can be reduced to logic.

It seems less obvious how to reconcile this position with the others, as long as properties of natural numbers are postulated rather than derived (even though logicists usually also recognize an axiom of infinity).

If we reject logicism, what should take its place? We are tempted to follow Lawvere (1967, 1969*a*) and adopt the view that the growth of mathematics should be guided by various categorical slogans (see Part 0) or the more widely held view that category theory underlines the general principles common to different areas of mathematics. For example, as regards Part II, the free topos came to our attention when we tried to find a left adjoint to the forgetful functor from toposes to graphs, while the categorical notion of projectivity, which had proved useful in algebra, showed unexpected applications in logic. The categorical viewpoint thus treats logic less as a foundation than as part of mathematics, which is also what Brouwer had in mind.

1 Intuitionistic type theory

A *type theory* is a formal theory given by the following data:

(*a*) a class of *types*, including a special type Ω;

(*b*) a class of *terms* of each type, including countably many variables of each type;

(*c*) for each finite set X of variables a binary relation $\vdash_X$ of *entailment* between terms of type Ω all free variables of which are elements of X.

These data are subject to the following conditions:

(*a*) The class of types is closed under the inductive clauses:
(*a*1) $1, N$ and Ω are types ('basic' types);
(*a*2) if A and B are types, so are $A \times B$ and PA.

We allow the possibility of additional types besides those specified by clauses (*a*1) and (*a*2), and even identifications between types. We interpret N as the type of natural numbers, 1 as a one-element type and Ω as the type of truth values or propositions

(*b*) The class of terms is freely generated from certain basic terms by certain operations including the following. Among the terms of type A are countably many variables $x_1^A, x_2^A, \ldots$. Usually we shall not quote a variable by name, but merely say: 'let x be a variable of type A' or even 'let $x \in A$'. The set of terms also contains a number of specific terms and is closed under certain operations.
(*b*1) $*$ is a term of type 1;
(*b*2) if a is a term of type A and b is a term of type B, then $\langle a, b \rangle$ is a term of type $A \times B$;
(*b*3) if a is a term of type A and α is a term of type PA, then $a \in \alpha$ is a term of type Ω;
(*b*4) if $\varphi(x)$ is a term of type Ω (possibly containing the free variable x of type A), then $\{x \in A \mid \varphi(x)\}$ is a term of type PA not containing a free occurrence of x;
(*b*5) 0 is a term of type N;
(*b*6) if n is a term of type N so is Sn;

(*b*7) $\top$ and $\bot$ are terms of type Ω;
(*b*8) if p and q are terms of type Ω, so are $p \wedge q$, $p \vee q$ and $p \Rightarrow q$;
(*b*9) if $\varphi(x)$ is a term of type Ω (possibly containing the free variable x of type A), then $\forall_{x \in A}\varphi(x)$ and $\exists_{x \in A}\varphi(x)$ are terms of type Ω not containing free occurrences of x.

We allow the possibility of additional basic terms and term forming operations besides those specified by clauses (*b*1) to (*b*9). The notions of *free* variable, *bound* variable and *closed* term are defined as usual (see below). We summarize the formation of terms in the following list, in addition to which there are variables of each type:

1	N	PA	$A \times B$	Ω
$*$	0	$\{x \in A \mid \varphi(x)\}$	$\langle a, b \rangle$	$\top, \bot$
	Sn			$p \wedge q, p \vee q$
				$p \Rightarrow q$
				$\forall_{x \in A}\varphi(x), \exists_{x \in A}\varphi(x)$
				$a \in \alpha$

where the subterms, $n, \varphi(x), a, b$ and α have the appropriate types:

N	Ω	A	B	PA
n	$\varphi(x)$	a	b	α
	p			
	q			

Thus $S(-)$, $\{x \in A \mid -\}$, $\langle -, - \rangle$ etc. are term forming operations. Terms of type Ω are also called *formulas*.

The term $\bot$ and the term forming operations $(-) \vee (-)$ and $\exists_{x \in A}(-)$ may be eliminated by means of the following definitions:

$$\bot \equiv \forall_{t \in \Omega} t,$$
$$p \vee q \equiv \forall_{t \in \Omega}(((p \Rightarrow t) \wedge (q \Rightarrow t)) \Rightarrow t),$$
$$\exists_{x \in A}\varphi(x) \equiv \forall_{t \in \Omega}(\forall_{x \in A}(\varphi(x) \Rightarrow t) \Rightarrow t).$$

One may also define a number of widely used term forming operations:

$$\neg p \equiv p \Rightarrow \bot,$$
$$p \Leftrightarrow q \equiv (p \Rightarrow q) \wedge (q \Rightarrow p),$$
$$a = a' \equiv \forall_{u \in PA}(a \in u \Leftrightarrow a' \in u),$$
$$\{a\} \equiv \{x \in A \mid a = x\},$$

$$\exists!_{x\in A}\varphi(x) \equiv \exists_{x'\in A}\{x\in A\,|\,\varphi(x)\} = \{x'\},$$

$$\{\langle x, y\rangle\in A\times B\,|\,\varphi(x,y)\} \equiv \{z\in A\times B\,|\,\exists_{x\in A}\exists_{y\in B}(z=\langle x,y\rangle \wedge \varphi(z))\},$$

$$\alpha\subseteq\beta\equiv\forall_{x\in A}(x\in\alpha\Rightarrow x\in B),$$

where α and β are terms of type PA.
The definition of $=$ given above may be ascribed to Leibniz.

The variable x of $\varphi(x)$ is said to be *bound* in $\{x\in A\,|\,\varphi(x)\}$ and $\forall_{x\in A}\varphi(x)$, hence also in $\exists_{x\in A}\varphi(x)$ and $\exists!_{x\in A}\varphi(x)$. An occurrence of a variable is said to be *free* when it is not bound.

(*c*) Entailment satisfies the following *axioms* and *rules of inference*:

1. Structural rules

1.1. $p \vdash_X p$;

1.2. $\dfrac{p\vdash_X q \quad q\vdash_X r}{p\vdash_X r}$;

1.3. $\dfrac{p\vdash_X q}{p\vdash_{X\cup\{y\}} q}$;

1.4. $\dfrac{\varphi(y)\vdash_{X\cup\{y\}}\psi(y)}{\varphi(b)\vdash_X\psi(b)}$,

if y is a variable of type B and b a term of type B with no free occurrences of variables other than elements of X, it being assumed that b is substitutable for x, that is, no free variable in b becomes bound in $\varphi(b)$ or $\psi(b)$.

2. Logical rules

2.1. $p\vdash_X \top$;

2.1′. $\bot\vdash_X p$;

2.2. $r\vdash_X p\wedge q$ iff $r\vdash_X p$ and $r\vdash_X q$;

2.2′. $p\vee q\vdash_X r$ iff $p\vdash_X r$ and $q\vdash_X r$;

2.3. $p\vdash_X q\Rightarrow r$ iff $p\wedge q\vdash_X r$;

2.4. $p\vdash_X \forall_{y\in B}\psi(y)$ iff $p\vdash_{X\cup\{y\}}\psi(y)$;

2.4′. $\exists_{y\in B}\psi(y)\vdash_X p$ iff $\psi(y)\vdash_{X\cup\{y\}} p$.

The rules 2.1′, 2.2′ and 2.4′ can be derived if $\bot$, $\vee$ and $\exists$ are defined as above.

We write $\vdash$ for $\vdash_{\varnothing}$, that is, for $\vdash_X$ when X is the empty set. The reason for the subscript X on the entailment symbol becomes apparent when we look at the following 'proof tree'

$$\dfrac{\dfrac{\overset{(1.1)}{\forall_{x\in A}\varphi(x)\vdash\forall_{x\in A}\varphi(x)}}{\forall_{x\in A}\varphi(x)\vdash_{x}\varphi(x)}\,(2.4)\qquad\dfrac{\overset{(1.1)}{\exists_{x\in A}\varphi(x)\vdash\exists_{x\in A}\varphi(x)}}{\varphi(x)\vdash_{x}\exists_{x\in A}\varphi(x)}\,(2.4')}{\dfrac{\forall_{x\in A}\varphi(x)\vdash_{x}\exists_{x\in A}\varphi(x)}{\forall_{x\in A}\varphi(x)\vdash\exists_{x\in A}\varphi(x)}\,(1.4)}\,(1.2)$$

where the last step is justified by replacing every free occurrence of the variable x (there are none) by the closed term a of type A, *provided* there is such a closed term. Had we not insisted on the subscripts, we could have deduced this in any case, even when A is an empty type, that is, when there are no closed terms of type A. We then would have been able to infer from the fact that all unicorns have horns that some unicorns have horns. Aristotle would have avoided this conclusion by denying that all unicorns have horns. Some modern authors avoid it by abandoning the transitivity of entailment!

In what follows, we write $\vdash_X p$ for $\top\vdash_X p$.

3. Extralogical axioms

Comprehension:

3.1. $\vdash_X \forall_{x\in A}(x\in\{x\in A\,|\,\varphi(x)\}\Leftrightarrow\varphi(x))$.

Extensionality:

3.2. $\vdash\forall_{u\in PA}\forall_{v\in PA}(\forall_{x\in A}(x\in u\Leftrightarrow x\in v)\Rightarrow u=v)$,

3.3. $\vdash\forall_{s\in\Omega}\forall_{t\in\Omega}((s\Leftrightarrow t)\Rightarrow s=t)$.

Products:

3.4. $\vdash\forall_{z\in 1}z=*$,

3.5. $\vdash\forall_{z\in A\times B}\exists_{x\in A}\exists_{y\in B}z=\langle x,y\rangle$,

3.6. $\vdash\forall_{x\in A}\forall_{x'\in A}\forall_{y\in B}\forall_{y'\in B}(\langle x,y\rangle=\langle x',y'\rangle\Rightarrow(x=x'\wedge y=y'))$.

Peano axioms:

3.7. $\vdash\forall_{x\in N}(Sx=0\Rightarrow\bot)$;

3.8. $\vdash\forall_{x\in N}\forall_{y\in N}(Sx=Sy\Rightarrow x=y)$,

3.9. $\vdash\forall_{u\in PN}((0\in u\wedge\forall_{x\in N}(x\in u\Rightarrow Sx\in u))\Rightarrow\forall_{y\in N}y\in u)$.

The type theories described so far are intuitionistic. A *classical type theory* satisfies an additional axiom:

$$\forall_{t\in\Omega}(t\vee\neg t)$$

or, equivalently,

$$\forall_{t\in\Omega}(\neg\neg t \Rightarrow t).$$

Example 1.1. *Pure type theory* is the type theory in which there are no types or terms other than those defined inductively by the above closure rules, there are no non-trivial identifications between types, and $\vdash_X$ is the smallest binary relation between terms satisfying the stated axioms and rules of inference.

In pure type theory there are closed terms of each type, hence subscripts on the entailment symbol may be omitted. Indeed, we have the following closed terms of the indicated type:

$$\frac{1 \quad N \quad \Omega \quad PA \quad A\times B}{* \quad 0 \quad \top \quad \{x\in A \mid \bot\} \quad \langle a,b\rangle},$$

it being assumed that a and b are already known to be of types A and B respectively.

We shall prove later that pure type theory has the important

Existence Property. If $\vdash \exists_{x\in A}\varphi(x)$ then $\vdash \varphi(a)$ for some closed term a of type A.

It is possible to consider some variants of our formulation of type theory which are logically equivalent, but for which the existence property does not hold. For example, one might replace the term forming operation $\{x\in A\mid -\}$ together with the above comprehension scheme by the axiom scheme

$$\vdash \exists_{u\in PA}\forall_{x\in A}(x\in u \Leftrightarrow \varphi(x)),$$

but then it would be impossible to witness this statement by a term of the language. Similarly one could replace the term $*$ of type 1 by the axiom

$$\vdash \exists_{z\in 1}\top,$$

which could then no longer be witnessed.

One may formulate a seemingly stronger version of type theory where instead of $PA = \Omega^A$ one requires B^A for arbitrary types A and B. This would necessitate replacing $\{x\in A\mid \varphi(x)\}$ by $\lambda_{x\in A}\varphi(x)$ when $\varphi(x)$ is of type B.

Example 1.2. Given a graph $\mathscr{G}$, the type theory *generated* by $\mathscr{G}$ is defined as follows. Its types are generated inductively by the type forming operations $(-)\times(-)$ and $P(-)$ from the basic types $1, N, \Omega$ and the vertices of $\mathscr{G}$ (which now count as basic types). Its terms are generated inductively from the basic terms x_i^A, $*$, $0, \top$, $\bot$ by the old term forming operations $\langle -,-\rangle$, $(-)\in(-)$, $\{x\in A\mid -\}$, etc. together with the new term forming operations for each arrow $f: A\to B$ of $\mathscr{G}$: if $a\in A$ then $fa\in B$. Finally, its axioms and rules of inference are precisely those which follow from (*c*) and no others. Note that, if *G* is the empty graph, Example 1.2 reduces to Example 1.1.

Exercises

1. If $\bot$, $\vee$ and $\exists$ are defined, deduce the logical axioms and rules 2.1′, 2.2′ and 2.4′.
2. Prove the equivalence of the axioms $\forall_{t\in\Omega}(t \vee \neg t)$ and $\forall_{t\in\Omega}(\neg\neg t \Rightarrow t)$.
3. Prove that in a classical type theory

   $\neg\forall_{x\in A}\varphi(x) \vdash_X \exists_{x\in A}\neg\varphi(x)$.

4. Describe a strong version of type theory with exponential types B^A as a typed λ-calculus with an entailment relation.

2 Type theory based on equality

We shall consider another variant of type theory here in which all logical symbols are defined in terms of equality. This is less intuitive than the version presented in Section 1, but has a practical advantage in facilitating interpretation in certain categories, as we shall see.

Types are subject to the same closure conditions as in Section 1, but the closure conditions for terms, aside from assuring that there are countably many variables of each type, are now summarized more briefly as follows:

$$\frac{1 \quad N \qquad\quad PA \qquad\quad A\times B \quad \Omega}{\ast \quad \begin{matrix}0\\ Sn\end{matrix} \quad \{x\in A \mid \varphi(x)\} \quad \langle a,b\rangle \quad \begin{matrix}a\in\alpha\\ a=a'\end{matrix}},$$

where it is assumed that

$$\frac{N \quad\; \Omega \quad\; A \quad\; B \;\; PA}{n \quad \varphi(x) \quad \begin{matrix}a\\ a'\end{matrix} \quad b \;\; \alpha}.$$

We then introduce the old symbols $\top$, $\wedge$, $\Rightarrow$ and $\forall$ by definitions as follows:

$$\begin{aligned}
\top &\equiv \ast = \ast,\\
p\wedge q &\equiv \langle p,q\rangle = \langle \top,\top\rangle,\\
p\Rightarrow q &\equiv p\wedge q = p,\\
\forall_{x\in A}\varphi(x) &\equiv \{x\in A \mid \varphi(x)\} = \{x\in A \mid \top\}.
\end{aligned}$$

We already saw in Section 1 how to define $\perp$, $\vee$ and $\exists$ in terms of these.

It is convenient here to view *entailment* $\vdash_X$ not as a binary relation between formulas, but as a relation between finite sets of formulas and formulas. Thus we write

$$\Gamma \vdash_X q$$

to say that q is entailed by $\Gamma = \{p_1, \ldots, p_n\}$, where $n \geqslant 0$. It is assumed that all variables occurring freely in $p_1, \ldots, p_n$ and q are elements of X. Such a notion of entailment (without the subscript X) was first used by Gentzen. As is customary, we may also omit the braces $\{\}$ in Γ and write more briefly

$$p_1, \ldots, p_n \vdash_X q.$$

When $n = 0$, that is, $\Gamma = \varnothing$, this may also be written $\vdash_X q$. Furthermore, when $X = \varnothing$, the subscript X may be omitted.

$\Gamma \vdash_X q$ also has a conventional meaning when Γ is infinite. It then asserts that $\Gamma' \vdash_X q$ for some finite subset Γ' of Γ.

It remains to lay down the axioms and rules of inference for the new entailment relation.

1. Structural rules

1.1. $p \vdash_X p$;

1.2. $$\frac{\Gamma \vdash_X p \quad \Gamma \cup \{p\} \vdash_X q}{\Gamma \vdash_X q};$$

1.3. $$\frac{\Gamma \vdash_X q}{\Gamma \cup \{p\} \vdash_X q};$$

1.4. $$\frac{\Gamma \vdash_X q}{\Gamma \vdash_{X \cup \{y\}} q};$$

1.5. $$\frac{\Gamma(y) \vdash_{X \cup \{y\}} \varphi(y)}{\Gamma(b) \vdash_X \varphi(b)},$$

where it is assumed that b may be substituted for y in $\varphi(y)$ and $\Gamma(y) = \{\varphi_1(y), \ldots, \varphi_n(y)\}$, that is, no occurrence of a free variable in b becomes bound in $\varphi_i(b)$ or $\varphi(b)$.

2. Pure equality rules

1.2. $\vdash_X a = a$;

2.2. $a = b, \quad \varphi(a) \vdash_X \varphi(b)$,

where it is assumed that a and b may be substituted for x in $\varphi(x)$;

2.3. $$\frac{\Gamma, p \vdash_X q \quad \Gamma, q \vdash_X p}{\Gamma \vdash_X p = q}.$$

3. Other axioms and rules.

3.1. $\langle a,b\rangle = \langle c,d\rangle \vdash_X a = c$;

3.2. $\langle a,b\rangle = \langle c,d\rangle \vdash_X b = d$;

3.3. $\vdash_X (x \in \{x \in A \mid \varphi(x)\}) = \varphi(x)$,

where it is assumed that x is an element of X;

3.4. $$\frac{\Gamma \vdash_{X \cup \{x\}} \varphi(x) = x \in \alpha}{\Gamma \vdash_X \{x \in A \mid \varphi(x)\} = \alpha},$$

where it is assumed that x is not free in Γ;

3.5. $\vdash_z z = *$,

where it is assumed that z is of type 1;

3.6. $$\frac{\Gamma, z = \langle x, y\rangle \vdash_{X \cup \{x,y,z\}} \varphi(z)}{\Gamma \vdash_{X \cup \{z\}} \varphi(z)},$$

where it is assumed that x and y are not free in Γ or $\varphi(z)$;

3.7. $Sx = 0 \vdash_{X \cup \{x\}} p$;

3.8. $Sx = Sy \vdash_{\{x,y\}} x = y$;

3.9. $$\frac{\Gamma \vdash_X \varphi(0) \quad \Gamma, \varphi(x) \vdash_{X \cup \{x\}} \varphi(Sx)}{\Gamma \vdash_{X \cup \{x\}} \varphi(x)}$$

where it is assumed that x is not free in Γ.

We wish to compare the traditional (old) type theories of Section 1 with the (new) type theories based on equality of Section 2.

Proposition 2.1. Every traditional type theory $\mathscr{O}$ gives rise to a type theory $N(\mathscr{O})$ based on equality, provided $=$ is defined by Leibniz' rule and $p_1, \ldots, p_n \vdash_X q$ is taken to mean $p_1 \wedge \ldots \wedge p_n \vdash_X q$, in particular, $\varnothing \vdash_X q$ means $\top \vdash q$.

Proof. Types and variables have not changed. It is an easy exercise to check that the new axioms and rules of inference are theorems and derived rules of inference in the old system. We shall just give a sample proof of new rule 3.6, which perhaps requires some explanation. To simplify the argument, we take Γ and X empty.

Suppose

$$z = \langle x, y \rangle \vdash_{\{x,y,z\}} \varphi(z).$$

Then

$$\exists_{x \in A} \exists_{y \in B} z = \langle x, y \rangle \vdash_{z} \varphi(z),$$

by old 2.4′ twice. But

$$\vdash_{z} \exists_{x \in A} \exists_{y \in B} z = \langle x, y \rangle,$$

by old 3.5 and 2.4. Therefore

$$\vdash_{z} \varphi(z),$$

by old 1.2.

Proposition 2.2. Every type theory $\mathcal{N}$ based on equality gives rise to a traditional type theory $O(\mathcal{N})$, provided $\top$, $\wedge$, $\Rightarrow$ and $\forall$ are defined in terms of $=$ and $p \vdash_{X} q$ means $\{p\} \vdash_{X} q$.

Proof. Types and variables do not change. It is an exercise, but not quite a routine one, to verify that the old axioms and rules of inference are theorems and derived rules in the new system. We shall content ourselves with two samples here, leaving some of the rules of equality to Lemma 2.3 below.

As our first example, we shall establish the old 2.2:

$$r \vdash_{X} p \wedge q \quad \text{if and only if } r \vdash_{X} p \text{ and } r \vdash_{X} q.$$

For simplicity, we shall omit the subscript.

Assume that

$$r \vdash p \wedge q.$$

This means

$$r \vdash \langle p, q \rangle = \langle \top, \top \rangle.$$

Now, by new 3.1 and 1.3,

$$r, \langle p, q \rangle = \langle \top, \top \rangle \vdash p = \top.$$

Hence, by new 1.2,

$$r \vdash p = \top.$$

But, by Lemma 2.3 below,

$$p = \top \vdash p.$$

Therefore, by new 1.3 and 1.2,

$$r \vdash p.$$

Similarly,

$$r \vdash q.$$

Conversely, suppose

$$r \vdash p \quad \text{and} \quad r \vdash q,$$

we wish to show that $r \vdash p \wedge q$, that is, $r \vdash \langle p, q \rangle = \langle \top, \top \rangle$. First we argue as follows, for convenience in tree form:

$$\dfrac{\dfrac{\overset{\text{(given)}}{r \vdash p} \qquad \overset{\text{(Lemma 2.3)}}{p \vdash p = \top}}{r \vdash p = \top} \qquad \overset{(2.2)}{p = \top \vdash \langle p, q \rangle = \langle \top, q \rangle}}{r \vdash \langle p, q \rangle = \langle \top, q \rangle}.$$

(Transitivity of entailment follows of course from 1.3 and 1.2.) Similarly

$$r \vdash \langle \top, q \rangle = \langle \top, \top \rangle.$$

Using transitivity of equality (see Lemma 2.3 below), we infer the desired result.

As a second example, we shall derive the old 3.5, assuming that the old logical rules have already been established.

We begin by citing a special case of the old 1.1: if z is a variable of type $A \times B$,

$$\exists_{x \in A} \exists_{y \in B} z = \langle x, y \rangle \vdash_{z} \exists_{x \in A} \exists_{y \in B} z = \langle x, y \rangle.$$

Using the old 2.4′ twice, we infer that

$$z = \langle x, y \rangle \vdash_{\{x, y, z\}} \exists_{x \in A} \exists_{y \in B} z = \langle x, y \rangle.$$

In view of the new rule 3.6 and transitivity of entailment we deduce

$$\vdash_{z} \exists_{x \in A} \exists_{y \in B} z = \langle x, y \rangle.$$

Using the old 2.4, we obtain the desired result. Incidentally, it appears from this proof that we might as well have taken Γ empty in 3.6.

It remains to state and prove the promised lemma.

Lemma 2.3. The following derived rules of equality hold in any type theory based on equality:

(i) $a = b \vdash_{X} b = a$;

(ii) $a = b, \varphi(b) \vdash_{X} \varphi(a)$;

(iii) $a = b, b = c \vdash_{X} a = c$;

(iv) $p = \top \vdash_{X} p$;

(v) $p \vdash_{X} p = \top$.

Proof. For simplicity we take X empty. Specializing 2.2 to the case $\varphi(x) \equiv x = a$, we obtain

$$a = b,\ a = a \vdash b = a.$$

Citing 1.1 and using 1.3 and 1.2, we infer (i).

From (i) and 1.3 we get

$$a = b,\ \varphi(b) \vdash b = a,$$

and from 2.2 and 1.3 we get

$$a = b,\ \varphi(b), b = a \vdash \varphi(a).$$

Therefore, taking $\Gamma \equiv \{a = b, \varphi(b)\}$, $p \equiv b = a$ and $q \equiv \varphi(a)$, we use 1.2 to infer $\Gamma \vdash q$, that is, (ii).

(iii) is a special case of (ii) with $\varphi(x) \equiv x = c$.

Next, we use 2.1 to infer $\vdash * = *$, that is, $\vdash \top$. Now by (ii),

$$p = \top,\ \top \vdash p.$$

Hence, by 1.3 and 1.2, (iv) follows.

Finally, from $\vdash \top$ and 1.3 we have $p, p \vdash \top$. Also, from 1.1 and 1.3 we have $p, \top \vdash p$.

Hence, by 2.3 taking $\Gamma \equiv \{p\}$ and $q \equiv \top$, we get (v).

Up to now, terms were generated freely by certain term forming operations in both old and new type theories. From now on, we shall usually identify terms if they are provably equal.

Theorem 2.4. Traditional type theories and type theories based on equality are equivalent in the sense that

$$ON(\mathcal{O}) = \mathcal{O}, \quad NO(\mathcal{N}) = \mathcal{N}$$

for all traditional $\mathcal{O}$ and type theories based on equality $\mathcal{N}$.

Proof. Given an old type theory $\mathcal{O}$, we form the old type theory $\mathcal{O}' = ON(\mathcal{O})$. On the face of it, the term forming operations in $\mathcal{O}'$ are different from those in $\mathcal{O}$. For example, $p \wedge q$ in $\mathcal{O}'$ is $\langle p, q \rangle = \langle \top, \top \rangle$ in $\mathcal{O}$, where $=$ is defined by Leibniz's rule. However in $\mathcal{O}$ one can prove $\langle p, q \rangle = \langle \top, \top \rangle \vdash_X p \wedge q$ and conversely, hence $\langle p, q \rangle = \langle \top, \top \rangle$ and $p \wedge q$ are provably equal by old 3.3. In this way one shows that the term forming operations in $\mathcal{O}'$ are really the same as those in $\mathcal{O}$, if attention is paid to provable equality between formulas in $\mathcal{O}$. Moreover, $p \vdash_X q$ in $\mathcal{O}'$ if and only if $\{p\} \vdash_X q$ in $N(\mathcal{O})$, that is, $p \vdash_X q$ in $\mathcal{O}$. Thus $\mathcal{O}$ and $\mathcal{O}' = ON(\mathcal{O})$ are the same theories. In a similar manner one shows that for any new type theory $\mathcal{N}, \mathcal{N}$ and $NO(\mathcal{N})$ are the same. Later we shall consider morphisms between type theories called 'translations', so that type theories, traditional or based on equality, will form a category. The assignments N and O of propositions 2.1 and 2.2 will then become functors and Theorem 2.4 will assert that the category of traditional type theories is isomorphic to the category of type theories based on equality.

Type theory based on equality also has a seemingly stronger version where instead of $PA = \Omega^A$ one requires B^A for all types A and B.

Exercises

1. Prove Proposition 2.1.
2. Complete the proof of Proposition 2.2.
3. Prove that in any traditional type theory the definitions of $\top$, $\wedge$, $\Rightarrow$ and $\forall$ in terms of $=$ become theorems.
4. Prove that in any type theory based on equality Leibniz's rule is a theorem.
5. Describe a strong version of type theory with exponential types B^A as a λ-calculus with equality and entailment.

3 The internal language of a topos

In this section we shall associate with every topos $\mathscr{T}$ a type theory $L(\mathscr{T})$, its so-called internal language.

An (elementary) *topos* $\mathscr{T}$ is a cartesian closed category in which the subobject functor is representable. What this means is that there is given an object Ω, called the *subobject classifier* and a natural isomorphism

$$\mathrm{Sub} \cong \mathrm{Hom}(-, \Omega).$$

More precisely, it means that there is given an arrow $\top: 1 \to \Omega$ such that

(i) for every arrow $h: A \to \Omega$ an equalizer of h and $\top \bigcirc_A: A \to 1 \to \Omega$ exists, call it a *kernel* of h and write

$$\ker h: \mathrm{Ker}\, h \to A;$$

(ii) for every monomorphism* $m: B \to A$ there is a unique arrow $\mathrm{char}\, m$: $A \to \Omega$, called its *characteristic* morphism, such that m is a kernel of $\mathrm{char}\, m$.

The statement that $m: B \to A$ is a kernel of $h: A \to \Omega$, or that h is the characteristic morphism of m, may also be expressed by saying that the following square is a pullback:

* An arrow m in a category $\mathscr{C}$ is a *monomorphism* if $ma = mb$ implies $a = b$. An *epimorphism* is a monomorphism in $\mathscr{C}^{\mathrm{op}}$.

$$\begin{array}{ccc} B & \xrightarrow{\bigcirc_B} & 1 \\ {\scriptstyle m}\big\downarrow & & \big\downarrow{\scriptstyle \top} \\ A & \xrightarrow[h]{} & \Omega \end{array}$$

While characteristic morphisms are unique, kernels are only unique up to isomorphism. Thus

$$\text{char}(\ker h) \cdot\!=\!\cdot h, \quad \ker(\text{char}\, m) \cong m.$$

The reader will notice that we denote equality of arrows in a topos by the symbol $\cdot\!=\!\cdot$; this is to distinguish it from the internal equality to be introduced later.

From now on we shall include in the definition of a topos the requirement that it possess a natural numbers object in the sense of Lawvere. This means that there is a diagram

$$1 \xrightarrow{0} N \xrightarrow{S} N$$

initial in the category of all diagrams

$$1 \xrightarrow{a} A \xrightarrow{f} A,$$

that is, for every such diagram there is a unique arrow $g: N \to A$ such that

$$g0 \cdot\!=\!\cdot a, \quad gS \cdot\!=\!\cdot fg.$$

It is possible to give an alternative definition of a topos which is a little more economical: instead of requiring B^A for all B and A, it suffices to require this only when $B = \Omega$ (see Section 13). We write PA for Ω^A and think of this as the power set of A. However, for now, we shall insist that a topos is cartesian closed. We repeat:

Definition 3.1. A *topos* is a cartesian closed category with a subobject classifier and a natural numbers object.

Many examples of toposes will be found in sections 9 and 10.

For future reference we shall state and prove two immediate consequences of Definition 3.1.

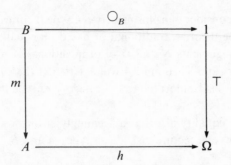

Proposition 3.2. Let $\delta_B: B \to \Omega$ be the characteristic morphism of the

monomorphism $\langle 1_B, 1_B\rangle: B \to B \times B$. For any two arrows $f, g: A \rightrightarrows B$,

$$f \cdot = \cdot g \quad \text{if and only if } \delta_B\langle f, g\rangle \cdot = \cdot \top \bigcirc_A.$$

Proof. One implication is immediate, since $\delta_B\langle f, f\rangle \cdot = \cdot \delta_B\langle 1_B, 1_B\rangle f \cdot = \cdot \top \bigcirc_B f \cdot = \cdot \top \bigcirc_A$. To prove the other implication, assume $\delta_B\langle f, g\rangle \cdot = \cdot \top \bigcirc_A$. Now the following square is a pullback:

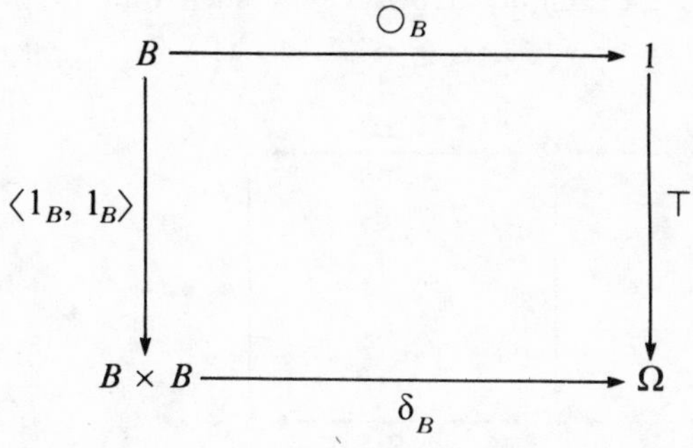

Therefore, there exists a unique arrow $k: A \to B$ such that

$$\langle f, g\rangle \cdot = \cdot \langle 1_B, 1_B\rangle k \cdot = \cdot \langle k, k\rangle.$$

Applying $\pi_{B,B'}$ and $\pi'_{B,B}$, to this, we obtain

$$f \cdot = \cdot k \cdot = \cdot g.$$

The second consequence of Definition 3.1 we have in mind is a categorical version of a scheme that is used in the definition of primitive recursive functions.

Proposition 3.3. (Recursion scheme). Given any object A in a cartesian closed category with natural numbers object and arrows $a: 1 \to A$ and $h: N \times A \to A$, there is a unique arrow $g: N \to A$ such that

$$g0 \cdot = \cdot a, \quad gS \cdot = \cdot h\langle 1_N, g\rangle.$$

Proof. In view of the universal property of the natural numbers object, there is a unique arrow $\langle f, g\rangle: N \to N \times A$ such that the following diagram commutes:

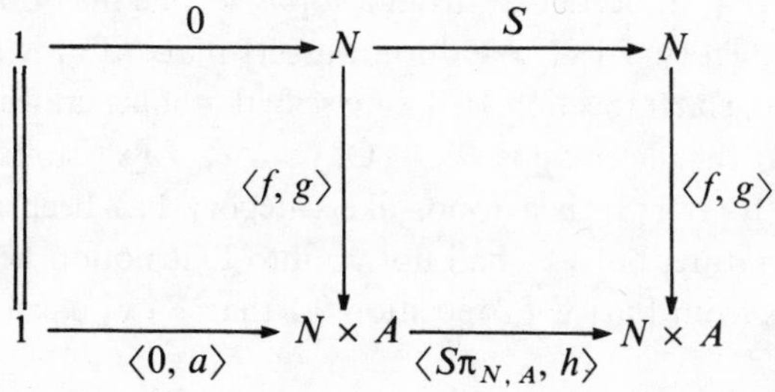

This translates into the following four equations:

(i) $f0 \cdot = \cdot 0,$

(ii) $fS \cdot = \cdot Sf,$

(iii) $g0 \cdot = \cdot a,$

(iv) $gS \cdot = \cdot h\langle f, g\rangle.$

Now the identity is the unique arrow $N \to N$ such that the following diagram commutes:

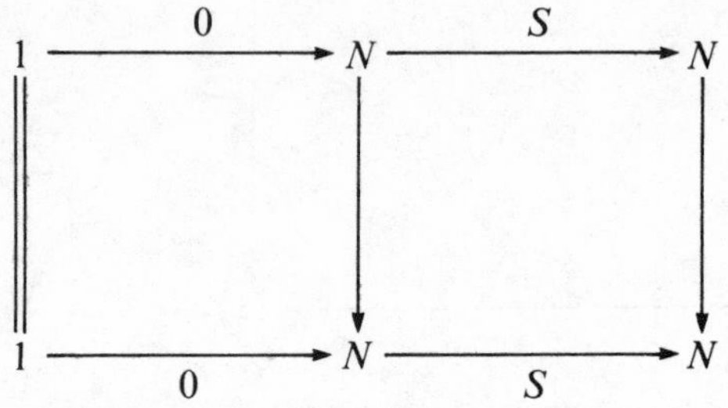

Therefore $f \cdot = \cdot 1_N$, in view of (i) and (ii). Then (iv) becomes $gS \cdot = \cdot h\langle 1_N, g\rangle$, and (iii) and (iv) are the required equations.

Corollary 3.4. In any cartesian closed category with natural numbers object, $1 \xrightarrow{0} N \xleftarrow{S} N$ is a coproduct diagram, so that $N + 1 \cong N$.

Proof. Suppose we are given $1 \xrightarrow{a} A \xleftarrow{k} N$. Put $h \cdot = \cdot k\pi_{N,A}$ and apply Proposition 3.3 to obtain a unique arrow $g: N \to A$ such that $g0 \cdot = \cdot a$ and $gS \cdot = \cdot h\langle 1_N, g\rangle \cdot = \cdot k\pi_{N,A}\langle 1_N, g\rangle \cdot = \cdot k.$

Note that g is usually written as $[a, k]$ and that the existence and uniqueness of g may be stated equationally by $[a, k]0 \cdot = \cdot a$, $[a, k]S \cdot = \cdot k$, $[g0, gS] \cdot = \cdot g$, for all $g: N \to A$ (see Part I, Section 8).

At this point it is useful to recall from Part I, Section 5 that to any cartesian closed category $\mathscr{A}$ one may adjoin an indeterminate arrow $x: 1 \to A$, where A is any object of $\mathscr{A}$, to obtain the polynomial cartesian closed category $\mathscr{A}[x]$. In particular, from a topos $\mathscr{T}$ one may obtain the cartesian closed category $\mathscr{T}[x]$. We do not assert that $\mathscr{T}[x]$ is a topos, although we shall see later (Section 16, Exercise 2) that it generates a topos $\mathscr{T}(x)$ equivalent to the slice topos $\mathscr{T}/A$. Of course, $\mathscr{T}[x]$ shares many properties with a topos. (Such a topos-like category has been called a 'dogma' in the literature, but we shall not go into that notion here.) For example, it follows from Part I, Proposition 9.1 that $\mathscr{T}[x]$ has a natural numbers object.

We also recall from Part I, Section 6 the important property of *functional completeness* of cartesian closed categories. In particular, this property implies that for every polynomial $\varphi(x): 1 \to B$ in $\mathscr{T}[x]$, depending on the indeterminate $x: 1 \to A$, there is a unique arrow $f: A \to B$ in $\mathscr{T}$ such that

$$\varphi(x) \cdot \underset{x}{=} \cdot fx.$$

Here $\cdot \underset{x}{=} \cdot$ is the equality relation in $\mathscr{T}[x]$. Another way of putting this is that there is a unique arrow $g: 1 \to B^A$ in $\mathscr{T}$ such that

$$\varphi(x) \cdot \underset{x}{=} \cdot g^{\ulcorner} x \equiv \varepsilon_{B,A}\langle g, x \rangle.$$

One also writes

$$g \equiv \lambda_{x \in A}\varphi(x).$$

Definition 3.5. The *internal language* $L(\mathscr{T})$ of a topos $\mathscr{T}$ has as types the objects of $\mathscr{T}$. It is understood that the type Ω is the object Ω, the type N is the object N, etc. It has as terms of type A in the variables x_i of type A_i $(i = 1, \ldots, n)$ polynomial expressions $\varphi(x_1, \ldots, x_n): 1 \to A$ in the indeterminate arrows $x_i: 1 \to A_i$ over $\mathscr{T}$. In particular,

variables of type A are indeterminate arrows $1 \to A$,

$*$ is $1 \to 1$,

0 is $0: 1 \to N$,

Sn is $1 \xrightarrow{n} N \xrightarrow{S} N$,

$\langle a, b \rangle$ is $1 \xrightarrow{\langle a, b \rangle} A \times B$,

$a = a'$ is $1 \xrightarrow{\langle a, a' \rangle} A \times A \xrightarrow{\delta_A} \Omega$, where $\delta_A \equiv \mathrm{char}\langle 1_A, 1_A \rangle$,

$a \in \alpha$ is $1 \xrightarrow{\langle \alpha, a \rangle} PA \times A \xrightarrow{\varepsilon_A} \Omega$, where $\varepsilon_A \equiv \varepsilon_{\Omega, A}$,

$\{x \in A \mid \varphi(x)\}$ is $\lambda_{x \in A}\varphi(x)$, the unique arrow $\alpha: 1 \to PA$ such that $x \in \alpha \cdot \underset{X \cup \{x\}}{=} \cdot \varphi(x)$.

It is understood in the above that n, a, b, a', α and $\varphi(x)$ have already been suitably interpreted as arrows. A more careful description of the terms in $L(\mathscr{T})$ would proceed as in Example 10.6.

Furthermore, if $X = \{x_1, \ldots, x_m\}$,

$$\varphi_1(X), \ldots, \varphi_n(X) \vdash_X \varphi_{n+1}(X)$$

means that, for all objects C of $\mathscr{T}$ and all arrows $h: C \to A$,

$$\text{if } f_i h \cdot = \cdot \top \bigcirc_C \quad (i = 1, \ldots, n) \text{ then } f_{n+1} h \cdot = \cdot \top \bigcirc_C,$$

where f_i $(i = 1, \ldots, n+1)$ is defined by functional completeness as follows:

$$f_i\langle x_1, \ldots, x_m\rangle \cdot{=}\cdot \varphi_i(X).$$

In other words, putting $A \equiv A_1 \times A_2 \times \cdots \times A_m$ (association on the left),

$$\ulcorner f_i \urcorner \cdot{=}\cdot \lambda_{x \in A} \varphi_i(\pi_{A,A_1} x, \ldots, \pi_{A,A_n} x).$$

It is easily verified that the internal language of a topos is indeed a type theory based on equality, except perhaps for the Peano axioms. We shall return to this point later.

The reader will recall that we have written $\cdot{=}\cdot$ for equality of arrows in the topos $\mathscr{T}$ (*external equality*) to distinguish it from $=$, which is here the equality symbol in $L(\mathscr{T})$ (*internal equality*).

In particular, $\vdash p$ in $L(\mathscr{T})$ means that $p \cdot{=}\cdot \top$ as arrows in $\mathscr{T}$. It will be convenient to have yet another notation for this, namely $\mathscr{T} \models p$, saying that the topos $\mathscr{T}$ *satisfies* the proposition p or that p *holds* in $\mathscr{T}$. This is particularly useful when p is a closed formula of pure type theory, which may be viewed as a kind of sublanguage of $L(\mathscr{T})$. In fact we thus have an interpretation of pure type theory in any topos $\mathscr{T}$. Similarly, $\vdash_X \psi(x)$ means $\psi(x) \cdot \underset{x}{=} \cdot \top$.

More generally, what is the meaning of $\varphi(X) \vdash_X \psi(X)$ according to the above definition? To simplify the discussion, take $X = \{x\}$, where $x: 1 \to A$. In view of functional completeness, we may write

$$\varphi(x) \cdot \underset{x}{=} \cdot fx, \quad \psi(x) \cdot \underset{x}{=} \cdot gx,$$

where f and g are uniquely determined arrows $A \to \Omega$. According to Definition 3.5, $fx \vdash_x gx$ means: for all $h: C \to A$, $fh \cdot{=}\cdot \top \bigcirc_C$ implies $gh \cdot{=}\cdot \top \bigcirc_C$. In particular, if we take g as a kernel of f, this condition asserts that a kernel of f is contained in a kernel of g.

In particular, if $fx \vdash_x gx$ and $gx \vdash_x fx$, we may infer that f and g have the same kernels. Since characteristic morphisms are unique, this implies that $f \cdot{=}\cdot g$. We have thus established, in the case $X = \{x\}$, that the following rule holds in the internal language of any topos:

$$\frac{\varphi(X) \vdash_X \psi(X) \quad \psi(X) \vdash_X \varphi(X)}{\varphi(X) \cdot \underset{X}{=} \cdot \psi(X)}.$$

Exercises

1. Show that in a topos $f_1 x, \ldots, f_n x \vdash gx$ means that $\ker f_1 \cap \ldots \cap \ker f_n \subseteq \ker g$.

2. Verify that the internal language of a topos satisfies all the axioms and rules of inference of type theory based on equality other than Peano's axioms (which will be discussed in Section 4).

4 Peano's rules in a topos

We shall prove that Peano's rules hold in any topos, thus completing the proof of the following:

Theorem 4.1. The internal language of a topos is a type theory based on equality.

For the rest of the proof of Theorem 4.1, the reader is referred back to Exercise 2 of Section 3.

Peano's first rule. For example, if X is empty, this may be written thus:

P1. $\delta_N\langle S, 0\bigcirc_N\rangle x \vdash_x fx,$

where $f: N \to \Omega$ is any arrow. (Usually f is the arrow $\bot\bigcirc_N: N \to 1 \to \Omega$.)

According to Definition 3.5, this means: for all objects C and all arrows $h: C \to N$, if

(i) $\delta_N\langle S, 0\bigcirc_N\rangle h \cdot = \cdot \top\bigcirc_C,$

then

(ii) $fh \cdot = \cdot \top\bigcirc_C.$

Assume (i), which, by Proposition 3.2, may be written

$$Sh \cdot = \cdot 0\bigcirc_N h \cdot = \cdot 0\bigcirc_C.$$

Now, by Corollary 3.4, there is a commutative diagram:

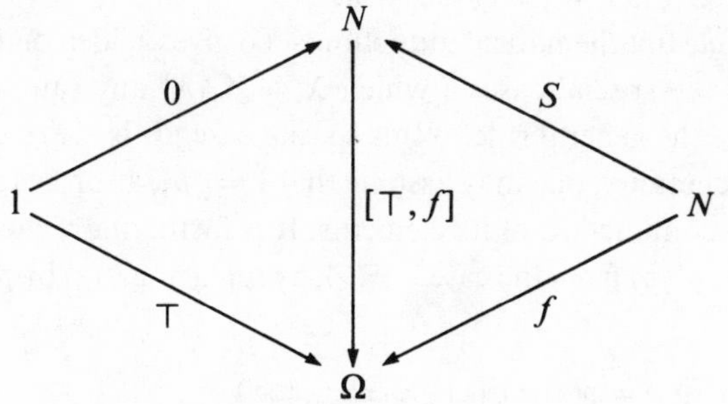

Thus

$$[\top, f]0 \cdot = \cdot \top, \quad [\top, f]S \cdot = \cdot f,$$

hence

$$fh \cdot = \cdot [\top, f]Sh \cdot = \cdot [\top, f]0\bigcirc_C \cdot = \cdot \top \bigcirc_C,$$

which is (ii).

Peano's second rule. This may be written

P2. $\delta_N\langle S\pi_{N,N'}, S\pi'_{N,N'}\rangle\langle x, y\rangle \vdash_{\{x,y\}} \delta_N\langle x, y\rangle.$

According to Definition 3.5, this means: for all objects C and all arrows $\langle h, k\rangle: C \to N \times N$,

$$\text{if } \delta_N\langle Sh, Sk\rangle \cdot = \cdot \top \bigcirc_C \quad \text{then} \quad \delta_N\langle h, k\rangle \cdot = \cdot \top \bigcirc_C.$$

In view of Proposition 3.2, this implication may be rendered as follows:

$$\text{if } Sh \cdot = \cdot Sk \quad \text{then} \quad h \cdot = \cdot k.$$

In other words, P2 asserts that S is a monomorphism.

By Corollary 3.4, we have the following commutative diagram:

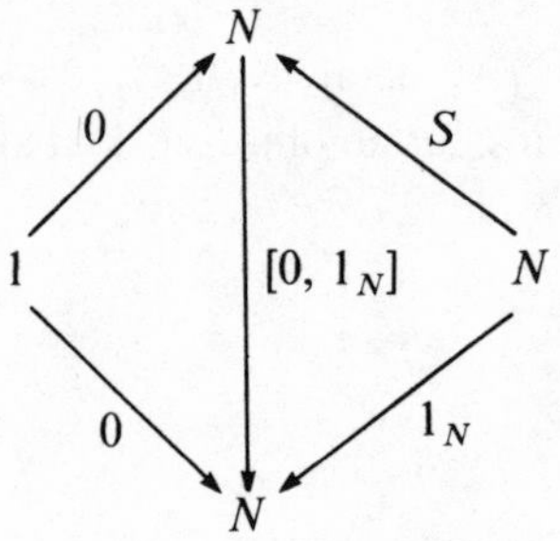

Thus

$$[0, 1_N]0 \cdot = \cdot 0, \quad [0, 1_N]S \cdot = \cdot 1_N.$$

$[0, 1_N]$ is of course the usual predecessor arrow. Assuming that $Sh \cdot = \cdot Sk$, we then have $h \cdot = \cdot [0, 1_N]Sh \cdot = \cdot [0, 1_N]Sk \cdot = \cdot k$.

Peano's third rule (mathematical induction). To give an idea of its proof, we shall look at the special case in which $X = \varnothing$. (At any rate, it is not difficult to reduce the general rule to this special case.) It is also clear that, without loss in generality, one may assume that $\Gamma = \{p\}$, as one can always replace Γ by the conjunction of its elements. It is furthermore possible to deduce the case $\Gamma = \{p\}$ from the case $\Gamma = \varnothing$, by replacing $\varphi(x)$ by $p \Rightarrow \varphi(x)$, as follows:

$$\dfrac{\dfrac{\dfrac{\dfrac{p \vdash \varphi(0)}{\vdash p \Rightarrow \varphi(0)} \qquad \dfrac{\dfrac{p, p \Rightarrow \varphi(x) \vdash_x \varphi(x) \quad \varphi(x) \vdash_x \varphi(Sx)}{p, p \Rightarrow \varphi(x) \vdash_x \varphi(Sx)}}{p \Rightarrow \varphi(x) \vdash_x p \Rightarrow \varphi(Sx)}}{\vdash_x p \Rightarrow \varphi(x)}}{p \vdash_x \varphi(x)}.$$

It therefore remains to prove Peano's third rule in the following form:

$$\text{P3.}\qquad \frac{\vdash \varphi(0) \quad \varphi(x) \vdash_{x} \varphi(Sx)}{\vdash_{x} \varphi(x)}.$$

Before showing that P3 holds in $L(\mathscr{T})$, we wish to introduce a symbol for the binary relation 'less than' between natural numbers. Apply Proposition 3.3 to the situation $a\colon 1 \to PN, h\colon N \times PN \to PN$, where

$$a \cdot = \cdot \{x \in N \mid \bot\},$$
$$h\langle x, u\rangle \cdot \underset{\{x,u\}}{=} \cdot \{x\} \cup u \cdot \underset{\{x,u\}}{=} \cdot \ \{y \in N \mid y = x \vee y \in u\}.$$

By Proposition 3.3, there is a unique arrow $g\colon N \to PN$ such that

$$g0 \cdot = \cdot \{x \in N \mid \bot\}, \quad gSx \cdot \underset{x}{=} \cdot \{y \in N \mid y = x \vee y \in gx\}.$$

Now define

$$m < n \equiv m \in gn$$

for terms m, n of type N. Then we have

(i) $\qquad x < 0 \cdot \underset{x}{=} \cdot \bot,$

(ii) $\qquad y < Sx \cdot \underset{\{x,y\}}{=} \cdot y = x \vee y < x.$

Let us now return to P3. We are given that $\vdash \varphi(0)$ and $\varphi(x) \vdash_{x} \varphi(Sx)$ and wish to prove that $\vdash_{x} \varphi(x)$. To do so, we pass from ordinary induction to 'course of values induction'. Define the arrow $f\colon N \to \Omega$ by

$$fx \cdot \underset{x}{=} \cdot \forall_{y \in N}(y < Sx \Rightarrow \varphi(y)).$$

Then

$$\begin{aligned} f0 &\cdot = \cdot \forall_{y \in N}(y < S0 \Rightarrow \varphi(y)) \\ &\cdot = \cdot \forall_{y \in N}((y = 0 \vee y < 0) \Rightarrow \varphi(y)) && \text{by (ii)} \\ &\cdot = \cdot \forall_{y \in N}(y = 0 \Rightarrow \varphi(y)) && \text{by (i)} \\ &\cdot = \cdot \varphi(0) \cdot = \cdot \top && \text{(given).} \end{aligned}$$

Also

$$\begin{aligned} fSx &\cdot \underset{x}{=} \cdot \forall_{y \in N}(y < S(Sx) \Rightarrow \varphi(y)) \\ &\cdot \underset{x}{=} \cdot \forall_{y \in N}((y = Sx \vee y < Sx) \Rightarrow \varphi(y)) && \text{by (ii)} \\ &\cdot \underset{x}{=} \cdot \varphi(Sx) \wedge fx \cdot \underset{x}{=} \cdot fx, \end{aligned}$$

because

$$\begin{aligned} fx &\cdot \underset{x}{=} \cdot \forall_{y \in N}(y < Sx \Rightarrow \varphi(y)) \\ &\cdot \underset{x}{=} \cdot \forall_{y \in N}((y = x \vee y < x) \Rightarrow \varphi(y)) && \text{by (ii)} \end{aligned}$$

and

$$\varphi(x) \vdash_{x} \varphi(Sx) \qquad\qquad \text{(given).}$$

Thus

$$f0 \cdot = \cdot \top, \quad fS \cdot = \cdot f.$$

Now, by Lawvere's definition of the natural numbers object, there is a unique arrow $N \to \Omega$ to make the following diagram commute:

$$\begin{array}{ccccc} 1 & \xrightarrow{\;0\;} & N & \xrightarrow{\;S\;} & N \\ \| & & \downarrow & & \downarrow \\ 1 & \xrightarrow[\;\top\;]{} & \Omega & \xrightarrow[\;1_\Omega\;]{} & \Omega \end{array}$$

As we have just shown, the arrow $f: N \to \Omega$ will do so, but so will the arrow $\top \bigcirc_N: N \to 1 \to \Omega$. Therefore $f \cdot = \cdot \top \bigcirc_N$, that is, $fx \cdot \underset{x}{=} \cdot \top$. Since $fx \underset{x}{\vDash} \varphi(x)$, also $\varphi(x) \cdot \underset{x}{=} \cdot \top$, that is, $\underset{x}{\vDash} \varphi(x)$. The proof of P3 is now complete.

Exercise

Show how to define $\leqslant$ using the recursion scheme.

5 The internal language at work

The internal language of a topos $\mathscr{T}$ may be put to work to express various properties of arrows internally or even to carry out certain categorical constructions.

The following result is sometimes summarized by the slogan: *external equality in a topos is equivalent to (provable) internal equality.*

Proposition 5.1 For any arrows $f, g: A \rightrightarrows B$ in a topos $\mathscr{T}$, $f \cdot = \cdot g$ if and only if

$$\mathscr{T} \vDash \forall_{x \in A} fx = gx.$$

Proof. That external equality implies internal equality is evident. We shall prove the converse.

That in $L(\mathscr{T})$ we have $\vdash \forall_{x \in A} fx = gx$ means $\underset{x}{\vDash} fx = gx$, by Rule 2.4 of Section 1, that is, $fx = gx \cdot \underset{x}{=} \cdot \top$, that is, $\delta_B \langle f, g \rangle \cdot = \cdot \top \bigcirc_A$, by functional completeness. Therefore $f \cdot = \cdot g$, by Proposition 3.2.

The following lemma turns out to be useful.

Lemma 5.2. Given a monomorphism $m: B \to C$ and an arrow $f: A \to C$ in the topos $\mathscr{T}$ such that

$$\mathscr{T} \vDash \forall_{x \in A} \exists_{y \in B} my = fx,$$

there is a unique arrow $g: A \to B$ such that $mg \cdot = \cdot f$.

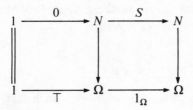

Proof. Since $\mathscr{T}$ is a topos, m is a kernel of its characteristic morphism $h: C \to \Omega$. In particular

$$hm \cdot = \cdot \top \bigcirc_C m \cdot = \cdot \top \bigcirc_B .$$

Now clearly

$$my = fx \vdash_{\{x,y\}} hfx = hmy,$$

hence

$$my = fx \vdash_{\{x,y\}} hfx = \top .$$

In view of Lemma 2.3, we infer that

$$my = fx \vdash_{\{x,y\}} hfx,$$

therefore

$$\exists_{y \in B} my = fx \vdash_x hfx,$$

by the usual rule of existential specification (see Section 1, rule 2.4′). As we are given that

$$\vdash_x \exists_{y \in B} my = fx,$$

it follows that

$$\vdash_x hfx.$$

According to our interpretation of $\vdash_x$ in $\mathscr{T}$ (see Definition 3.5), this means that, for all $k: D \to A$,

$$hfk \cdot = \cdot \top \bigcirc_D .$$

So, in particular,

$$hf \cdot = \cdot \top \bigcirc_A \cdot = \cdot \top \bigcirc_C f.$$

Since m is an equalizer of h and $\top \bigcirc_C$, it follows that $f \cdot = \cdot mg$ for a unique $g: A \to B$.

We recall that in a topos, as in any cartesian closed category, there is a one-to-one correspondence between arrows $h: C \to \Omega$ and arrows $\ulcorner h \urcorner : 1 \to \Omega^C = PC$. In fact, by Part I, Section 3 (3.3), we have

$$h \cdot = \cdot \ulcorner h \urcorner^{\int} \cdot = \cdot \varepsilon_{\Omega,C} \langle \ulcorner h \urcorner \bigcirc_C, 1_C \rangle .$$

Now let $z: 1 \to C$ be an indeterminate arrow, then

$$hz \cdot \underset{z}{=} \cdot \varepsilon_{\Omega,C} \langle \ulcorner h \urcorner, z \rangle \cdot \underset{z}{=} \cdot z \in \ulcorner h \urcorner ,$$

according to the definition of $\in$ in the internal language (see Definition 3.5). We may express this as follows:

***Lemma* 5.3.** In any topos, if $h: C \to \Omega$, then $\ulcorner h \urcorner : 1 \to PC$ is given by

$$\ulcorner h \urcorner \cdot = \cdot \{ z \in C \mid hz \}.$$

We are now ready to express characteristic morphisms in the internal language.

Proposition 5.4. If $m: B \to C$ is a monomorphism in the topos $\mathscr{T}$, then

$$\ulcorner \text{char}\, m \urcorner \cdot = \cdot \{z \in C \mid \exists_{y \in B} my = z\}.$$

Proof. Let $h: C \to \Omega$ be such that

$$\ulcorner h \urcorner \cdot = \cdot \{z \in C \mid \exists_{y \in B} my = z\}.$$

Thus, for an indeterminate arrow $z: 1 \to C$,

$$hz \cdot \underset{z}{=} \cdot \exists_{y \in B} my = z,$$

in view of Lemma 5.3. It is easily checked that $hmy \cdot \underset{y}{=} \cdot \top$, hence that $hm \cdot = \cdot \top \bigcirc_B$. We shall prove that m is a kernel of h. Replacing z by fx, where $x: 1 \to A$ is another indeterminate, we get

$$hfx \cdot \underset{x}{=} \cdot \exists_{y \in B} my = fx.$$

Suppose now that $hf \cdot = \cdot \top \bigcirc_A$, that is,

$$hfx \cdot \underset{x}{=} \cdot \top,$$

then we may infer that

$$\top \cdot \underset{x}{=} \cdot \exists_{y \in B} my = fx.$$

According to our interpretation of $\vdash_x$ in Definition 3.5, this implies

$$\vdash_x \exists_{y \in B} my = fx$$

in $L(\mathscr{T})$, that is

$$\mathscr{T} \models \forall_{x \in A} \exists_{y \in B} my = fx.$$

By Lemma 5.2, there is a unique arrow $g: A \to B$ such that $mg \cdot = \cdot f$. Thus m is a kernel of h, that is, h is the characteristic morphism of m, as was to be proved.

It is easily checked that $\langle 1_A, 1_A \rangle: A \to A \times A$ is a monomorphism. In fact, this was presumed in Proposition 3.2 when δ_A was defined by

$$\delta_A \equiv \text{char} \langle 1_A, 1_A \rangle.$$

Corollary 5.5. In any topos

$$\ulcorner \delta_A \urcorner \cdot = \cdot \{z \in A \times A \mid \exists_{x \in A} \langle x, x \rangle = z\}.$$

Proof. This is an immediate consequence of Proposition 5.4. It is an easy exercise to check that also

$$\ulcorner \delta_A \urcorner \cdot = \cdot \{z \in A \times A \mid \pi_{A,A} z = \pi'_{A,A} z\}.$$

But, recalling the general definition of $\{\langle x, y \rangle \in A \times B \mid \varphi(x, y)\}$ from

Section 1, we may also write Corollary 5.5 as

$$\ulcorner\delta_A\urcorner \cdot\!=\!\cdot \{\langle x, y\rangle \in A \times B \mid x = y\}$$

or even as

$$\ulcorner\delta_A\urcorner \cdot\!=\!\cdot \{\langle x, x\rangle \in A \times A \mid \top\}.$$

Lemma 5.6. If $h^*: A \to PB$ corresponds to $h: A \times B \to \Omega$ in a topos,

$$h^*x \cdot\!\underset{x}{=}\!\cdot \{y \in B \mid h\langle x, y\rangle\}.$$

Proof. h^* is the unique arrow $A \to PB$ such that $\varepsilon_B(h^* \times 1_B) \cdot\!=\!\cdot h$, where $\varepsilon_B \equiv \varepsilon_{\Omega,B}$. Let $x: 1 \to A$ and $y: 1 \to B$ be indeterminate arrows, then

$$\begin{aligned} y \in h^*x &\underset{\{x,y\}}{\cdot\!=\!\cdot} \varepsilon_B\langle h^*x, y\rangle \\ &\underset{\{x,y\}}{\cdot\!=\!\cdot} \varepsilon_B(h^* \times 1_B)\langle x, y\rangle \\ &\underset{\{x,y\}}{\cdot\!=\!\cdot} h\langle x, y\rangle. \end{aligned}$$

The result now follows.

If we define the *singleton morphism* $\iota_A \equiv \delta_A^*: 1 \to PA$, we have immediately:

Corollary 5.7. $\iota_A x \cdot\!\underset{x}{=}\!\cdot \{x' \in A \mid x = x'\} \equiv \{x\}$.

Proposition 5.8. In any topos the singleton morphism $\iota_A: A \to PA$ is a monomorphism.

Proof. Suppose $f, g: B \rightrightarrows A$ are such that $\iota_A f \cdot\!=\!\cdot \iota_A g$. By Proposition 5.1,

$$\vdash \forall_{y \in B} \iota_A f y = \iota_A g y,$$

that is, by Corollary 5.7,

$$\vdash_{y} \{fy\} = \{gy\}.$$

It now follows from the usual properties of equality that $\vdash_{y} fy = gy$, that is, that $f \cdot\!=\!\cdot g$, by Proposition 5.1.

The following result is sometimes summarized by saying that *description holds in a topos.*

Theorem 5.9. Suppose that

$$\mathscr{T} \models \forall_{x \in A} \exists!_{y \in B} \varphi(x, y),$$

then there is a unique arrow $g: A \to B$ such that

$$\mathscr{T} \models \forall_{x \in A} \varphi(x, gx);$$

in fact,

$$y = gx \underset{\{x,y\}}{\cdot\!=\!\cdot} \varphi(x, y).$$

Proof. We recall from Proposition 5.8 that the singleton morphism ι_B: $B \to PB$ is a monomorphism.

In view of functional completeness, we may write

$$\{y \in B \mid \varphi(x, y)\} \cdot \underset{x}{=} \cdot fx$$

for a unique arrow $f: A \to PB$. The assumption of the theorem may then be written:

$$\mathcal{T} \models \forall_{x \in A} \exists_{y \in B} \iota_B y = fx.$$

Applying Lemma 5.2, we obtain a unique arrow g: $A \to B$ such that $\iota_B g \cdot = \cdot f$, hence

$$\mathcal{T} \models \forall_{x \in A} \{gx\} = \{y \in B \mid \varphi(x, y)\}.$$

Thus $\vdash_{x} \varphi(x, gx)$ and therefore $\vdash \forall_{x \in A} \varphi(x, gx)$ in $L(\mathcal{T})$.

Exercises

1. With any object C in a topos there is associated a subobject $m_C: \tilde{C} \to PC$ of PC with

   $$\ulcorner \text{char}\, m_C \urcorner \cdot = \cdot \{w \in PC \mid \forall_{z \in C} (z \in w \Rightarrow w = \{z\})\}.$$

   (*a*) Show that there is an arrow $n_C: C \to \tilde{C}$ such that $m_C n_C \cdot = \cdot \iota_C$, the singleton morphism.

   (*b*) If m: $A \to B$ is a monomorphism and $f: A \to C$ any arrow, show that there is a unique arrow g: $B \to \tilde{C}$ extending f, that is, such that $gm \cdot = \cdot n_C f$ forms a pullback. (Hint: First let h: $B \to PC$ be such that

   $$hy \cdot \underset{y}{=} \cdot \{z \in C \mid \exists_{x \in A} (y = mx \wedge fx = z)\},$$

   then find the unique g such that $h \cdot = \cdot m_C g$.)

   (*c*) A *partial map* from B to C is a pair of arrows (m: $A \to B, f: A \to C$), where m is a monomorphism. Given two partial maps from B to C, say the above and (m': $A' \to B, f': A' \to C$), an *isomorphism* between them is an isomorphism $\sigma: A \cong A'$ such that $m'\sigma \cdot = \cdot m$ and $f'\sigma \cdot = \cdot f$. Let Part (B, C) be the set of isomorphism classes of partial maps from B to C; show that Part $(-, C)$ is a contravariant functor (by pulling back).

   (*d*) Show that in a topos Part$(-, C)$ is representable; in fact, Part$(-, C) \cong$ Hom$(-, \tilde{C})$. $\tilde{C}$ is called a *partial map classifier* of C.

   (*e*) Obtain the definition of the subobject classifier as a special case of this construction.

2. If $A + B$ is a coproduct of objects A and B in a topos, clearly there is a monomorphism $m_{A,B}: A + B \to P(A + B) \cong PA \times PB$. Show that a coproduct of A and B may indeed be constructed by taking $\ulcorner \text{char}\, m_{A,B} \urcorner \equiv \{\langle u, v \rangle \in PA \times PB \mid (\exists! u \wedge v = \varnothing_B) \vee (u = \varnothing_A \wedge \exists! v)\}$, where $\exists! u \equiv \exists!_{x \in A} x \in u, \varnothing_A \equiv \{x \in A \mid \bot\}$. The canonical injections $\kappa_{A,B}$:

$A \to A + B$ and $\kappa'_{A,B}: B \to A + B$ are given by $m_{A,B}\kappa_{A,B}x \cdot \underset{x}{=} \cdot \langle \{x\}, \varnothing_B \rangle$, $m_{A,B}\kappa'_{A,B}y \cdot \underset{y}{=} \cdot \langle \varnothing_A, \{y\} \rangle$. If $f: A \to C, g: B \to C$, use Theorem 5.9 to obtain a unique arrow $[f, g]: A + B \to C$ such that $[f, g]\kappa_{A,B} \cdot = \cdot f, [f, g]\kappa'_{A,B} \cdot = \cdot g$. Show that the uniqueness may also be expressed equationally by $[h\kappa_{A,B}, h\kappa'_{A,B}] \cdot = \cdot h$ for all $h: A + B \to C$.

3. In any topos $\mathscr{T}$, for $\alpha, \beta: 1 \to PA$, we may define $\alpha \leqslant \beta$ to mean $\mathscr{T} \vDash \alpha \subseteq \beta$. Show that, for each object A of $\mathscr{T}$, $\mathrm{Hom}(1, PA) \cong \mathrm{Sub}(A)$ is a Heyting algebra, that is, a poset which, regarded as a category, is bicartesian closed.

6 The internal language at work II

In this section we shall use the internal language to discuss monomorphisms, epimorphisms, injectives and projectives in a topos. First we note that monomorphisms and epimorphisms may be described exactly as in the category of sets.

Proposition 6.1. In a topos $\mathscr{T}$, $f: A \to B$ is a *monomorphism* if and only if

(*a*) $\mathscr{T} \vDash \forall_{x \in A} \forall_{x' \in A}(fx = fx' \Rightarrow x = x')$,

an *epimorphism* if and only if

(*b*) $\mathscr{T} \vDash \forall_{y \in B} \exists_{x \in A} fx = y$,

an *isomorphism* if and only if

(*c*) $\mathscr{T} \vDash \forall_{y \in B} \exists !_{x \in A} fx = y$.

Proof. Suppose (*a*) holds and $g, h: C \rightrightarrows A$ are such that $fg \cdot = \cdot fh$. Then clearly,

$$\mathscr{T} \vDash \forall_{z \in C} fgz = fhz.$$

Hence, in view of (*a*),

$$\mathscr{T} \vDash \forall_{z \in C} gz = hz.$$

Then by Proposition 5.1, $g \cdot = \cdot h$, and so f is a monomorphism.

Conversely, suppose f is a monomorphism. Let $\langle g, h \rangle: C \to A \times A$ be a kernel of $k: A \times A \to \Omega$, where

$$\ulcorner k \urcorner \equiv \{\langle x, x' \rangle \in A \times A \mid fx = fx'\}.$$

Then $k\langle g, h \rangle \cdot = \cdot \top \bigcirc_C$, hence

$$k\langle gz, hz \rangle \cdot \underset{z}{=} \cdot \top,$$

from which it follows that

$$\vdash_z \langle gz, hz \rangle \in \ulcorner k \urcorner,$$

that is,

$$\vdash_z fgz = ghz.$$

By Proposition 5.1, we infer $fg \cdot = \cdot fh$, hence $g \cdot = \cdot h$. Thus the kernel of k is $\langle g, g \rangle$ and so, by Proposition 5.4,

$$\ulcorner k \urcorner \cdot = \cdot \{\langle x, x' \rangle \in A \times A \mid \exists_{z \in C} \langle g, g \rangle z = \langle x, x' \rangle\}.$$

It now follows that

$$\begin{aligned} fx = fx' &\vdash_{\{x,x'\}} \langle x, x' \rangle \in \ulcorner k \urcorner \\ &\vdash_{\{x,x'\}} \exists_{z \in C} \langle gz, gz \rangle = \langle x, x' \rangle \\ &\vdash_{\{x,x'\}} x = x'. \end{aligned}$$

hence that condition (*a*) holds.

Next, assume (*b*) and suppose $g, h: B \rightrightarrows C$ are such that $gf \cdot = \cdot hf$, hence

$$\mathscr{T} \models \forall_{x \in A} gfx = hfx.$$

From this and (*b*) we infer by ordinary logic that

$$\mathscr{T} \models \forall_{y \in B} gy = hy,$$

hence, by Proposition 5.1, that $g \cdot = \cdot h$, and so f is an epimorphism.

Conversely, suppose f is an epimorphism. By functional completeness, there is a unique arrow $g: B \to \Omega$ such that

$$gy \cdot \underset{y}{\equiv} \cdot \exists_{x \in A} fx = y.$$

In particular,

$$gfx \cdot \underset{x}{\equiv} \cdot \top \cdot \underset{x}{\equiv} \cdot \top \bigcirc_B fx,$$

so that

$$gf \cdot = \cdot \top \bigcirc_B f,$$

again by functional completeness. Since f is an epimorphism, $g \cdot = \cdot \top \bigcirc_B$, hence

$$gy \cdot \underset{y}{\equiv} \cdot \top,$$

that is

$$\vdash_y \exists_{x \in A} fx = y,$$

from which (*b*) follows.

Finally, assume (*c*). Apply Theorem 5.9 to $\varphi(y, x) \equiv fx = y$. Thus, there is a unique arrow $g: B \to A$ such that

$$x = gy \underset{\{x,y\}}{\cdot = \cdot} fx = y.$$

From this it easily follows that g is the inverse of f, and so f is an isomorphism.

Conversely, assume that f is an isomorphism with inverse g. Then

$$\mathcal{T} \models \forall_{y \in B} fgy = y \wedge \forall_{x \in A} gfx = x,$$

from which (*c*) easily follows.

The following allows us to recognize and construct equalizers and pullbacks.

Proposition 6.2. (*a*) $m: C \to A$ is an equalizer of $f, g: A \rightrightarrows B$ if and only if

$$\ulcorner \text{char}\, m \urcorner \cdot = \cdot \{x \in A \mid fx = gx\},$$

that is,

$$fx = gx \cdot \underset{x}{\equiv} \cdot \exists_{z \in C} mz = x.$$

(*b*) $C \xrightarrow{m} A_1 \times A_2 \overset{\pi}{\underset{\pi'}{\rightrightarrows}} \begin{matrix} A_1 \\ A_2 \end{matrix}$ is a pullback of $A_1 \xrightarrow{f} B \xleftarrow{g} A_2$ if and only if

$$\ulcorner \text{char}\, m \urcorner \cdot = \cdot \{\langle x_1, x_2 \rangle \in A_1 \times A_2 \mid fx_1 = gx_2\},$$

that is

$$fx_1 = gx_2 \underset{\{x_1, x_2\}}{\cdot = \cdot} \exists_{z \in C} mz = \langle x_1, x_2 \rangle.$$

Proof. Let $\ulcorner \text{char}\, m \urcorner$ be defined as in (*a*). We must first show that $fm \cdot = \cdot gm$. By Proposition 5.4,

$$\ulcorner \text{char}\, m \urcorner \cdot = \cdot \{x \in A \mid \exists_{z \in C} mz = x\},$$

hence

$$\exists_{z' \in C} mz' = x \cdot \underset{x}{\equiv} \cdot fx = gx.$$

Replacing x by mz, we obtain

$$\vdash_z fmz = gmz,$$

hence $fm \cdot = \cdot gm$, by Proposition 5.1.

Next suppose $h: D \to A$ is such that $fh \cdot = \cdot gh$. Then, clearly

$$\vdash_t fht = ght,$$

that is,

$$\vdash_t ht \in \ulcorner \text{char}\, m \urcorner,$$

that is,

$$\vdash_{\tau} (\mathrm{char}\, m)ht,$$

that is,

$$(\mathrm{char}\, m)ht \cdot\underset{\tau}{=}\cdot \top \cdot\underset{\tau}{=}\cdot \top \bigcirc_D t.$$

Thus, by functional completeness,

$$(\mathrm{char}\, m)h \cdot{=}\cdot \top \bigcirc_D,$$

hence there is a unique $k: D \to C$ such that $mk \cdot{=}\cdot h$. This shows that m is an equalizer of (f, g).

If m' is any equalizer of (f, g), then $m' \cong m$, hence $\mathrm{char}\, m' = \mathrm{char}\, m$, which explains the 'only if' part of the statement. This completes the proof of (*a*).

As to (*b*), it suffices to point out that a pullback of (f, g) is given by an equalizer of $(f\pi_{A_1,A_2}, g\pi'_{A_1,A_2})$.

A monomorphism (epimorphism) is called *regular* if it is an equalizer (coequalizer) of a pair of arrows.

Lemma 6.3. In a topos all monomorphisms and epimorphisms are regular.

Proof. For monomorphisms this follows from the definition of a topos, according to which every monomorphism $m: A \to B$ is an equalizer of $(\mathrm{char}\, m, \top \bigcirc_B)$.

Now suppose $e: A \to B$ is an epimorphism. We shall prove that it is a coequalizer of its kernel pair $C \underset{g}{\overset{f}{\rightrightarrows}} A$, which may of course be constructed as a pullback:

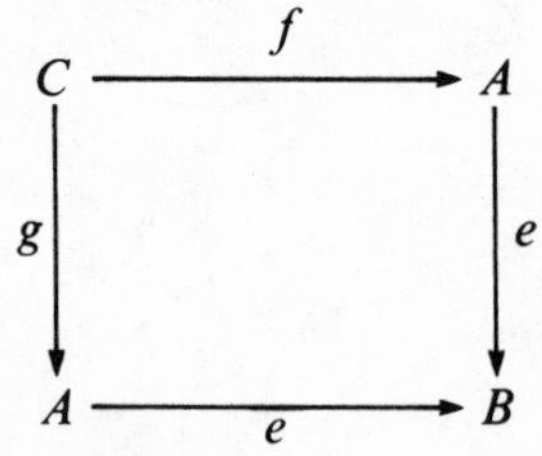

According to Proposition 6.2, we may take $f \cdot{=}\cdot \pi_{A,A} m$ and $g \cdot{=}\cdot \pi'_{A,A} m$, where $m: C \to A \times A$ is determined by its characteristic morphism

$$\ulcorner \mathrm{char}\, m \urcorner \cdot{=}\cdot \{\langle x, x'\rangle \in A \times A \mid ex = ex'\},$$

that is, by

$$ex = ex' \underset{\{x,x'\}}{\cdot{=}\cdot} \exists_{z\in C}\, mz = \langle x, x'\rangle.$$

Suppose $h: A \to D$ is such that $h\pi_{A,A}m \cdot = \cdot h\pi'_{A,A}m$. This is easily seen to imply that

$$\mathscr{T} \models \forall_{x\in C}\forall_{x'\in C}(ex = ex' \Rightarrow hx = hx').$$

We seek a unique $k: B \to D$ such that $ke\cdot = \cdot h$. Indeed, let

$$\varphi(y, t) \equiv \exists_{x\in A}(t = hx \wedge ex = y).$$

Using Proposition 6.1 and the above, it is an easy exercise in logic to prove that

$$\mathscr{T} \models \forall_{y\in B}\exists!_{t\in D}\varphi(y, t).$$

It then follows from Theorem 5.9 that there is a unique arrow $k: B \to D$ such that $ke\cdot = \cdot h$, as was to be proved.

An *injective* in a category is an object I such that for every (regular) monomorphism $m: B \to A$ and every arrow $f: B \to I$ there is an arrow $g: A \to I$ such that $gm\cdot = \cdot f$.

Dually, a *projective* is an object P such that for every (regular) epimorphism $e: A \to B$ and every arrow $f: P \to B$ there is an arrow $g: P \to A$ such that $eg\cdot = \cdot f$.

There is no unanimity in the literature whether to insist on the adjective 'regular' in these definitions. Fortunately, this does not matter in a topos, in view of Lemma 6.3.

Proposition 6.4. For any object C in a topos $\mathscr{T}$, $PC = \Omega^C$ is injective. In particular, $\Omega \cong P1$ is injective.

Proof. Given a monomorphism $m: B \to A$ and an arrow $f: B \to PC$, define $g: A \to PC$ by functional completeness thus:

$$gx \cdot \underset{x}{=} \cdot \{z \in C \mid \exists_{y\in B}(my = x \wedge z \in fy)\}.$$

Then

$$gmy \cdot \underset{y}{=} \cdot \{z \in C \mid \exists_{y'\in B}(my' = my \wedge z \in fy')\}.$$

By Proposition 6.1,

$$\mathscr{T} \models \forall_{y\in B}\forall_{y'\in B}(my' = my \Rightarrow y = y'),$$

hence

$$gmy \cdot \underset{y}{=} \cdot \{z \in C \mid z \in fy\} \cdot \underset{y}{=} \cdot fy,$$

and so $gm\cdot = \cdot f$, by functional completeness.

Lemma 6.5. In a topos pullbacks preserve epimorphisms. This means: if

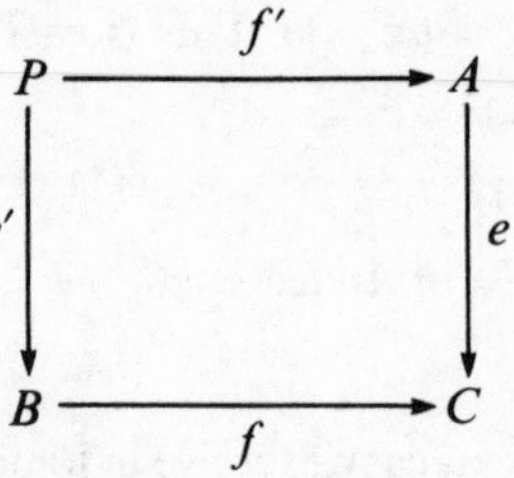

is a pullback and e is an epimorphism, then so is e'.

Proof. In view of Proposition 6.2, the pullback may be constructed by letting $m: P \to A \times B$ be an equalizer of $e\pi_{A,B}$ and $f\pi'_{A,B}$ and by setting $f' \cdot = \cdot \pi_{A,B}m$ and $e' \cdot = \cdot \pi'_{A,B}m$. Note that

$$\ulcorner \mathrm{char}\, m \urcorner \cdot = \cdot \{\langle x, y\rangle \in A \times B \mid ex = fy\}.$$

By Proposition 6.1,

$$\mathscr{T} \vDash \forall_{z\in C} \exists_{x\in A} ex = z,$$

hence, in particular,

$$\mathscr{T} \vDash \forall_{y\in B} \exists_{x\in A} ex = fy.$$

Therefore

$$\mathscr{T} \vDash \forall_{y\in B} \exists_{x\in A} \langle x, y\rangle \in \ulcorner \mathrm{char}\, m \urcorner,$$

that is, by Proposition 5.4,

$$\mathscr{T} \vDash \forall_{y\in B} \exists_{x\in A} \exists_{t\in P} mt = \langle x, y\rangle.$$

After interchanging the two existential quantifiers, we obtain from this

$$\mathscr{T} \vDash \forall_{y\in B} \exists_{t\in P} \pi'_{A,B} mt = y,$$

and so, by Proposition 6.1, $e' \cdot = \cdot \pi'_{A,B}m$ is an epimorphism.

Proposition 6.7. If C is an object in a topos $\mathscr{T}$, the following statements are equivalent:

(i) C is projective;

(ii) all epimorphisms $e: A \to C$ split (that is, there exists $m: C \to A$ such that $em \cdot = \cdot 1_C$);

(iii) from

$$\mathscr{T} \vDash \forall_{z\in C} \exists_{x\in A} \varphi(z, x)$$

we may infer that, for some arrow $f: C \to A$,

$$\mathscr{T} \vDash \forall_{z\in C} \varphi(z, fz).$$

Proof. That (i) implies (ii) is immediate. That (ii) implies (i) depends on

Proposition 6.6. Indeed, assume (ii) and let $e: A \to B$ be an epimorphism, $f: C \to B$ any arrow. Form the pullback

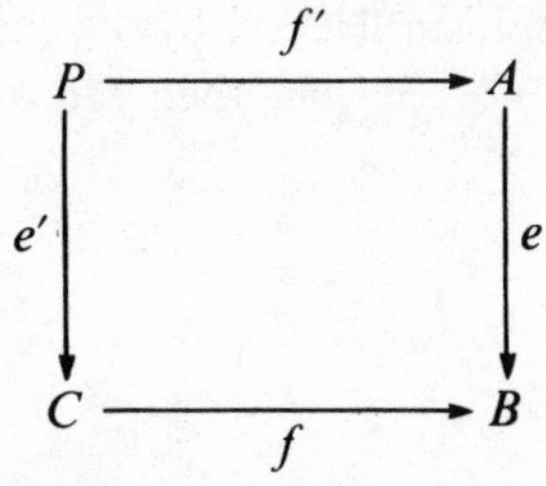

By Proposition 6.6, e' is an epimorphism. Hence, by (ii), there is an arrow $m': C \to P$ such that $e'm' \cdot = \cdot 1_C$. Therefore

$$ef'm' \cdot = \cdot fe'm' \cdot = \cdot f,$$

and so (i) holds.

To show that (iii) implies (ii), we let $\varphi(z, x) \equiv z = ex$, where e is an epimorphism. Then, by Proposition 6.1,

$$\mathscr{T} \models \forall_{z \in C} \exists_{x \in A} \varphi(z, x).$$

By (iii), there exists $f: C \to A$ such that

$$\mathscr{T} \models \forall_{z \in C} \varphi(z, fz),$$

that is,

$$\mathscr{T} \models \forall_{z \in C} z = efz,$$

that is, by Proposition 5.1, $ef \cdot = \cdot 1_C$, which shows (ii).

To show that (ii) implies (iii), let $\varphi(z, x)$ be any formula of $L(\mathscr{T})$ in the variables $z \in C$ and $x \in A$ and suppose

$$\mathscr{T} \models \forall_{z \in C} \exists_{x \in A} \varphi(z, z).$$

Let $n: B \to C \times A$ be a kernel of char n defined by

$$\ulcorner \mathrm{char}\, n \urcorner \cdot = \cdot \{\langle z, x \rangle \in C \times A \mid \varphi(z, x)\},$$

that is, in view of Proposition 5.4,

$$\exists_{y \in B} ny = \langle z, x \rangle \underset{\{z,x\}}{\cdot = \cdot} \varphi(z, x).$$

Therefore

$$\mathscr{T} \models \forall_{z \in C} \exists_{x \in A} \exists_{y \in B} ny = \langle z, x \rangle.$$

Interchanging existential quantifiers, we deduce

$$\mathscr{T} \models \forall_{z \in C} \exists_{y \in B} \pi_{C,A} ny = z.$$

Now define $e: B \to C$ by $e \cdot = \cdot \pi_{C,A} n$. Then

$$\mathscr{T} \models \forall_{z \in C} \exists_{y \in B} ey = z.$$

Thus, by Proposition 6.1, e is an epimorphism. Hence, by (ii), there is an arrow $m: C \to B$ such that $em \cdot = \cdot 1_C$. Put $f \cdot = \cdot \pi'_{C,A} nm$, then $f: C \to A$ and

$$\begin{aligned} \langle z, fz \rangle \cdot \underset{z}{=} \cdot & \langle emz, \pi' nmz \rangle \\ \cdot \underset{z}{=} \cdot & \langle \pi nmz, \pi' nmz \rangle \\ \cdot \underset{y}{=} \cdot & nmz. \end{aligned}$$

Hence

$$\begin{aligned} \varphi(z, fz) \cdot \underset{z}{=} \cdot & \exists_{y \in D} ny = \langle z, fz \rangle \\ \cdot = \cdot & \top, \end{aligned}$$

that is,

$$\mathscr{T} \models \forall_{z \in C} \varphi(z, fz),$$

and so (iii) holds.

Exercises

1. Give a direct proof of Proposition 6.2 (*a*).
2. Give a proof that Ω is injective without using $\Omega \cong P1$.
3. Show that, for any object C of a topos, the partial map classifier $\tilde{C}$ defined in the exercise of Section 5 is injective.
4. (Higgs). If $f: \Omega \to \Omega$ is a monomorphism in a topos, prove that $f^2 \cdot = \cdot 1_\Omega$. (Hint suggested by Ščedrov: First prove the following in the internal language:

   (i) $\vdash \forall_{x \in \Omega}(fx \Rightarrow (f\top \Leftrightarrow p))$,
   (ii) $\vdash \forall_{x \in \Omega}(f^2 x \Rightarrow x)$,
   (iii) $\vdash \forall_{x \in \Omega}(fx \Rightarrow f^3 x)$.

   Then conclude that $\vdash \forall_{x \in \Omega}(f^2 x = x)$.) (See Ščedrov 1984b.)

7 Choice and the Boolean axiom

If $\mathfrak{L}$ is a type theory, so is $\mathfrak{L}(z)$, where z is a variable of type C, say, regarded as a *parameter*. To be precise, $\mathfrak{L}(z)$ has the same terms as $\mathfrak{L}$, but open terms of $\mathfrak{L}$ with no free variables other than z are viewed as closed terms in $\mathfrak{L}(z)$. Moreover, the entailment relation $\vdash_X$ of $\mathfrak{L}(z)$ is the entailment relation $\vdash_{X \cup \{z\}}$ of $\mathfrak{L}$, assuming $z \notin X$.

Definition 7.1. A type theory satisfies the *rule of choice* if from

$$\vdash \forall_{x\in A}(x\in\alpha \Rightarrow \exists_{y\in B}\varphi(x, y)),$$

where A is any type, α any closed term of type PA and $\varphi(x, y)$ any formula with no free variables other than x and y, one may infer that

$$\vdash \forall_{x\in A}(x\in\alpha \Rightarrow \exists!_{y\in B}\psi(x, y))$$

for some formula $\psi(x, y)$ such that

$$\psi(x, y) \vdash_{\{x,y\}} \varphi(x, y).$$

This rule can be stated in a more transparent way if there are sufficiently many function symbols in $\mathfrak{L}$, see Proposition 7.5 below.

Lemma 7.2. If $\mathfrak{L}$ satisfies the rule of choice, then so does $\mathfrak{L}(z)$, where z is any variable of type C say.

Proof. Suppose

$$\vdash_z \exists_{x\in A}(x\in\alpha(z) \Rightarrow \exists_{y\in B}\varphi(z, x, y)),$$

where α and φ depend on the parameter z. Then

$$\vdash \forall_{\langle z,x\rangle\in C\times A}(\langle z, x\rangle\in\alpha' \Rightarrow \exists_{y\in B}\varphi(z,x, y)),$$

where

$$\alpha' \equiv \{\langle z, x\rangle\in C\times A \mid x\in\alpha(z)\}.$$

If $\mathfrak{L}$ satisfies the rule of choice, we can find $\psi(z, x, y)$ where

$$\psi(z, x, y) \vdash_{\{z,x,y\}} \varphi(z, x, y)$$

such that

$$\vdash \forall_{\langle z,x\rangle\in C\times A}(\langle z, x\rangle\in\alpha' \Rightarrow \exists!_{y\in B}\psi(z, x, y)),$$

that is

$$\vdash_z \forall_{x\in A}(x\in\alpha(z) \Rightarrow \exists!_{y\in B}\psi(z, x, y)).$$

Proposition 7.3. If $\mathfrak{L}$ satisfies the rule of choice and p is any closed formula then $\vdash p \vee \neg p$.

Proof. We define

$$\alpha_p \equiv \{x\in\Omega \mid x \vee (\neg x \wedge p)\},$$

$$\beta_p \equiv \{x\in\Omega \mid \neg x \vee (x \wedge p)\},$$

$$\gamma_p \equiv \{\alpha_p, \beta_p\} \equiv \{y\in P\Omega \mid y = \alpha_p \vee y = \beta_p\}.$$

We assert that

(1) $\vdash \forall_{y \in P\Omega}(y \in \gamma_p \Rightarrow \exists_{x \in \Omega} x \in y)$.

Indeed, we argue informally thus: if $y \in \gamma_p$ then either $y = \alpha_p$ or $y = \beta_p$, so either $\top \in y$ or $\bot \in y$. By the rule of choice, we can then find a formula $\psi(y, x)$ such that

(2) $\psi(y, x) \vdash_{\{x,y\}} x \in y$,

(3) $\vdash \forall_{y \in P\Omega}(y \in \gamma_p \Rightarrow \exists!_{x \in \Omega} \psi(y, x))$.

Arguing informally again, we have $\alpha_p \in \gamma_p$ and $\beta_p \in \gamma_p$, hence, by (3), $\exists!_{x \in \Omega} \psi(\alpha_p, x)$ and $\exists!_{x \in \Omega} \psi(\beta_p, x)$. Say $\psi(\alpha_p, x_1)$ and $\psi(\beta_p, x_2)$, so, by (2), $x_1 \in \alpha_p$ and $x_2 \in \beta_p$. Therefore

$$(x_1 \vee (\neg x_1 \wedge p)) \wedge (\neg x_2 \vee (x_2 \wedge p)),$$

hence $p \vee (x_1 \wedge \neg x_2)$. The result now follows: the first disjunct is p; the second implies $\neg p$; for if p holds, $\alpha_p = \{x \in \Omega \mid x \vee \neg x\} = \beta_p$, hence $x_1 = x_2$, which contradicts $x_1 \wedge \neg x_2$. Thus $p \vee \neg p$.

***Corollary* 7.4.** If $\mathfrak{L}$ satisfies the rule of choice, then $\mathfrak{L}$ is classical, that is $\vdash \forall_{x \in \Omega}(x \vee \neg x)$.

Proof. By Lemma 7.2, $\mathfrak{L}(x)$ satisfies the rule of choice. Hence, by Proposition 7.3, $\vdash_{\bar{x}} x \vee \neg x$.

We should point out that in the language $L(\mathscr{T})$ of a topos $\mathscr{T}$ the rule of choice takes a somewhat simpler form.

***Proposition* 7.5.** The language $L(\mathscr{T})$ of a topos satisfies the rule of choice if and only if, for all formulas $\varphi(x, y)$, one may infer from $\mathscr{T} \vDash \forall_{x \in A} \exists_{y \in B} \varphi(x, y)$ that $\mathscr{T} \vDash \forall_{x \in A} \varphi(x, gx)$ for some arrow $g: A \to B$ in $\mathscr{T}$.

Proof. Assume the rule of choice as stated in Definition 7.1 with $\alpha \equiv \{x \in A \mid \top\}$ and suppose that $\mathscr{T} \vDash \forall_{x \in A} \exists_{y \in B} \varphi(x, y)$. Then we may infer that $\mathscr{T} \vDash \forall_{x \in A} \exists!_{y \in B} \psi(x, y)$ for some formula $\psi(x, y)$ such that $\psi(x, y) \vdash_{\{x,y\}} \varphi(x, y)$. In view of description (Theorem 5.9), there is a unique arrow $g: A \to B$ such that $\mathscr{T} \vDash \forall_{x \in A} \psi(x, gx)$, hence $\mathscr{T} \vDash \forall_{x \in A} \varphi(x, gx)$.

Conversely, assume the simpler condition of Proposition 7.5 and suppose that $\mathscr{T} \vDash \forall_{x \in A}(x \in \alpha \Rightarrow \exists_{y \in B} \varphi(x, y))$. Let $m: A' \to A$ be a kernel of $\alpha^{\int}: A \to \Omega$ and let $x': 1 \to A'$ be an indeterminate arrow. Then $\vdash_{\bar{x}'} mx' \in \alpha$, hence $\vdash_{\bar{x}'} \exists_{y \in B} \varphi(mx', y)$. By assumption, there is an arrow $g: A' \to B$ so that $\vdash_{\bar{x}'} \varphi(mx', gx')$. Take $\psi(x, y) \equiv \exists_{x' \in A}(x = mx' \wedge y = gx' \wedge \varphi(x, y))$. It is now easily shown that $\mathscr{T} \vDash \forall_{x \in A}(x \in \alpha \Rightarrow \exists!_{y \in B} \psi(x, y))$, by observing that $x \in \alpha$ means $x \in \ulcorner \mathrm{char}\, m \urcorner$, that is, $\exists_{x' \in A'} x = mx'$, in view of Proposition 5.4.

Moreover, from $\psi(x, y)$ one may clearly infer that $\varphi(x, y)$, so that $L(\mathscr{T})$ satisfies the rule of choice according to Definition 7.1.

According to Lawvere, a topos *has choice* if all epimorphisms split. In view of Proposition 6.7, this means that all objects are projective, or, equivalently, that rule (iii) of Proposition 6.7 holds for all objects C and A. In view of Proposition 7.5, this is evidently the same as saying that $L(\mathscr{T})$ satisfies the rule of choice. Thus we immediately have the following consequence of Corollary 7.4.

Corollary *7.6.* If a topos has choice then it is Boolean, that is, it satisfies $\forall_{t\in\Omega}(t \vee \neg t)$.

This result is due to Diaconescu, whose proof was quite different.

As long as we are discussing the Boolean axiom, we may point out the following alternative characterization.

Proposition *7.7.* A topos is Boolean if and only if $1 \xrightarrow{\top} \Omega \xleftarrow{\bot} 1$ is a coproduct diagram.

Proof. Assume that $1 \xrightarrow{\top} \Omega \xleftarrow{\bot} 1$ is a coproduct diagram in the topos $\mathscr{T}$.

Let $\neg : \Omega \to \Omega$ be defined by $\neg x$ for an indeterminate arrow $x: 1 \to \Omega$. Then $\neg : \Omega \to \Omega$ is clearly the unique arrow such that the following triangles commute:

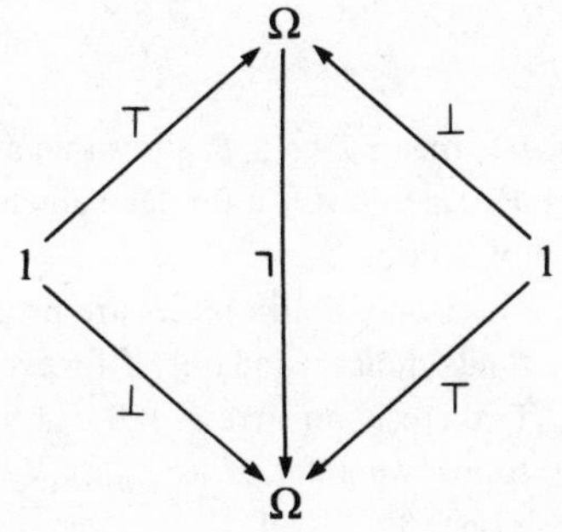

Therefore $\neg\neg \cdot = \cdot 1_\Omega$. From this it follows that, for an indeterminate $t: 1 \to \Omega$, $\neg\neg t \cdot \underset{t}{=} \cdot t$, hence

$$\mathscr{T} \models \forall_{t\in\Omega} \neg\neg t \Rightarrow t.$$

Conversely, assume that $\mathscr{T}$ is Boolean. Given $1 \xrightarrow{a} A \xleftarrow{b} 1$, we aim to show that there exists a unique arrow $f: \Omega \to A$ such that $f\top \cdot = \cdot a$ and $f\bot \cdot = \cdot b$.

Consider the formula $\varphi(t, x)$, with variables t of type Ω and x of type A, given by

$$\varphi(t, x) \equiv (t \wedge x = a) \vee (\neg t \wedge x = b).$$

Since

$$\mathscr{T} \vDash \forall_{t\in\Omega}(t \vee \neg t),$$

we easily infer that

$$\mathscr{T} \vDash \forall_{t\in\Omega}\exists!_{x\in A}\varphi(t, x).$$

Since $\mathscr{T}$ has description, there is a unique arrow $f:\Omega \to A$ such that $\mathscr{T} \vDash \forall_{t\in\Omega}\varphi(t, ft)$. Thus in the language $L(\mathscr{T})$.

$$\vdash_t (t \wedge ft = a) \vee (\neg t \wedge ft = b).$$

Substituting $\top$ and $\bot$ for t, we obtain $\vdash f\top = a$ and $\vdash f\bot = b$, hence $f\top \cdot\!=\!\cdot a$ and $f\bot \cdot\!=\!\cdot b$, by Proposition 3.2.

It remains to show that these last two equations determine f uniquely. They clearly imply

$$t \vdash_t ft = a, \quad \neg t \vdash_t ft = b,$$

hence

$$t \vee \neg t \vdash_t (t \wedge ft = a) \vee (\neg t \wedge ft = b).$$

Since $\vdash_t t \vee \neg t$, we infer $\vdash_t \varphi(t, ft)$, that is $\mathscr{T} \vDash \forall_{t\in\Omega}\varphi(t, ft)$, which determines f uniquely, as already mentioned.

Exercises

1. For $\alpha, \beta \in \mathrm{Hom}(1, PA)$ define $\alpha \leqslant b$ to mean $\mathscr{T} \vDash \alpha \subseteq \beta$. Show that a topos $\mathscr{T}$ is Boolean if and only if $\mathrm{Hom}(1, PA) \cong \mathrm{Sub}\, A$ is a Boolean algebra for all objects A of $\mathscr{L}$. (See Exercise 3 of Section 5.)
2. Prove that a topos has choice if and only if all objects are projective.
3. (Lawvere)—Prove that a topos $\mathscr{T}$ has choice if and only if, for every arrow $f: A \to B$ such that $\mathscr{T} \vDash \forall_{y\in B}\exists_{x\in A}\top$, there is an arrow $g: B \to A$ such that $fgf \cdot\!=\!\cdot f$ (Hint: In a Boolean topos we infer $\mathscr{T} \vDash \forall_{y\in B}\exists_{x\in A}\forall_{x'\in A}(fx' = y \Rightarrow fx = y)$. Now apply Proposition 7.5.)

8 Topos semantics

Let $\mathscr{T}$ be a topos and $L(\mathscr{T})$ its internal language. In this section we examine the provability predicate $\vdash$ for $L(\mathscr{T})$ in more detail.

Definition 8.1. If $\varphi(x)$ is a formula of $L(\mathscr{T})$ in the variable x of type A, then

$\varphi(x) \cdot \underset{x}{=} \cdot fx$ for a uniquely determined arrow $f: A \to \Omega$. For an arrow $a: C \to A$ in $\mathscr{T}$ one writes $\varphi(a) \equiv fa$, by abuse of notation, and regards a as a *generalized element of A at stage C*. One also writes $C \Vdash \varphi(a)$ for $fa \cdot = \cdot \top \bigcirc_C$ and reads this as '$\varphi(a)$ *holds at stage C*' or '*C forces* $\varphi(a)$'.

More generally, in the presence of a parameter y of type B, if

$$\varphi(y, x) \cdot \underset{\{y,x\}}{=} \cdot g\langle y, x\rangle,$$

where $g: B \times A \to \Omega$, one writes

$$\varphi(y, a) \equiv g\langle y\bigcirc_C, a\rangle.$$

The following is an easy consequence of the above definition.

Proposition 8.2. (1) If $C \Vdash \varphi(a)$ and $h: D \to C$, then $D \Vdash \varphi(ah)$.

(2) $\vDash_x \varphi(x)$ in $L(\mathscr{T})$ if and only if, for all objects C and all generalized elements $a: C \to A$, $C \Vdash \varphi(a)$.

(3) $C \Vdash \varphi(a)$ if and only if $\vdash \forall_{z \in C} \varphi(az)$.

(4) If $h: D \to C$ is an epimorphism and $D \Vdash \varphi(ah)$, then $C \Vdash \varphi(a)$.

Proof. For example, we shall show (3). According to Definition 8.1, $C \Vdash \varphi(a)$ means $fa \cdot = \cdot \top \bigcirc_C$, that is, $faz \cdot \underset{z}{=} \cdot \top$, which translates into $\vDash_z \varphi(az)$, that is, $\vdash \forall_{z \in C} \varphi(az)$.

By (2) of Proposition 8.2, 'truth' in a topos is equivalent to truth at all stages and for all generalized elements. Actually, the stages may be restricted to a generating set of $\mathscr{T}$. A set $\mathscr{C}$ of objects of $\mathscr{T}$ is called a *generating set* if, for any two arrows $f, g: A \rightrightarrows B$, $f \cdot = \cdot g$, if and only if, for all C in $\mathscr{C}$ and all $h: C \to A$, $fh \cdot = \cdot gh$. We shall state this formally:

Proposition 8.3. If $\mathscr{C}$ is a generating set of objects of $\mathscr{T}$, then $\vDash_x \varphi(x)$ in $L(\mathscr{T})$ if and only if, for all objects C in $\mathscr{C}$ and all generalized elements a of A at stage C, $C \Vdash \varphi(a)$.

Proof. Suppose $C \Vdash \varphi(a)$ for all $C \in \mathscr{C}$ and all $a: C \to A$. Then $fa \cdot = \cdot \top \bigcirc_C \cdot = \cdot \top \bigcirc_A a$ for all $C \in \mathscr{C}$ and $a: C \to A$, and so $f \cdot = \cdot \top \bigcirc_A$, that is, $\vDash_x \varphi(x)$. The converse is evident.

Proposition 8.3 may be exploited to reduce the notion of 'truth' in $L(\mathscr{T})$ to an inductive definition of '$\varphi(a)$ holds at stage C' in a way familiar to logicians. We shall here regard $L(\mathscr{T})$ as a type theory in the sense of Section 1 of Part II, not as a type theory based on equality. Atomic formulas $\varphi(x)$ in an indeterminate x of type A are the following: x itself, in case $A = \Omega$; $b(x) \in \beta(x)$, where $b(x)$ has type B and $\beta(x)$ has type PB; $\top$ and $\bot$. Compound formulas are constructed with the help of the connectives $\wedge$, $\vee$ and $\Rightarrow$ and the quantifiers $\forall_{y\in B}$ and $\exists_{y\in B}$.

The following theorem should be called 'Beth–Kripke–Joyal semantics'.

Theorem 8.4. Given that $a: C \to A$, then:

(0) $C \Vdash a$ iff $a \cdot = \cdot \top \bigcirc_C$, in case $A = \Omega$;

(1) $C \Vdash b(a) \in \beta(a)$ iff $\varepsilon_B \langle \beta(a), b(a) \rangle \cdot = \cdot \top \bigcirc_C$;

(2) $C \Vdash \top$ always;

(3) $C \Vdash \bot$ iff C is an initial object of $\mathcal{T}$;

(4) $C \Vdash \varphi(a) \wedge \psi(a)$ iff $C \Vdash \varphi(a)$ and $C \Vdash \psi(a)$;

(5) $C \Vdash \varphi(a) \vee \psi(a)$ iff there is an epimorphism $[k, l]$: $D + E \to C$ such that $D \Vdash \varphi(ak)$ and $E \Vdash \psi(al)$;

(6) $C \Vdash \varphi(a) \Rightarrow \psi(a)$ iff, for all $h: D \to C$, if $D \Vdash \varphi(ah)$ then $D \Vdash \psi(ah)$,

(7) $C \Vdash \forall_{y\in B} \psi(y, a)$ iff, for all $h: D \to C$ and all $b: D \to B$, $D \Vdash \psi(b, ah)$;

(8) $C \Vdash \exists_{y\in B} \psi(y, a)$ iff there is an epimorphism $h: D \to C$ and an arrow $b: D \to B$ such that $D \Vdash \psi(b, ah)$.

Although the symbols $\neg$ and $=$ are defined symbols, we add the following clauses in the same spirit:

(9) $C \Vdash \neg\varphi(a)$ iff, for all $h: D \to C$, if $D \Vdash \varphi(a)$ then $D \cong 0$;

(10) $C \Vdash b(a) = b'(a)$ iff $b(a) \cdot = \cdot b'(a)$.

Proof. Most of the above clauses may be left as exercises; we shall only prove (3), (5) and (8).

(3) Suppose $C \Vdash \bot$. By Proposition 8.2, this means $\vdash \forall_{z\in C} \bot$, whence it follows that $\vdash \forall_{z\in C} \exists!_{x\in A} \bot$. By description (see Theorem 5.9), there exists a unique arrow $f: C \to A$ such that $\vdash \forall_{z\in C} \bot$. Since this is the case, C is an initial object.

Conversely, suppose C is initial. Then the two arrows $\top \bigcirc_C, \bot \bigcirc_C : C \rightrightarrows A$ must coincide, hence $\vdash \forall_{z\in C} (\top = \bot)$, that is, $\vdash \forall_{z\in C} \bot$.

(5) Suppose $C \Vdash \varphi(a) \vee \psi(a)$, that is, by Proposition 8.2, $\vdash_z faz \vee gaz$, where $\varphi(x) \cdot \underset{x}{=} \cdot fx$ and $\psi(x) \cdot \underset{x}{=} \cdot gx$. Let $k \equiv \ker(fa)$, $l \equiv \ker(ga)$. Then, by Lemma 5.3,

$$\ulcorner \mathrm{char}\, k \urcorner \cdot = \cdot \{z \in C \mid faz\}, \quad \ulcorner \mathrm{char}\, l \urcorner \cdot = \cdot \{z \in C \mid gaz\}.$$

Let $D \equiv \mathrm{Ker}(fa)$ be the source of k, $E \equiv \mathrm{Ker}(ga)$ the source of l. By Proposition 5.4,

$$faz \cdot \underset{z}{=} \cdot \exists_{s\in D} z = ks,$$

and similarly for g. Therefore

$$\vdash_z (\exists_{s\in D} z = ks \vee \exists_{t\in E} z = lt),$$

hence

$$\models_z \exists_{u\in D+E} z = [k, l]u;$$

for, in the first case, $u = \kappa_{D,E}s$ and, in the second case, $u = \kappa'_{D,E}t$. By Proposition 6.1, $[k, l]: D + E \to C$ is an epimorphism. Moreover, $fak \cdot\!=\!\cdot \top \bigcirc_D$, hence $\vdash \forall_{s\in D} faks$, that is, $D \Vdash \varphi(ak)$. Similarly one obtains $E \Vdash \psi(al)$.

Conversely, suppose $[k, l]: D + E \to C$ is an epimorphism such that $D \Vdash \varphi(ak)$ and $E \Vdash \psi(al)$. Thus, in $L(\mathscr{T})$,

(i) $\vdash \forall_{z\in C} \exists_{u\in D+E} [k, l]u = z$,
(ii) $\vdash \forall_{s\in D} \varphi(aks)$, $\quad \vdash \forall_{t\in E} \psi(alt)$.

Now, by Lemma 8.5 below,

$$\vdash \forall_{u\in D+E} (\exists_{s\in D} u = \kappa_{D,E}s \vee \exists_{t\in E} u = \kappa'_{D,E}t).$$

Hence, by (i),

$$\vdash \forall_{z\in C} (\exists_{s\in D} z = ks \vee \exists_{t\in E} z = lt),$$

using $[k, l]\kappa_{D,E} \cdot\!=\!\cdot k$, etc. Therefore, by (ii),

$$\vdash \forall_{z\in C} (\varphi(az) \vee \psi(az)).$$

(8) Suppose $C \Vdash \exists_{y\in B} \psi(y, a)$, where $a: C \to A$. In view of Proposition 8.2, this means that in $L(\mathscr{T})$

(i) $\vdash \forall_{z\in C} \exists_{y\in B} \psi(y, az)$.

Let $g: B \times A \to \Omega$ be the unique arrow such that

(ii) $\psi(y, x) \underset{\{y,x\}}{\cdot\!=\!\cdot} g\langle y, x\rangle$

and write $n: D \to C \times B$ for the kernel of

$$C \times B \xrightarrow{\langle \pi'_{C,B}, a\pi_{C,B}\rangle} B \times A \xrightarrow{g} \Omega.$$

Thus

$$\ulcorner \mathrm{char}\, n \urcorner \cdot\!=\!\cdot \{\langle z, y\rangle \in C \times B \mid g\langle y, az\rangle\},$$

that is, by Proposition 5.4,

(iii) $\exists_{s\in D} ns = \langle z, y\rangle \underset{\{z,y\}}{\cdot\!=\!\cdot} g\langle y, az\rangle$.

From (i), (ii) and (iii) we infer

$$\vdash \forall_{z\in C} \exists_{y\in B} \exists_{s\in D} (ns = \langle z, y\rangle),$$

hence

$$\vdash \forall_{z\in C} \exists_{s\in D} \pi_{C,B} ns = z.$$

Thus $h \equiv \pi_{C,B}n$ is an epimorphism, by Proposition 6.1. Moreover, if $b \equiv \pi'_{C,B}n$, we have

$$\begin{aligned}\psi(bs, ahs) &\cdot\underset{s}{=}\cdot g\langle bs, ahs\rangle \\ &\cdot\underset{s}{=}\cdot g\langle \pi'_{C,B}, a\pi_{C,B}\rangle ns \\ &\cdot\underset{s}{=}\cdot \top,\end{aligned}$$

hence

$$\vdash \forall_{s\in D}\psi(bs, ahs),$$

that is $D \Vdash \psi(b, ah)$, by Proposition 8.2.

Conversely, assume that $h: D \to C$ is an epimorphism and that $b: D \to B$ is such that $D \Vdash \psi(b, ah)$, that is, $\vdash \forall_{s\in D}\psi(bs, ahs)$. Now, by Proposition 6.1, $\vdash \forall_{z\in C}\exists_{s\in D} hs = z$, hence $\vdash \forall_{z\in C}\exists_{s\in D}\psi(bs, az)$. It follows that $\vdash \forall_{z\in C}\exists_{y\in B}\psi(y, az)$, that is, $C \Vdash \exists_{y\in B}\psi(y, a)$.

It remains to prove the following:

Lemma 8.5. In $L(\mathscr{T})$,

$$\vdash \forall_{z\in A+B}(\exists_{x\in A} z = \kappa_{A,B}x \vee \exists_{y\in B} z = \kappa'_{A,B}y).$$

Proof. We refer the reader to Section 5, Exercise 2. Using the particular construction of $A + B$ given there, we want to show that

$$\vdash \forall_{z\in A+B}(\exists_{x\in A} m_{A,B}z = \langle\{x\}, \varnothing_B\rangle \vee \exists_{y\in B} m_{A,B}z = \langle\varnothing_A, \{y\}\rangle).$$

Indeed, $m_{A,B}: A + B \to PA \times PB$ was there defined by

$$\begin{aligned}\ulcorner\mathrm{char}\, m_{A,B}\urcorner \equiv \{\langle u, v\rangle \in PA \times PB \mid (\exists! u \wedge v = \varnothing_B) \\ \vee (u = \varnothing_A \wedge \exists! v)\},\end{aligned}$$

which is easily seen to be the same as

$$\{\omega \in PA \times PB \mid \exists_{x\in A}\omega = \langle\{x\}, \varnothing_B\rangle \vee \exists_{y\in B}\omega = \langle\varnothing_A, \{y\}\rangle\}.$$

Take an indeterminate arrow $z: 1 \to A + B$, then $\vdash_z m_{A,B}z \in \ulcorner\mathrm{char}\, m_{A,B}\urcorner$ and the result follows.

Sometimes clauses (5) and (8) in Theorem 8.4 can be strengthened, as was done in the original semantics of Kripke. We recall from Proposition 6.7 that an object C in a topos is *projective* if and only if all epimorphisms $e: D \to C$ split. We shall call an object C *indecomposable* if, for all arrows $k: D \to C$, $l: E \to C$ such that $[k, l]: D + E \to C$ is an epimorphism, either k or l is an epimorphism.

Proposition 8.6. Given that $a: C \to A$, then:

(5′) if C is indecomposable, then $C \Vdash \varphi(a) \vee \psi(a)$ iff $C \Vdash \varphi(a)$ or $C \Vdash \psi(a)$;

(8′) If C is projective, then $C \Vdash \exists_{y \in B} \psi(y, a)$ iff there is an arrow $b' \colon C \to B$ such that $C \Vdash \psi(b', a)$.

Proof. (5′) Assume that C is indecomposable and that $C \Vdash \varphi(a) \vee \psi(a)$. By (5) of Theorem 8.4, there is an epimorphism $[k, l] \colon D + E \to C$ such that $D \Vdash \varphi(ak)$ and $E \Vdash \psi(al)$. Since C is indecomposable, either k or l is an epimorphism. Therefore, by Proposition 8.2(4), either $C \Vdash \varphi(a)$ or $C \Vdash \psi(a)$.

(8′) Assume that C is projective and that $C \Vdash \exists_{y \in B} \psi(y, a)$. By (8) of Theorem 8.4, there is an epimorphism $h \colon D \to C$ and an arrow $b \colon D \to B$ such that $D \Vdash \psi(b, ah)$. Since C is projective, h splits, so there is an arrow $k \colon C \to D$ such that $hk \cdot = \cdot 1_C$. Now, by Proposition 8.2(1), $C \Vdash \psi(bk, ahk)$. Writing $b' \equiv bk$, we have $C \Vdash \psi(b', a)$.

Exercises

1. Complete the proofs of Proposition 8.2 and Theorem 8.4.

2. Prove that an object C is indecomposable if and only if it is not the union of two proper subobjects.

3. Given an object C in a topos $\mathscr{T}$, form a type theory $\mathfrak{L}_C$ whose closed terms of type A are the arrows $C \to A$ of $\mathscr{T}$, that is, elements of type A at stage C. In the notation of Definition 8.1, show that $\vdash \varphi(a)$ in $\mathfrak{L}_C$ if and only if $C \Vdash \varphi(a)$.

4. Given an object C in a topos $\mathscr{T}$, construct a topos $\mathscr{T}/C$ whose objects are arrows $A \to C$ and relate $L(\mathscr{T}/C)$ to $\mathfrak{L}_C$ above. (See Part I, Section 7, Exercise 2 and also Exercise 2 of Section 16 below.

5. In clauses (6) and (7) of Theorem 8.4, show that D need only be taken in the generating set $\mathscr{C}$ of objects of $\mathscr{T}$.

9 Topos semantics in functor categories

Example 9.1. Special examples of toposes are functor categories $\mathbf{Sets}^{\mathscr{C}}$, where $\mathscr{C}$ is a small category. Before discussing the semantics of such toposes, we briefly review some of their pertinent structure.

Limits and colimits are constructed objectwise. For example, the terminal object 1 is the functor defined on objects C of $\mathscr{C}$ by $1(C) \equiv \{*\}$.

Exponentiation is defined with the help of Yoneda's Lemma. For any object A of $\mathscr{C}$, we write

$$h^A \equiv \mathrm{Hom}_{\mathscr{C}}(A, -).$$

If $F, G: \mathscr{C} \to \mathbf{Sets}$, we ought to have

$$G^F(A) \cong \mathrm{Nat}(h^A, G^F) \cong \mathrm{Nat}(h^A \times F, G),$$

so we are led to define

$$G^F(A) \equiv \mathrm{Nat}(h^A \times F, G).$$

Moreover, for $f: A \to B$ and C in $\mathscr{C}$, $G^F(f): G^F(A) \to G^F(B)$ is defined by

$$G^F(f)(t)(C)(g, c) \equiv t(C)(gf, c),$$

for any $t \in G^F(A)$, $g: B \to C$ and $c \in F(C)$.

The evaluation functor $\varepsilon_{G,F}: G^F \times F \to G$ is defined by

$$\varepsilon_{G,F}(C)(t, c) \equiv t(C)(1_C, c)$$

for any $t \in G^F(C)$ and $c \in F(C)$.

The exponential adjunction

$$\frac{t: H \times F \to G}{t^*: H \to G^F}$$

is defined by

$$t^*(A)(a)(C)(h, c) \equiv t(C)(H(h)(a), c),$$

for any objects A and C in $\mathscr{C}$, $h: A \to C$, $a \in H(A)$ and $c \in F(C)$.

The reader will be able to satisfy himself that $\mathbf{Sets}^{\mathscr{C}}$ is a cartesian closed category. To obtain the topos structure, we first observe that Yoneda's Lemma requires

$$\Omega(A) \cong \mathrm{Nat}(h^A, \Omega) \cong \mathrm{Sub}(h^A),$$

so we are led to define $\Omega(A)$ as the set of all subfunctors of h^A. Such a subfunctor is essentially given by a collection S of arrows with source A satisfying:

$$\text{if } h \in S \text{ then } gh \in S,$$

for all arrows g whose source is the target of h. S is called an *A-sieve*, although an algebraist may think of S as a *left ideal* of arrows with source A. If $f: A \to B$, $\Omega(f): \Omega(A) \to \Omega(B)$ is given by

$$\Omega(f)(S) \equiv \bigcup_{C \in \mathscr{C}} \{g: B \to C \mid gf \in S\}.$$

Trivial examples of A-sieves are the empty set $\varnothing$ and the set $A/\mathscr{C}$ of all arrows with source A.

$\top: 1 \to \Omega$ is the natural transformation defined by

$$\top(A)(*) \equiv A/\mathscr{C}.$$

If $m: F \to G$ is a monomorphism in $\mathbf{Sets}^{\mathscr{C}}$, $\operatorname{char} m: G \to \Omega$ is the natural transformation defined by

$$(\text{char}\, m)(A)(a) \equiv \bigcup_{B \in \mathscr{C}} \{f: A \to B \mid \exists_{b \in F(B)} G(f)(a) \cdot = \cdot m(B)(b)\},$$

for all objects A of $\mathscr{C}$ and all $a \in G(A)$. In particular, equality is given by

$$\delta_F(A)(a, a') \equiv \bigcup_{B \in \mathscr{C}} \{f: A \to B \mid F(f)(a) \cdot = \cdot F(f)(a')\},$$

for all a, $a' \in F(A)$.

If $t: F \to \Omega$ is any natural transformation we define the subfunctor $\text{Ker}\, t$ of F on objects A of $\mathscr{C}$ by

$$(\text{Ker}\, t)(A) \equiv \{a \in F(A) \mid t(A)(a) \cdot = \cdot A/\mathscr{C}\},$$

while $(\ker t)(A)$ is taken to be the inclusion. It is now easily verified that

$$\text{char}(\ker t) \cdot = \cdot t, \quad \ker(\text{char}\, m) \cong m.$$

Finally, the constant functor $N(A) \equiv \mathbb{N}$, $N(A \to B) \equiv 1_{\mathbb{N}}$ is a natural numbers object.

In order to apply the clauses of Theorem 8.4 and Proposition 8.6 to the semantics of $\mathbf{Sets}^{\mathscr{C}}$, we observe the following:

Proposition 9.2. The representable functors $h^C \equiv \text{Hom}_{\mathscr{C}}(C, -)$, with C ranging over the objects of $\mathscr{C}$, form a generating set for $\mathbf{Sets}^{\mathscr{C}}$. Moreover, each representable functor h^C is indecomposable and projective.

Proof. Suppose $t, t': F \rightrightarrows G$ are two distinct natural transformations, then there is an object A of $\mathscr{C}$ and an element $a \in F(A)$ such that $t(A)(a) \neq t'(A)(a)$. Now, by Yoneda's Lemma, $F(A) \cong \text{Nat}(h^A, F)$, so a corresponds to a natural transformation $\hat{a}: h^A \to F$. More precisely, $\hat{a}(B): \text{Hom}(A, B) \to F(B)$ is defined by

$$\hat{a}(B)(f) \equiv F(f)(a).$$

In particular,

$$\hat{a}(A)(1_A) = F(1_A)(a) = a.$$

Now

$$t(A)(a) = t(A)\hat{a}(A)(1_A) = (t \circ \hat{a})(A)(1_A),$$

so $t \circ \hat{a} \neq t' \circ \hat{a}$. (Here $\circ$ denotes composition of natural transformations.) This shows that the representable functors h^A form a generating set.

Why is h^A indecomposable? Suppose $[k, l]: F + G \to h^A$ is an epimorphism, we claim that k or l is an epimorphism. Since epimorphisms are defined pointwise, we know, in particular, that $[k(A), l(A)]: F(A) + G(A) \to \text{Hom}(A, A)$ is surjective. Therefore 1_A must be in the image of $k(A)$ or $l(A)$, say the former. Then $k(A)(a) = 1_A$ for some $a \in F(A)$. Hence

$$(k \circ \hat{a})(B)(f) \cdot = \cdot k(B)\hat{a}(B)(f) \cdot = \cdot k(B)F(f)(a) \cdot = \cdot \text{Hom}(A, f)k(A)(a)$$

$$\cdot\!=\!\cdot \mathrm{Hom}(A,f)(1_A)$$
$$\cdot\!=\!\cdot f,$$

so $k \circ \hat{a} = 1_{h^A}$, which makes k an epimorphism.

Finally, to show that h^A is projective, we suppose that the natural transformation $k: F \to h^A$ is an epimorphism and show that it splits. Now $k(A): F(A) \to \mathrm{Hom}(A, A)$ is a surjection, so there is an element $a \in F(A)$ such that $k(A)(a) \cdot\!=\!\cdot 1_A$. As above, we deduce that $k \circ \hat{a} = 1_{h^A}$, so that k splits.

The proof of Proposition 9.2 is now complete.

To rewrite the clauses of Theorem 8.4 and Proposition 8.6 for $\mathbf{Sets}^{\mathscr{C}}$, we shall write 'stage C' for 'stage h^C'. We recall that, in view of Yoneda's Lemma, an arrow $c: h^C \to F$ is uniquely determined by an element $\check{c} \in F(C)$, where

$$c(A)(f) \equiv F(f)(\check{c}),$$

for any $f: C \to A$ in $\mathscr{C}$, and therefore

$$\check{c} \equiv c(C)(1_C).$$

Proposition 9.3. Given $c: h^C \to F$, then

(0) $C \Vdash c$ iff $\check{c} = C/\mathscr{C}$, in case $F = \Omega$;

(1) $C \Vdash b \in \beta$ if $\check{\beta}(C)(1_C, \check{b}) = C/\mathscr{C}$;

(2) $C \Vdash \top$ always;

(3) $C \Vdash \bot$ never;

(4) $C \Vdash \varphi(c) \wedge \psi(c)$ iff $C \Vdash \varphi(c)$ and $C \Vdash \psi(c)$;

(5′) $C \Vdash \varphi(c) \vee \psi(c)$ iff $C \Vdash \varphi(c)$ or $C \Vdash \psi(c)$;

(6) $C \Vdash \varphi(c) \Rightarrow \psi(c)$ iff, for all $k: C \to D$, if $D \Vdash \varphi(c_k)$ then $D \Vdash \psi(c_k)$, where $c_k \equiv ch^k$, $h^k: h^D \to h^C$;

(7) $C \Vdash \forall_{y \in G} \psi(y, c)$ iff, for all $k: C \to D$ and all $b: h^D \to G$, $D \Vdash \psi(b, c_k)$, where $c_k \equiv ch^k$;

(8′) $C \Vdash \exists_{y \in G} \psi(y, c)$ iff $C \Vdash \psi(b, c)$ for some $b: h^C \to G$.

Proof. (0) If $\check{c} \in \Omega(C)$, then $\check{c}$ is a C-sieve. Moreover, by Theorem 8.4, $C \Vdash c$ if and only if $c \cdot\!=\!\cdot \top \bigcirc_C$, which easily translates into the assertion that $\check{c} = \top(C)(*) = C/\mathscr{C}$.

(1) Here we assume that $\check{b} \equiv \check{b}(c)$ is an element of $G(C)$ and that $\check{\beta} \equiv \check{\beta}(c)$ is an element of $(PG)(C) \equiv \Omega^G(C)$. Thus $b: h^C \to G$ and $\beta: h^C \to PG$ are natural transformations and

$$b \in \beta \cdot\!=\!\cdot \varepsilon_G \circ \langle \beta, b \rangle$$

is obtained by composing two natural transformations, the first being $\varepsilon_G \equiv \varepsilon_{G,\Omega}$. Taking any object A and any arrow $f: C \to A$, one easily calculates that

$$(b \in \beta)(A)(f) = \check{\beta}(A)(f, G(f)(\check{b})).$$

Therefore,

$$(b \in \beta)\check{\ }\cdot = \cdot \check{\beta}(C)(1_C, b),$$

from which the result follows.

(2) is immediate.

(3) We know from Theorem 8.4 that $h^C \Vdash \bot$ if and only if h^C is initial in $\mathbf{Sets}^{\mathscr{C}}$, which implies, in particular, that $\mathrm{Hom}(C, C) = h^C(C)$ is initial in **Sets**, that is, empty. This cannot be because it contains 1_C.

(4) is immediate.

(5′) follows from Proposition 8.6 and the fact, established in Proposition 9.2, that h^C is indecomposable.

(6) The reader will easily verify that Theorem 8.4(6) is still true if D is restricted to be an object of a generating set for $\mathscr{T}$. In case $\mathscr{T} = \mathbf{Sets}^{\mathscr{C}}$, we replace $a: C \to A$ by $c: h^C \to F$ and $h: D \to C$ by $h^k: h^D \to h^C$, where $k: C \to D$ is any arrow in $\mathscr{C}$. Thus (6) should read:

$C \Vdash \varphi(c) \Rightarrow \psi(c)$ iff, for all $k: C \to D$ in $\mathscr{C}$, if $D \Vdash \varphi(ch^k)$ then $D \Vdash \psi(ch^k)$.

An easy calculation shows that if $c_k \equiv ch^k$, then $\check{c}_k \equiv F(k)(\check{c}) \in F(D)$ is what has become of $\check{c} \in F(C)$ in changing states from C to D along k.

(7) In Theorem 8.4(7) likewise D need only be taken in a generating set. In $\mathbf{Sets}^{\mathscr{C}}$, this becomes:

$$\text{for all } c: h^C \to F, \quad C \Vdash \forall_{y \in G} \psi(y, c)$$

iff, for all $k: C \to D$ and all $b: h^D \to G$, $D \Vdash \psi(b, ch^k)$.

(8′) follows from Proposition 8.6 and the fact, established in Proposition 9.2, that h^C is projective.

The proof of Proposition 9.3 is now complete.

Remark 9.4. In (6) and (7) above, the element c_k of F at stage D is what has become of the element c of F at stage C in passing along $k: C \to D$.

Example 9.5. Kripke models. We now look at a special case of Example 9.1 in which $\mathscr{C}$ is a preordered set regarded as a category. For objects (= elements) A, B of $\mathscr{C}$, $\mathrm{Hom}(A, B)$ has precisely one element if $A \leqslant B$, it is empty otherwise. The objects of $\mathbf{Sets}^{\mathscr{C}}$ may be regarded as $\mathscr{C}$-indexed families of sets $F = \{F(A) | A \in \mathscr{C}\}$ such that, whenever $A \leqslant B$, there is a 'transition mapping' $F_{AB}: F(A) \to F(B)$ satisfying

$$F_{AA} = 1_{F(A)}, \quad F_{BC} F_{AB} = F_{AC}.$$

A morphism $t: F \to G$ is a family of mappings $\{t(A): F(A) \to G(A) | A \in \mathscr{C}\}$ such that, for all $B \geqslant A$, the following square commutes:

$$\begin{array}{ccc} F(A) & \xrightarrow{t(A)} & G(A) \\ {\scriptstyle F_{AB}}\downarrow & & \downarrow{\scriptstyle G_{AB}} \\ F(B) & \xrightarrow[t(B)]{} & G(B) \end{array}$$

Limits and colimits are defined elementwise.

Exponentials are given thus: for $F, G: \mathscr{C} \to \mathbf{Sets}$,

$$G^F(A) \equiv \mathrm{Nat}(h^A \times F, G)$$

is essentially the set of all families $\{t(B): F(B) \to G(B) \mid B \geqslant A\}$ such that, for all $C \geqslant B$, $G_{BC}t(B) = t(C)F_{BC}$. We may interpret such a family as an arrow $F_A \to G_A$, where F_A is the restriction of F to $\{B \in \mathscr{C} \mid B \geqslant A\}$. We note that, if $A \leqslant B$, then $G^F(A) \subseteq G^F(B)$. Evaluation $\varepsilon_{G,F}: G^F \times F \to G$ is then given by

$$\varepsilon_{G,F}(C)(t, c) \equiv t(C)(c),$$

for any $t \in G^F(C)$ and $c \in F(C)$.

To describe the topos structure of $\mathbf{Sets}^{\mathscr{C}}$ when $\mathscr{C}$ is a preordered set we let

$$(A) \equiv \{B \in \mathscr{C} \mid B \geqslant A\},$$

$$\Omega(A) \equiv \{\mathscr{S} \subseteq (A) \mid \forall_{B,C \in \mathscr{C}}((B \in \mathscr{S} \wedge C \geqslant B) \Rightarrow C \in \mathscr{S}\}.$$

The two extreme cases are $\mathscr{S} = \varnothing$ and $\mathscr{S} = \{B \in \mathscr{C} \mid B \geqslant A\} \equiv (A)$. If $A \leqslant B$, the transition mapping $\Omega_{AB}: \Omega(A) \to \Omega(B)$ is given by

$$\Omega_{AB}(\mathscr{S}) \equiv \mathscr{S} \cap (B)$$

and $\top : 1 \to \Omega$ is given by

$$\top(A)(*) \equiv (A).$$

We let the reader work out characteristic morphisms and kernels.

The semantics for $\mathbf{Sets}^{\mathscr{C}}$ when $\mathscr{C}$ is a preordered set is essentially that proposed by Kripke for first order intuitionistic logic, now generalized to higher order. As in general functor categories, stages are representables h^C, usually just written as C, with the difference that now there is at most one transition $C \to D$, precisely when $C \leqslant D$.

The modifications of Proposition 9.3 in case $\mathscr{C}$ is a preordered set are obvious. For example:

(0) $C \Vdash c$ iff $\check{c} = (C)$, in case $F = \Omega$;

(1) $C \Vdash b \in \beta$ iff $\check{\beta}(C)(\check{b}) = (C)$;

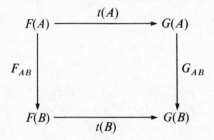

(6) $C \Vdash \varphi(c) \Rightarrow \psi(c)$ iff, whenever $C \leqslant D$, if $D \Vdash \varphi(c_D)$ then $D \Vdash \psi(c_D)$, where $\check{c}_D \equiv F_{CD}(\check{c})$.

Example 9.6. $\mathscr{M}$-sets. At the other extreme from Example 9.5 is the case of a functor category $\mathbf{Sets}^{\mathscr{C}}$ in which $\mathscr{C}$ has exactly one object C. $\mathscr{C}$ is then given by a monoid $\mathscr{M} = \mathrm{Hom}(C, C)$ and may be identified with $\mathscr{M}$. A functor $\mathscr{C} \to \mathbf{Sets}$ is essentially a set X with a monoid homomorphism $\mathscr{M} \to X^X$ or, equivalently, an $\mathscr{M}$-*set* (X, λ), that is, a set X with a mapping $\lambda: M \times X \to X$ (where M is the underlying set of $\mathscr{M}$), called the *action*, satisfying

$$\lambda(1, x) = x, \quad \lambda(mn, x) = \lambda(m, \lambda(n, x))$$

for all $x \in X$, m and $n \in M$. In particular, the sole representable functor h^C: $\mathscr{M} \to \mathbf{Sets}$ corresponds to the $\mathscr{M}$-set $\mathscr{M}$, the action $M \times M \to M$ being multiplication.

If two functors $\mathscr{M} \to \mathbf{Sets}$ are given by $\mathscr{M}$-sets (X, λ) and (Y, μ), a natural transformation between them is given by an $\mathscr{M}$-set homomorphism $f: (X, \lambda) \to (Y, \mu)$, that is, a mapping $f: X \to Y$ such that

$$f(\lambda(m, x)) = \mu(m, f(x))$$

for all $m \in M$ and $x \in X$.

Limits and colimits are calculated as in **Sets**, with the obvious 'induced' action. In particular, the empty $\mathscr{M}$-set is an initial object, while the terminal object is the set $\{*\}$ with trivial action.

To describe the cartesian closed structure, we may convert the definitions of Example 9.1 to $\mathscr{M}$-sets, but it seems easier to argue directly. Given $\mathscr{M}$-sets $A = (|A|, \lambda)$ and $B = (|B|, \mu)$, we want

$$|B^A| \cong \mathrm{Hom}(\mathscr{M}, B^A) \cong \mathrm{Hom}(\mathscr{M} \times A, B),$$

so we define $|B^A| \equiv \mathrm{Hom}(\mathscr{M} \times A, B)$ and endow it with the action θ given by

$$\theta(m, f)(n, a) \equiv f(nm, a),$$

for all $m, n \in M$, $f \in |B^A|$ and $a \in |A|$. Evaluation $\varepsilon_{B,A}$: $B^A \times A \to B$ is then given by

$$\varepsilon_{B,A}(f, a) \equiv f(1, a),$$

for all $a \in |A|$ and $f \in |B^A|$.

To describe the topos structure of $\mathbf{Sets}^{\mathscr{M}}$ let us first point out that, in the language of Example 9.1, $\Omega(C)$ is the set of all left ideals of $\mathscr{M}$. Examples of left ideals are: $\varnothing$, $\mathscr{M}$ and the principal left ideal $(m) \equiv \{nm \mid n \in M\}$ generated by $m \in M$. For any $m \in M \equiv \mathrm{Hom}(C, C)$, we define $\Omega(m): \Omega(C) \to \Omega(C)$ by stipulating, for any left ideal L of $\mathscr{M}$,

$$\Omega(m)(L) \equiv \{n \in M \mid nm \in L\}.$$

This makes $\Omega(C)$ into an $\mathscr{M}$-set with action $(m, L) \to \Omega(m)(L)$. In passing from $\mathbf{Sets}^{\mathscr{M}}$ to $\mathscr{M}$-sets, we should write Ω for $\Omega(C)$, so Ω is now the $\mathscr{M}$-set of left ideals of $\mathscr{M}$. Moreover, $\top\colon \{*\} \to \Omega$ is defined by $\top(*) \equiv \mathscr{M}$. Again we leave characteristic morphisms and kernels to the reader.

Before examining the topos semantics for $\mathscr{M}$-sets, we note that there is only one generator h^C, hence only one stage C, reference to which may as well be omitted. However, there are many transitions $C \to C$.

For any $\mathscr{M}$-set A, an arrow $a\colon \mathscr{M} \to A$ is determined by the element $\breve{a} \equiv a(1) \in |A|$. We shall present some selected clauses for the semantics of $\mathscr{M}$-sets, as modified from Proposition 9.3:

(0) $\Vdash a$ iff $\breve{a} = \mathscr{M}$, in case $A = \Omega$;
(1) $\Vdash b \in \beta$ iff $\breve{\beta}(1, \breve{b}) = \mathscr{M}$, where $\breve{b} \in |B|$ and $\breve{\beta} \in |\Omega^B|$;
(6) $\Vdash \varphi(a) \Rightarrow \psi(a)$ iff, for all $k \in M$, if $\Vdash \varphi(ka)$ then $\Vdash \psi(ka)$;
(7) $\Vdash \forall_{y \in B} \psi(y, a)$ iff, for all $k \in M$ and all $b \in |B|$, $\Vdash \psi(b, ka)$;
(8′) $\Vdash \exists_{y \in B} \psi(y, a)$ iff $\Vdash \psi(b, a)$ for some $\breve{b} \in |B|$.

Exercises

1. A type F in the internal language of $\mathbf{Sets}^{\mathscr{C}}$ is said to be *partially empty* if $F(A) = \varnothing$ for some object $A \in \mathscr{C}$. If this is so, prove that there are no closed terms of type F. Complete the proof that $\mathbf{Sets}^{\mathscr{C}}$ is a topos.

2. Fill in the missing details in the proofs of 9.3 and 9.5.

3. Prove that the topos of $\mathscr{M}$-sets is Boolean if and only if $\mathscr{M}$ is a group.

4. (Johnstone). Prove that the following statements are equivalent:
   (i) DeMorgan's law for $\mathscr{M}$-sets:

   $\mathbf{Sets}^{\mathscr{M}} \models \forall_{s \in \Omega} \forall_{t \in \Omega} (\neg(s \wedge t) \Rightarrow \neg s \vee \neg t)$.

   (ii) For any left ideals K, L of $\mathscr{M}$, if $K \cap L = \varnothing$ then $K = \varnothing$ or $L = \varnothing$.
   (iii) Left Ore condition for $\mathscr{M}$: for all $m, n \in M$, there are $u, v \in M$ such that $um \cdot = \cdot vn$.

5. Prove the following:
   (*a*) for an $\mathscr{M}$-set A, the arrows $1 \to A$ correspond to the fixed points under the action;
   (b) 1 is projective in $\mathbf{Sets}^{\mathscr{M}}$ if and only if $\mathscr{M}$ has an element 0 such that $0m = 0$ for all $m \in M$.

6. Prove that an arrow $p\colon 1 \to \Omega$ in $\mathbf{Sets}^{\mathscr{M}}$ is given by a left ideal $L \equiv p(C)(*)$ of $\mathscr{M}$ such that, for all $m \in M$, $\Omega(m)(L) = L$. Show that this can only happen

when $L = \varnothing$ or M, hence that $\mathrm{Hom}(1, \Omega)$ is a two-element Boolean algebra.

7. (Bradd Hart). Prove that in the topos of small graphs the subobject classifier Ω is a graph with two vertices and five arrows.

8. Let $\mathbf{n}$ be the n-element chain

$$\underset{0}{\cdot} \to \underset{1}{\cdot} \to \underset{2}{\cdot} \to \cdots \to \underset{n-1}{\cdot}$$

and consider the Kripke model $\mathbf{Set^n}$.

(*a*) Prove that $\mathrm{Hom}(1\ \Omega) \cong \mathbf{n} + \mathbf{1}$.

(*b*) Show that $\mathbf{Sets^n} \models \neg\neg\beta$, where $\beta \equiv \forall_{t \in \Omega}(t \vee \neg t)$ is the Boolean axiom. (Hint: check that $n-1 \Vdash \beta$ and $0 \models \neg\neg\beta$.)

(*c*) Generalize part of the above argument by showing that, if p is a terminal element of a pest $\mathscr{P}$, then $p \models \beta$ in $\mathbf{Sets}^{\mathscr{P}}$.

9. Let ω be the poset of natural numbers. Show that not $\mathbf{Sets}^{\omega} \models \neg\neg\beta$ (by checking that not $0 \Vdash \neg\neg\beta$) and infer that not $\vdash \neg\neg\beta$ in pure type theory $\mathfrak{L}_0$.

10 Sheaf categories and their semantics

We recall that a *topological space* $X = (|X|, \mathcal{O}(X))$ is given by a set $|X|$ of *points* and a lattice $\mathcal{O}(X)$ of *open* subsets of $|X|$ closed under finite intersection and arbitrary union. We shall often also write X for $|X|$. The lattice $\mathcal{O}(X)$ is a Heyting algebra with

$$1 = X, \quad 0 = \varnothing, \quad U \wedge V = U \cap V, \quad U \vee V = U \cup V$$

and, less obviously,

$$U \Rightarrow V = \mathrm{int}((X - U) \cup V),$$

this being the largest open set W such that $W \cap U \subseteq V$. In general $\mathcal{O}(X)$ is not a Boolean algebra, in fact

$$\neg\neg U = \mathrm{int}(X - \mathrm{int}(X - U))$$

is the interior of the closure of U.

Definition 10.1. A *presheaf* on X is a contravariant functor from $\mathcal{O}(X)$ to **Sets**, that is, a functor $F: \mathcal{O}(X)^{\mathrm{op}} \to \mathbf{Sets}$. Here $\mathcal{O}(X)$ is regarded as a category as in Example 9.5.

Because F is contravariant, the mapping $F_{UV}: F(U) \to F(V)$ applies whenever $V \subseteq U$. One calls $s \in F(U)$ a *section over* U and writes

$$F_{UV}(s) = s|_V,$$

which is called the *restriction* of s to V. According to Example 9.5, the presheaves on X form a topos.

Example 10.2. A presheaf F on X is called a *sheaf* provided it has the following property: for every *open covering*

$$U = \bigcup_{i \in I} U_i$$

of U and all *pairwise compatible* sections $\{s_i \in F(U_i)\}_{i \in I}$ such that

$$s_i|_{U_i \cap U_j} = s_j|_{U_i \cap U_j},$$

for all $i, j \in I$, there exists a unique $s \in F(U)$ such that $s|_{U_i} = s_i$, for all $i \in I$.

The reader may easily verify that this is the same as saying that

$$F(U) \to \prod_{i \in I} F(U_i) \rightrightarrows \prod_{i, j \in I} F(U_i \cap U_j)$$

is an equalizer diagram in **Sets**, where the mappings are the obvious ones induced by the restrictions.

The full subcategory of $\mathbf{Sets}^{\mathcal{O}(X)^{\mathrm{op}}}$ consisting of sheaves will be denoted by $\mathbf{Sh}(X)$. We shall see later that it is a topos. In the meantime we mention two special examples of sheaves on X:

(*a*) The sheaf of *continuous real valued functions* on X assigns to each open subset V of X the set

$$F(V) = \mathrm{Cont}(V, \mathbb{R})$$

of continuous functions from V to $\mathbb{R}$.

(*b*) Each open subset U of X determines a sheaf $F = h_U$, this being the contravariant representable functor given by

$$h_U(V) = \mathrm{Hom}(V, U) = \begin{cases} \{*\} & \text{if } V \subseteq U, \\ \varnothing & \text{otherwise.} \end{cases}$$

The reader will notice that we have taken the sole element of $\mathrm{Hom}(V, U)$ to be $*$ rather than incl_{VU}, the inclusion mapping of V into U.

In both these examples, $F(\varnothing)$ is a one-element set. To see that this is a consequence of the definition of 'sheaf' one must use the fact that the empty set admits a covering by an empty family!

We shall denote the category of topological spaces and continuous mappings by **top**, to distinguish it from the category **Top** of toposes and strict logical functors to be considered in Section 13. If $X \in \mathbf{top}$, the slice category $\mathbf{top}/X$ is called the category of *spaces over* X. Its objects are arrows $Y \to X$ in **top** and its arrows are commutative triangles:

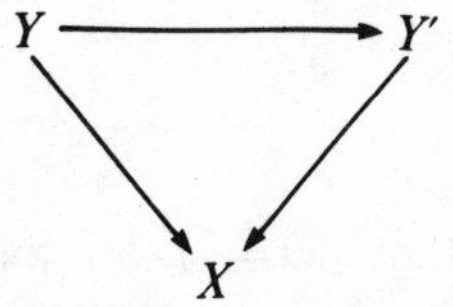

The category of spaces over X has a full subcategory whose objects are called *sheaf spaces*: they are *local homeomorphism* $p: Y \to X$, which means that for every point $y \in Y$ there is an open neighborhood V of y such that $p|_V$ is a homeomorphism.

Theorem 10.3. For any topological space X there is a pair of adjoint functors

$$\mathbf{Sets}^{\mathcal{O}(X)^{\mathrm{op}}} \underset{\Gamma}{\overset{L}{\rightleftarrows}} \mathbf{top}/X,$$

L left adjoint to Γ, which induce an equivalence of categories between $Sh(X)$ and the full subcategory of **top**/X consisting of sheaf spaces. Moreover, the former is a reflective subcategory with reflector ΓL and the latter a coreflective subcategory with coreflector $L\Gamma$.

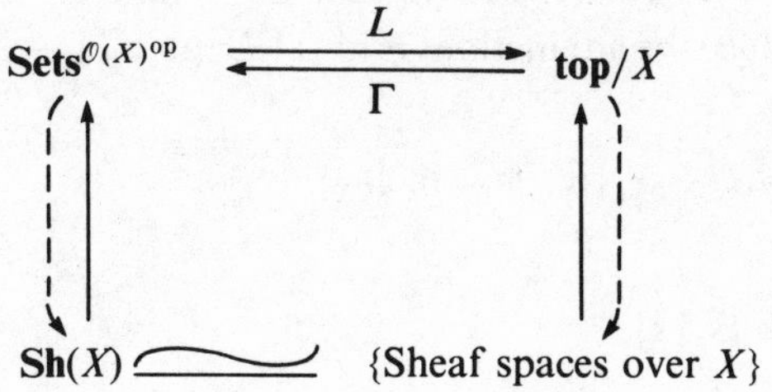

The reflector ΓL is called the *associated sheaf* functor.

Proof. We describe Γ first on objects $p: Y \to X$ of **top**/X:

$$\Gamma(p)(U) \equiv \{s: U \to Y \mid ps = \mathrm{incl}_{UX}\}.$$

A continuous mapping $s: U \to Y$ such that ps is the inclusion of U into X is called a *section* of p. If $V \subseteq U$, $\Gamma(p)(U) \to \Gamma(p)(V)$ is given by genuine restriction.Thus Γ is a functor. It is easily verified that $\Gamma(p)$ is a sheaf: the *sheaf of sections* of p.

We also describe L first on objects $F: \mathcal{O}(X)^{\mathrm{op}} \to \mathbf{Sets}$. The local homeomorphism $L(F): Y_F \to X$ will be defined presently. The underlying set

$$|Y_F| \equiv \dot{\bigcup_{x \in X}} F_x$$

is the disjoint union of *stalks* F_x, where

$$F_x \equiv \varinjlim_{x \in U} F(U) = \dot{\bigcup_{x \in U}} F(U)/ \underset{x}{\sim}.$$

Here, for $s \in F(U)$ and $t \in F(V)$, $s \underset{x}{\sim} t$ means that $s|_W = t|_W$ for some open $W \subseteq U \cap V$ such that $x \in W$. If $[s]_x$ denotes the equivalence class of s modulo $\underset{x}{\sim}$, we thus have

$$|Y_F| \equiv \{(x, [s]_x) | \exists_{U \in \mathcal{O}(X)} (x \in U \wedge s \in F(U))\}.$$

The topology on Y_F is generated by basic open sets of the form

$$V(s, U) \equiv \{(x, [s]_x) | x \in U\},$$

where $U \in \mathcal{O}(X)$ and $s \in F(U)$.

After these preliminaries, we define $L(F)$ as the projection:

$$L(F)(x, [s]_x) \equiv x.$$

This is continuous, because

$$L(F)^{-1}(U) = V(s, U).$$

It is in fact a local homeomorphism, because it maps $V(s, U)$ biuniquely onto U. Finally, it is easily verified that a natural transformation $F \to G$ induces an arrow $L(F) \to L(G)$ in **top**/X, so that L is a functor.

Next, we shall introduce the adjunctions η: id $\to \Gamma L$ and ε: $L\Gamma \to$ id. Recall that

$$\Gamma L(F)(U) = \{\sigma: U \to Y_F | L(F)\sigma = \mathrm{incl}_{UX}\}$$

and put

$$\eta(F)(U)(s)(x) \equiv (x, [s]_x),$$

for any $s \in F(U)$ and $x \in U$. Recall also that

$$|Y_{\Gamma(p)}| = \{(x, [s]_x) | \exists_{U \in \mathcal{O}(X)} (x \in U \wedge ps = \mathrm{incl}_{UX})\},$$
$$L\Gamma(p)(x, [s]_x) = x,$$

and put

$$\varepsilon(p)(x, [s]_x) \equiv s(x).$$

It is easily checked that $\varepsilon(p)$ is well-defined and continuous and that η and ε satisfy the appropriate identities assuring that L is left adjoint to Γ.

We know from general principles (see Part 0, Section 4) that Fix $\eta \simeq$ Fix ε, where

$$\mathrm{Fix}\,\eta \equiv \{F | \eta(F) \text{ is an isomorphism}\},$$
$$\mathrm{Fix}\,\varepsilon \equiv \{p | \varepsilon(p) \text{ is an isomorphism}\}.$$

We shall prove that Fix $\eta \simeq$ **Sh**(X) and that Fix $\varepsilon \simeq$ {sheaf spaces over X}, so that the equivalence asserted in Theorem 10.3 will follow.

A straightforward calculation shows that $s \mapsto \bar{s} \equiv \eta(F)(U)(s)$, where $\bar{s}(x) \equiv (x, [s]_x)$, is a biunique correspondence whenever F is a sheaf. Hence

$$\operatorname{Im} \mathrm{L} \subseteq \mathbf{Sh}(X) \subseteq \operatorname{Fix} \eta,$$

and so all three categories are equivalent. Thus $\eta\Gamma$ is a natural equivalence and $\mathbf{Sh}(X)$ is a reflective subcategory.

It is also easy to calculate that $(x, [s]_x) \mapsto \varepsilon(p)(x, [s]_x) \equiv s(x)$ is a biunique correspondence whenever p is a local homeomorphism. Hence

$$\operatorname{Im} \mathrm{L} \subseteq \{\text{sheaf spaces over } X\} \subseteq \operatorname{Fix} \varepsilon,$$

and so all three categories are equivalent. Thus εL is a natural equivalence and sheaf spaces over X form a coreflective subcategory.

The proof of Theorem 10.3 is now complete.

Proposition 10.4. $\mathbf{Sh}(X)$ is a topos.

Proof. We begin by pointing out that, as a reflective subcategory, $\mathbf{Sh}(X)$ is closed under limits, so these are constructed objectwise as for functor categories in general. Exponentiation also is constructed as for presheaves: given sheaves F and G, one wants

$$G^F(U) \cong \operatorname{nat}(h_U \times F, G);$$

but, since $\mathcal{O}(X)$ is a poset, we have

$$h_U(V) \equiv \begin{cases} \{*\} & \text{if } V \subseteq U, \\ \varnothing & \text{otherwise.} \end{cases}$$

Therefore, we define

$$G^F(U) \equiv \operatorname{nat}(F|_U, G|_U),$$

where $F|_U$ is the restriction of F to $\mathcal{O}(U)^{\mathrm{op}}$. This is easily shown to be a sheaf. Evaluation $\varepsilon_{G,F}\colon G^F \times F \to G$ is then also given as for presheaves by

$$\varepsilon_{G,F}(U)(t, s) \equiv t(U)(s),$$

where $t \in \operatorname{nat}(F|_U, G|_U)$ and $s \in F(U)$.

Clearly, the presheaf N is a sheaf. However, the subobject classifier differs from that for presheaves, which follows Example 9.5; instead we put

$$\Omega(U) \equiv \{V \in \mathcal{O}(X) \mid V \subseteq U\}.$$

If $U' \subseteq U$ we obtain the mapping $\Omega(U) \to \Omega(U')$ given by $V \mapsto V \cap U'$ for $V \in \Omega(U)$. The reader may verify that Ω is a sheaf. The arrow $\top\colon 1 \to \Omega$ is given by

$$\top(U)(*) \equiv U.$$

To check that we have made the right choice for the sheaf Ω, we could apply the associated sheaf functor to the presheaf Ω. However, it may be more instructive to verify instead that

$$\text{char}(\ker g) \cdot = \cdot g, \quad \ker(\text{char}\, m) \cong m,$$

for any arrow $g: F \to \Omega$ and any monomorphism $m: G \to F$, provided characteristic morphisms and kernels are defined appropriately.

To simplify matters, we shall assume that G is actually a subsheaf of F and that $m(U)$ is the inclusion mapping of $G(U)$ into $F(U)$ for each $U \in \mathcal{O}(X)$. We then put

$$(\text{char}\, m)(U)(s) \equiv \bigcup \{V \in \mathcal{O}(X) \mid V \subseteq U \wedge s|_V \in G(V)\}$$

for $U \in \mathcal{O}(X)$ and $s \in F(U)$. It is easily seen that this is the largest subset W of U such that $s|_W \in G(W)$. In particular, $\delta_F(U)(s,t)$ is then the largest subset W of U such that the restrictions of s and t agree on W. We furthermore put

$$(\text{Ker}\, g)(U) \equiv \{s \in F(U) \mid g(U)(s) = U\},$$

and let $(\ker g)(U)$ be the inclusion of $(\text{Ker}\, g)(U)$ into $F(U)$.

In particular, every subobject of 1 in $\mathbf{Sh}(X)$ is isomorphic to a kernel of an arrow $g: 1 \to \Omega$. Letting $U \equiv g(X)(*)$, we easily see that $g \equiv g_U$ is then given by

$$g_U(V)(*) \equiv U \cap V$$

for each $V \in \mathcal{O}(X)$. Hence

$$\begin{aligned}(\text{Ker}\, g_U)(V) &\equiv \{s \in 1(V) \mid g_U(V)(s) = V\} \\ &\equiv \{s \in \{*\} \mid U \cap V = V\} \\ &\equiv \begin{cases} \{*\} & \text{if } V \subseteq U, \\ \varnothing & \text{otherwise.} \end{cases}\end{aligned}$$

Thus

$$(\text{Ker}\ g_U)(V) = \text{Hom}(V, U) = h_U(V),$$

and so every subobject of 1 is isomorphic to h_U for some $U \in \mathcal{O}(X)$.

We observe that the subobjects h_U of 1 form a generating set for $\mathbf{Sh}(X)$, because they form one for $\mathbf{Sets}^{\mathcal{O}(X)^{\text{op}}}$. However, there is no reason why the representable sheaves h_U should be indecomposable or projective in $\mathbf{Sh}(X)$. Thus, we derive the semantics of $\mathbf{Sh}(X)$ from Theorem 8.4 without the help of Proposition 8.6. We write $U \Vdash \varphi(a)$ for $h_U \Vdash \varphi(a)$ and assume that $a: h_U \to F$ is determined by $\breve{a} \in F(U)$ as usual, where $\breve{a} \equiv a(U)(*)$ or $a(U)(1_U)$. (The reader will recall that we had written $*$ for the sole element of $\text{Hom}(V, U)$ in place of incl_{VU}.)

If $h_{V \subseteq U} \equiv \text{Hom}(-, \text{incl}_{VU})$ is the natural transformation $h_V \to h_U$ induced by the inclusion $V \subseteq U$ and if $a: h_U \to F$, we write $a|_V \equiv a \circ h_{V \subseteq U}$.

Proposition 10.5. (Sheaf semantics). Let F be a sheaf on X, U an open subset of X, $a: h_U \to F$. Then

(0) $U \Vdash a$ iff $\check{a} = U$, in case $F = \Omega$;

(1) $U \Vdash b \in \beta$ iff $\check{\beta}(U)(\check{b}) = U$, where $b: h_U \to F$ and $\beta: h_U \to \Omega^F$;

(2) $U \Vdash \top$ always;

(3) $U \Vdash \bot$ iff $U = \varnothing$;

(4) $U \Vdash \varphi(a) \wedge \psi(a)$ iff $U \Vdash \varphi(a)$ and $U \Vdash \psi(a)$;

(5) $U \Vdash \varphi(a) \vee \psi(a)$ iff $U = V \cup W$ for some V, $W \in \mathcal{O}(X)$ such that $V \Vdash \varphi(a|_V)$ and $W \Vdash \psi(a|_W)$;

(6) $U \Vdash \varphi(a) \Rightarrow \psi(a)$ iff, for all $V \subseteq U$, if $V \Vdash \varphi(a|_V)$ then $V \Vdash \psi(a|_V)$;

(7) $U \Vdash \forall_{y \in G} \psi(y, a)$ iff, for all $V \subseteq U$ and all $b: h_V \to G$, $V \Vdash \psi(b, a|_V)$;

(8) $U \Vdash \exists_{y \in G} \psi(y, a)$ iff there is an open covering $U = \bigcup_{i \in I} U_i$ and arrows $b_i: h_{U_i} \to G$ such that $U_i \Vdash \psi(b_i, a|_{U_i})$ for all $i \in I$.

Proof. We look at a few less obvious clauses.

(1) We recall that $b \in \beta: h_U \to \Omega$ is defined by

$$b \in \beta \equiv \varepsilon_F \circ \langle \beta, b \rangle,$$

hence

$$\begin{aligned}(b \in \beta)\check{} &= (b \in \beta)(U)(*) \\ &= \varepsilon_F(U)(\check{\beta}, \check{b}) \\ &= \check{\beta}(U)(\check{b}),\end{aligned}$$

in view of the definition of $\varepsilon_F \equiv \varepsilon_{\Omega,F}$ in the proof of Proposition 10.4. It follows that $b \in \beta \cdot = \cdot \top \circ \bigcirc_{h_U}$ if and only if $\check{\beta}(U)(\check{b}) = U$.

(3) $U \Vdash \bot$ means that $\bot \circ \bigcirc_U \cdot = \cdot \top \circ \bigcirc_U$. The subscript U here should really be h_U. Now $\bigcirc_U: h_U \to 1$, so $\bigcirc_U(V): \mathrm{Hom}(V, U) \to \{*\}$, and therefore $\top(V)\bigcirc_U(V)$: $\mathrm{Hom}(V, U) \to \Omega(V)$ sends $*$ to $\top(V)(*) \equiv V$. Similarly $\bot(V)\bigcirc_U(V)$ sends $*$ to $\bot(V)(*) \equiv \varnothing$. So $U \Vdash \bot$ asserts that, for all $V \subseteq U$, $V = \varnothing$, that is, that $U = \varnothing$.

Note that $\varnothing \Vdash \bot$ in $\mathbf{Sh}(X)$, but not in $\mathbf{Sets}^{\mathcal{O}(X)^{\mathrm{op}}}$.

(5) Suppose $U \Vdash \varphi(a) \vee \psi(a)$. By Theorem 8.4, there are arrows $k: D \to h_U$ and $l: E \to h_U$ such that $[k, l]: D + E \to h_U$ is an epimorphism and such that $D \Vdash \varphi(ak)$ and $E \Vdash \psi(al)$. Writing $\varphi(x) \cdot \underset{x}{=} \cdot fx$, we thus have $f \circ a \circ k \cdot = \cdot \top \circ \bigcirc_D$. We now factor k through its image. Being a subobject of 1, this may be taken to be of the form h_V. Recall that $h_{V \subseteq U}$ is the natural transformation $h_V \to h_U$ induced by the inclusion $V \subseteq U$, then

$$k \cdot = \cdot h_{V \subseteq U} \circ e,$$

where e is an epimorphism. Now

$$f \circ a \circ h_{V \subseteq U} \circ e \cdot = \cdot f \circ a \circ k \cdot = \cdot \top \circ \bigcirc_D \cdot = \cdot \top \circ \bigcirc_V \circ e,$$

and so

$$f \circ a \circ h_{V \subseteq U} \cdot = \cdot \top \circ \bigcirc_V,$$

that is, $V \Vdash \varphi(a|_V)$. Similarly, we obtain $W \Vdash \psi(a|_W)$.

It remains to show that $U = V \cup W$. We argue informally in the internal language of the topos of sheaves on X. Let $x \in h_U$, then $\exists_{w \in D+E}[k, l]w = x$. By Lemma 8.5, $\exists_{y \in D} w = \kappa_{D,E} y$ or $\exists_{z \in E} w = \kappa'_{D,E} z$, hence $ky = x$ or $lz = x$, say the former. Now $k \cdot = \cdot h_{V \subseteq U} \circ e$, so $h_{V \subseteq U}\, ey = x$ in the internal language. But $ey \in h_V$, hence $x = h_{V \subseteq U} y'$ for some $y' \in h_V$. Writing $t \equiv h_{V \subseteq V \cup W} y'$, we infer that $x = h_{V \cup W \subseteq U} t$. The same conclusion holds in case $lz = x$, so it holds in any case.

We have thus proved

$$\vdash \forall_{x \in h_U} \exists_{t \in h_{V \cup W}} x = h_{V \cup W \subseteq U} t.$$

Since $h_{V \cup W \subseteq U}$ is a monomorphism, we may replace $\exists$ by $\exists!$. Therefore, by description, there is a unique arrow $f: h_U \to h_{V \cup W}$ such that $h_{V \cup W \subseteq U} \circ f \cdot = \cdot 1_U$. By Yoneda's Lemma, we may write $f \cdot = \cdot h_g$, where $g \equiv f(U)(*)$. It easily follows that

$$\mathrm{incl}_{V \cup W \subseteq U} \circ g \cdot = \cdot 1_U.$$

Thus $\mathrm{incl}_{V \cup W \subseteq U}$ is surjective, and so $U = V \cup W$.

Conversely, suppose $U = V \cup W$ and $V \Vdash \varphi(a|_V)$ and $W \Vdash \psi(a|_W)$. We shall construct an epimorphism $[k, l]: h_V + h_W \to h_U$ such that $V \Vdash \varphi(ak)$ and $W \Vdash \psi(al)$. Indeed, taking $k \equiv h_{V \subseteq U}$ and $l \equiv h_{W \subseteq U}$, we have $a \circ k \cdot = \cdot a|_V$ and $a \circ l \cdot = \cdot a|_W$. To show that $[k, l]$ is an epimorphism, suppose $f, g: h_U \to G$ are such that $f \circ [k, l] \cdot = \cdot g \circ [k, l]$, that is, $f \circ k \cdot = \cdot g \circ k$ and $f \circ l \cdot = \cdot g \circ l$. We shall prove that $f \cdot = \cdot g$.

Consider the following commutative square:

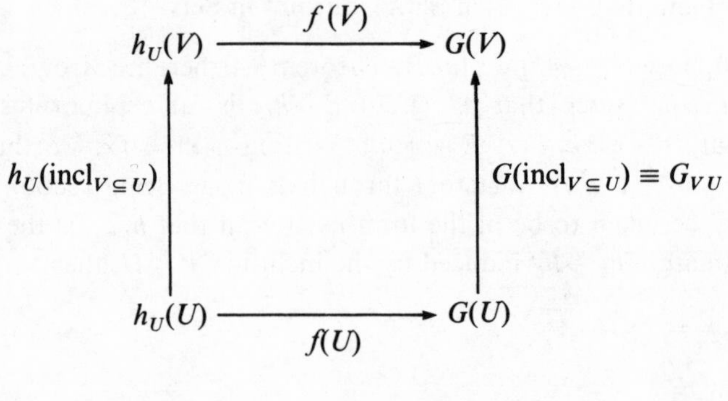

Writing $\check{f} \equiv f(U)(1_U)$ (at this point it is convenient not to replace 1_U by $*$), we see that

$$\begin{aligned}(f \circ k)(V)(1_V) \cdot = \cdot & f(V)(\mathrm{incl}_{V \subseteq U}) \\ \cdot = \cdot & G_{VU}(\check{f}).\end{aligned}$$

Therefore

$$G_{VU}(\check{f}) = G_{VU}(\check{g})$$

and similarly

$$G_{WU}(\check{f}) = G_{WU}(\check{g}).$$

Since G is a sheaf, we infer that $\check{f} = \check{g}$, hence that $f \cdot = \cdot g$.

(6) Suppose $U \Vdash \varphi(a) \Rightarrow \psi(a)$. Consider any open $V \subseteq U$, so $h_{V \subseteq U}$: $h_V \to h_U$. Writing $a|_V \equiv a h_{V \subseteq U}$, we use Theorem 8.4 to infer that if $V \Vdash \varphi(a_V)$ then $V \Vdash \psi(a_V)$.

Conversely, assume this implication. Suppose $k: D \to h_U$ and $D \Vdash \varphi(ak)$. We want to show that $D \Vdash \psi(ak)$, so that Theorem 8.4 will then yield $U \Vdash \varphi(a) \Rightarrow \psi(a)$. As in the proof of (5) above, we factor $k \cdot = \cdot e \circ h_{V \subseteq U}$. Now we are given that

$$f \circ a \circ h_{V \subseteq U} \circ e \cdot = \cdot \top \circ \bigcirc_D \cdot = \cdot \top \circ \bigcirc_V \circ e,$$

hence

$$f \circ a \circ h_{V \subseteq U} \cdot = \cdot \top \circ \bigcirc_V,$$

that is $V \Vdash \varphi(a_V)$. But then, by assumption, $V \Vdash \psi(a_V)$, from which it easily follows that $D \Vdash \psi(ak)$, as was to be shown.

The remaining clauses will be left as exercises to the reader.

Exercises

1. Prove that F is a sheaf on X if and only if $F(U) \to \prod_{i \in I} F(U_i) \rightrightarrows \prod_{i,j \in I} F(U_i \cap U_j)$ is an equalizer diagram for all coverings $U = \bigcup_{i \in I} U_i$ of open subsets U of X.
2. Prove that $\mathrm{Cont}(-, \mathbb{R})$ and $h_U \equiv \mathrm{Hom}(-, U)$ are sheaves on X.
3. Fill in the missing parts in the proofs of Theorem 10.3 and propositions 10.4 and 10.5
4. Prove that $\varnothing \Vdash \bot$ in $\mathbf{Sh}(X)$, but not in $\mathbf{Sets}^{\mathcal{O}(X)^{\mathrm{op}}}$.
5. Consider the sheaf $C \equiv \mathrm{Cont}(-, \mathbb{R})$ on X, U an open subset of X and $a: h_U \to C$. Show that $U \Vdash a \neq 0$ if and only if $\check{a} \equiv a(U)(1_U)$ vanishes on no nonempty open subset of U. Deduce that

   $\mathbf{Sh}(X) \vDash \forall_{x \in C}(\neg\neg x = 0 \Rightarrow x = 0)$.

6. Discuss the ring structure of C.

7. For any open subset U of X, let $Q(U)$ be the ring of all continuous real-valued functions defined on dense open subsets of U, with $f \in C(V)$ and $g \in C(W)$ identified when $f|_{V \cap W} = g|_{V \cap W}$, where V and W (hence also $V \cap W$) are dense open subsets of U. Show that Q is a sheaf and that

   $\mathbf{Sh}(X) \models \forall_{x \in Q}(\neg x = 0 \Rightarrow \exists_{y \in Q} x \cdot y = 1)$.

   (If $f \in C(V)$ and $g \in C(W)$, one defines $(f \cdot g)(x) \equiv f(x) \cdot g(x)$ for all $x \in V \cap W$.)

11 Three categories associated with a type theory

Given a type theory $\mathfrak{L}$, we first present a naive way of associating a category $D(\mathfrak{L})$ with $\mathfrak{L}$. Its objects are the types of $\mathfrak{L}$ and its arrows $A \to B$ are pairs $(x, \varphi(x))$, where x is a variable of type A and $\varphi(x)$ a term of type B with no free variables except possibly x.

We agree to identify arrows $(x, \varphi(x))$ and $(x', \psi(x'))$ if the terms $\varphi(x)$ and $\psi(x)$ are provably equal, that is, $\vdash_x \varphi(x) = \psi(x)$. Composition of arrows $(x, \varphi(x))$: $A \to B$ and $(y, \psi(y))$: $B \to C$ is given by $(x, \psi(\varphi(x)))$: $A \to C$. Moreover, the identity arrow $A \to A$ is of course (x, x).

The kind of category obtained in this way, but with a looser equality relation between arrows, has been described axiomatically and has been called a *dogma* in the literature. A dogma is not in general a topos, although it generates one. A dogma with the strong kind of equality advocated here may be faithfully embedded in a topos.

We next present a second way of associating a category $A(\mathfrak{L})$ with a type theory $\mathfrak{L}$. Its objects are closed terms α of type PA for any type A. We may think of α as denoting a *set* of type PA; but, taking a nominalistic position, we shall often regard α itself as a set. Equality between objects α and α' is defined to hold provided they have the same type PA and $\vdash \alpha = \alpha'$.

The arrows $f: \alpha \to \beta$ of $A(\mathfrak{L})$ are triples $(\alpha, |f|, \beta)$, where $|f|$ is a closed term of type $P(A \times B)$ and $\vdash |f| \subseteq \alpha \times \beta$. We often call $|f|$ the *graph* of f. We may think of f as denoting a *relation* between the sets denoted by α and β respectively; but, taking a nominalistic position, we shall often regard f as a relation between the sets α and β. Equality between relations $f, g: \alpha \rightrightarrows \beta$ is defined thus:

$$f \cdot\!=\!\cdot g \text{ means } \vdash |f| = |g|.$$

The identity relation 1_α: $\alpha \to \alpha$ is defined by

$$|1_\alpha| \equiv \{\langle x, x' \rangle \in A \times A \mid x = x'\}.$$

This is what has previously been denoted by $\ulcorner \delta_A \urcorner$. Composition of relations $f: \alpha \to \beta$ and $g: \beta \to \gamma$ is defined by

$$|gf| \equiv \{\langle x, z \rangle \in A \times C \mid \exists_{y \in B}(\langle x, y \rangle \in |f| \wedge \langle y, z \rangle \in |g|)\}.$$

It is easily seen that $A(\mathfrak{L})$ is a category. Such categories of 'sets and relations' have been described axiomatically and called *allegories* by Peter Freyd. We point out that with any relation $f: \alpha \to \beta$ one may associate its *converse* $f^{-1}: \beta \to \alpha$ defined by

$$|f^{-1}| \equiv \{\langle y, x \rangle \in B \times A \mid \langle x, y \rangle \in |f|\}.$$

Our final way of associating a category with a type theory concerns a subcategory $T(\mathfrak{L})$ of $A(\mathfrak{L})$, which later will be seem to be a topos. It has the same objects as $A(\mathfrak{L})$, namely sets. Its arrows are *functions*, that is relations $f: \alpha \to \beta$ satisfying

$$\vdash \forall_{x \in A}(x \in \alpha \Rightarrow \exists!_{y \in B} \langle x, y \rangle \in |f|).$$

Another way of expressing this is:

$$\vdash |1_\alpha| \subseteq |f^{-1}f|, \quad \vdash |ff^{-1}| \subseteq |1_\beta|.$$

The following lemma will be used in the next section.

Lemma 11.1. A function $f: \alpha \to \beta$ is a monomorphism if and only if $f^{-1}f \cdot = \cdot 1_\alpha$.

Proof. Assume $f^{-1}f \cdot = \cdot 1_\alpha$ and $fg \cdot = \cdot fh$, where g, $h: \gamma \rightrightarrows \alpha$. One immediately computes

$$g \cdot = \cdot f^{-1}fg \cdot = \cdot f^{-1}fh \cdot = \cdot h,$$

hence f is a monomorphism.

Conversely, suppose f is a monomorphism. Consider the set

$$\gamma \equiv |f^{-1}f| \cdot = \cdot \{\langle x, x' \rangle \in A \times A \mid \exists_{y \in B}(\langle x, y \rangle \in |f| \wedge \langle x', y \rangle \in |f|)\}.$$

Define g, $h: \gamma \rightrightarrows \alpha$ by

$$|g| \equiv \{\langle \langle x, x' \rangle, x \rangle \in (A \times A) \times A \mid \langle x, x' \rangle \in \gamma\},$$

$$|h| \equiv \{\langle \langle x, x' \rangle, x' \rangle \in (A \times A) \times A \mid \langle x, x' \rangle \in \gamma\}.$$

Then

$$|fg| \cdot = \cdot \{\langle \langle x, x' \rangle, y \rangle \in (A \times A) \times B \mid \langle x, y \rangle \in |f| \wedge \langle x, x' \rangle \in \gamma\},$$

$$|fh| \cdot = \cdot \{\langle \langle x, x' \rangle, y \rangle \in (A \times A) \times B \mid \langle x', y \rangle \in |f| \wedge \langle x, x' \rangle \in \gamma\}.$$

Since f is a function, it follows from the definition of γ that $fg \cdot = \cdot fh$, hence that $g \cdot = \cdot h$. Therefore $\vdash \gamma \subseteq |1_\alpha|$, that is, $\gamma \cdot = \cdot |1_\alpha|$.

Ultimately we shall prove that $T(\mathfrak{L})$ is a topos; but, in this section, we shall content ourselves with a weaker statement.

Proposition 11.2. $T(\mathfrak{L})$ is a cartesian closed category.

Proof. When necessary, we shall use bold face to distinguish objects in $T(\mathfrak{L})$ from types in $\mathfrak{L}$. The terminal object $\mathbf{1}$ of $T(\mathfrak{L})$ is defined by

$$\mathbf{1} \equiv \{*\},$$

while products are defined by

$$\alpha \times \beta \equiv \{\langle x, y\rangle \in A \times B \mid x \in \alpha \wedge y \in \beta\},$$

where α is of type PA and β of type PB. The arrows $\bigcirc_\alpha: \alpha \to \mathbf{1}$, $\pi_{\alpha,\beta}: \alpha \times \beta \to \alpha$ and $\pi'_{\alpha,\beta}: \alpha \times \beta \to \beta$ are defined thus:

$$|\bigcirc_\alpha| \equiv \alpha \times \{*\} \equiv \{\langle x, y\rangle \in A \times 1 \mid x \in \alpha\},$$

$$|\pi_{\alpha,\beta}| \equiv \{\langle\langle x, y\rangle, x\rangle \in (A \times B) \times A \mid x \in \alpha \wedge y \in \beta\},$$

$$|\pi'_{\pi,\beta}| \text{ similarly.}$$

Moreover, if $f: \gamma \to \alpha$ and $g: \gamma \to \beta$, we define $\langle f, g\rangle: \gamma \to \alpha \times \beta$ by

$$|\langle f, g\rangle| \equiv \{\langle z, \langle x, y\rangle\rangle \in C \times (A \times B) \mid \langle z, x\rangle \in |f| \wedge \langle z, y\rangle \in |g|\}.$$

It is a routine exercise to check that $T(\mathfrak{L})$ satisfies all the equations of a cartesian category.

Before continuing, we introduce a useful notation: Suppose α is a term of type PA, β is a term of type PB and ρ is a term of type $P(A \times B)$,

$$\rho: \alpha \to \beta \equiv \rho \subseteq \alpha \times \beta \wedge \forall_{x \in A}(x \in \alpha \Rightarrow \exists!_{y \in B} \langle x, y\rangle \in \rho).$$

It is important to distinguish the formula $\rho: \alpha \to \beta$ of $\mathfrak{L}$, in which ρ is a term of type $P(A \times B)$, from the statement $f: \alpha \to \beta$ about $T(\mathfrak{L})$, in which f is an arrow of $T(\mathfrak{L})$, which really means that $\vdash |f|: \alpha \to \beta$ in the notation just introduced.

We now define

$$\beta^\alpha \equiv \{w \in P(A \times B) \mid w: \alpha \to \beta\},$$

if α has type PA and β has type PB. We also define $\varepsilon_{\beta,\alpha}: \beta^\alpha \times \alpha \to \beta$ by

$$|\varepsilon_{\beta,\alpha}| \equiv \{\langle\langle w, x\rangle, y\rangle \in (P(A \times B) \times A) \times B \mid w: \alpha \to \beta \wedge \langle x, y\rangle \in w\}.$$

Moreover, if $h: \alpha \times \beta \to \gamma$, then $h^*: \alpha \to \gamma^\beta$ is defined thus:

$$|h^*| \equiv \{\langle x, w\rangle \in A \times P(B \times C) \mid x \in \alpha \wedge w: \beta \to \gamma \wedge$$
$$\forall_{y \in B}(y \in \beta \Rightarrow \exists_{z \in C}(\langle\langle x, y\rangle, z\rangle \in |h| \wedge \langle y, z\rangle \in w))\}.$$

It is again a routine exercise to show that the remaining equations of a cartesian closed category are satisfied.

In particular, if $\mathbf{\Omega} \equiv \{t \in \Omega | \top\}$,

$$\mathbf{\Omega}^{\alpha} \equiv \{w \in P(A \times \Omega) | w : \alpha \to \mathbf{\Omega}\}.$$

Instead of regarding $P\alpha$ as an abbreviation for $\mathbf{\Omega}^{\alpha}$, we may also define $P\alpha$ directly by

$$P\alpha \equiv \{w \in PA | w \subseteq \alpha\}.$$

The reader should note that then $P\alpha$ is no longer equal to $\mathbf{\Omega}^{\alpha}$, but only isomorphic to it, the proof of which fact we leave as an exercise. On the other hand, we then have

$$\vdash \beta^{\alpha} \subseteq P(\alpha \times \beta),$$

so that β^{α} becomes a subobject of $P(\alpha \times \beta)$ in $T(\mathfrak{L})$, in fact a 'canonical' subobject, in a sense that will be explained in Section 15.

Having defined $P\alpha$ directly, we do the same for $\varepsilon_{\alpha}: P\alpha \times \alpha \to \mathbf{\Omega}$ rather than regarding it as an abbreviation for $\varepsilon_{\mathbf{\Omega},\alpha}$. Thus

$$|\varepsilon_{\alpha}| \equiv \{\langle\langle u, x\rangle, t\rangle \in (PA \times A) \times \Omega | x \in \alpha \wedge u \subseteq \alpha \wedge t = (x \in u)\}.$$

Moreover, if $h: \beta \times \alpha \to \mathbf{\Omega}$, we obtain $h^*: \beta \to P\alpha$ as follows:

$$|h^*| \equiv \{\langle y, u\rangle \in B \times PA | u = \{x \in A | \langle\langle y, x\rangle, \top\rangle \in |h|\}\}.$$

Strictly speaking, this is not the same as the $h^*: \beta \to \mathbf{\Omega}^{\alpha}$ introduced earlier, but we use the same notation anyway.

Exercises

1. Prove that an arrow $f: \alpha \to \beta$ in $T(\mathfrak{L})$ is an epimorphism if and only if $ff^{-1} \cdot = \cdot 1_{\beta}$.
2. Check that the equations of a cartesian closed category are satisfied in $T(\mathfrak{L})$.
3. Show that there are functors $D(\mathfrak{L}) \to A(\mathfrak{L}) \to T(\mathfrak{L})$.
4. Prove directly that $P\alpha$ as defined above is isomorphic to $\mathbf{\Omega}^{\alpha}$.
5. Show that, for $h: \beta \times \alpha \to \mathbf{\Omega}$ and $k: \beta \to P\alpha$, $\varepsilon_{\alpha}(h^* \times 1_{\alpha}) \cdot = \cdot h, (\varepsilon_{\alpha}(k \times 1_{\alpha}))^* \cdot = \cdot k.$

12 The topos generated by a type theory

We shall show that the category $T(\mathfrak{L})$ associated with a type theory $\mathfrak{L}$ is a topos. In view of Proposition 11.2, it remains only to produce a subobject classifier and a natural numbers object.

The subobject classifier $\mathbf{\Omega}$ and the canonical arrow $\mathbf{T}:1\to\mathbf{\Omega}$ (to be distinguished from the term $\top$ of type Ω in $\mathfrak{L}$) are defined as follows:

$$\mathbf{\Omega}\equiv\{t\in\Omega\,|\,\top\},$$
$$|\mathbf{T}|\equiv\{\langle *,\top\rangle\}.$$

Lemma 12.1. Every arrow $h:\alpha\to\mathbf{\Omega}$ in $T(\mathfrak{L})$ has a kernel $\ker h:\operatorname{Ker} h\to\alpha$ that is, an equalizer of h and $\mathbf{T}\bigcirc_\alpha$, so that the following square is a pullback:

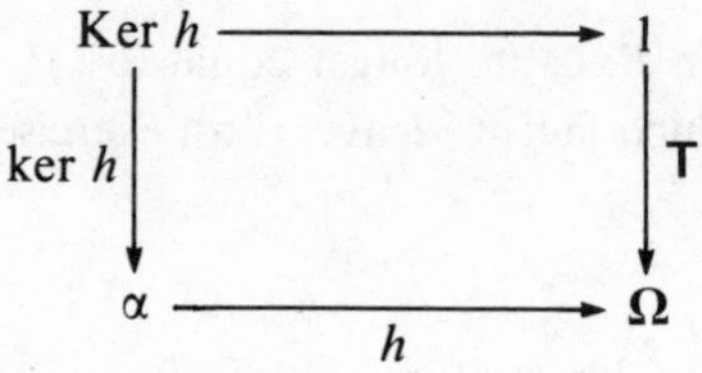

We define

$$\operatorname{Ker} h\equiv\{x\in A\,|\,\langle x,\top\rangle\in|h|\},$$
$$|\ker h|\equiv\{\langle x,x\rangle\in A\times A\,|\,\langle x,\top\rangle\in|h|\}.$$

Proof. First note that the square commutes. Suppose $f:\beta\to\alpha$ is such that $hf\cdot=\cdot\,\mathbf{T}\bigcirc_\beta$. Then

$$|hf|\cdot=\cdot\{\langle y,\top\rangle\in B\times\Omega\,|\,y\in\beta\}.$$

Define $g:\beta\to\operatorname{Ker} h$ by $|g|\equiv|f|$, then g is easily seen to be the unique arrow for which $(\ker h)g\cdot=\cdot f$.

Proposition 12.2. $(\mathbf{\Omega},\mathbf{T})$ is a subobject classifier of $T(\mathfrak{L})$.

Proof. If $m:\beta\to\alpha$ is a monomorphism in $T(\mathfrak{L})$, we define its characteristic morphism $\operatorname{char} m:\alpha\to\mathbf{\Omega}$ by

$$|\operatorname{char} m|\equiv\{\langle x,t\rangle\in A\times\Omega\,|\,t=(\exists_{y\in B}\langle y,x\rangle\in|m|)\}.$$

It remains to check that

(i) $\operatorname{char}(\ker h)\cdot=\cdot h$,
(ii) $\ker(\operatorname{char} m)\cong m$,

where $h:\alpha\to\mathbf{\Omega}$ and $m:\beta\to\alpha$ is a monomorphism.

Before proving this, let us establish a lemma, which will also be useful elsewhere. (See Lemma 20.3.).

Lemma 12.3. Given a formula $\varphi(t)$ of $\mathfrak{L}$ with a free variable t of type Ω such that

$(*)\qquad \vdash\exists!_{t\in\Omega}\varphi(t),$

then

$$(**) \qquad \vdash \forall_{t\in\Omega}(t = \varphi(\top) \Leftrightarrow \varphi(t)).$$

Proof. We argue informally. (*a*) Suppose $\varphi(t)$. Assuming t, we obtain $t = \top$, hence $\varphi(\top)$. Thus $t \Rightarrow \varphi(\top)$. Assuming $\varphi(\top)$, we obtain $\top = t$ from $(*)$, hence t. Thus $\varphi(\top) \Rightarrow t$. Therefore $\varphi(\top) \Leftrightarrow t$, hence $\varphi(\top) = t$, by propositional extensionality. This proves one of the implications in $(**)$.

(*b*) Conversely, suppose $t = \varphi(\top)$. By $(*)$, $\varphi(s)$ for some $s \in \Omega$. In view of (*a*), $t = \varphi(\top) = s$, hence $\varphi(t)$. This proves the other implication in $(**)$.

We now return to the proof of Proposition 12.2.

Proof of (i). We calculate

$$\begin{aligned} |\mathrm{char}(\ker h)| \cdot{=}\cdot & \{\langle x,t\rangle \in A \times \Omega \mid t = (\exists_{y\in A}\langle y,x\rangle \in |\ker h|)\} \\ \cdot{=}\cdot & \{\langle x,t\rangle \in A \times \Omega \mid t = (\exists_{y\in A}(\langle x,\top\rangle \in |h| \wedge y = x))\} \\ \cdot{=}\cdot & \{\langle x,t\rangle \in A \times \Omega \mid t = (\langle x,\top\rangle \in |h|)\} \\ \cdot{=}\cdot & \{\langle x,t\rangle \in A \times \Omega \mid \langle x,t\rangle \in |h|\} \cdot{=}\cdot |h|, \end{aligned}$$

by Lemma 12.3 with $\varphi(t) \equiv \langle x,t\rangle \in |h|$.

Proof of (ii). Suppose $m: \beta \to \alpha$ is a given monomorphism. We claim that $m \cong \ker(\mathrm{char}\, m)$. Indeed, define $g: \beta \to \mathrm{Ker}(\mathrm{char}\, m)$ by $|g| \equiv |m|$, as in the proof of Lemma 12.1, then

$$|g^{-1}g| \cdot{=}\cdot |m^{-1}m| \cdot{=}\cdot 1_\beta,$$

by Lemma 11.1.

On the other hand,

$$\begin{aligned} |gg^{-1}| \cdot{=}\cdot & |mm^{-1}| \\ \cdot{=}\cdot & \{\langle x,x\rangle \in A \times A \mid \exists_{y\in B}\langle y,x\rangle \in |m|\} \\ \cdot{=}\cdot & \{\langle x,x\rangle \in A \times A \mid \langle x,\top\rangle \in |\mathrm{char}\, m|\} \\ \cdot{=}\cdot & \{\langle x,x\rangle \in A \times A \mid x \in \mathrm{Ker}(\mathrm{char}\, m)\} \\ \cdot{=}\cdot & |1_{\mathrm{Ker}(\mathrm{char}\, m)}|. \end{aligned}$$

Therefore g is an isomorphism. The proof of Proposition 12.2 is now complete.

The natural numbers object $\mathbf{N}$ in $T(\mathfrak{L})$ with canonical arrows $\mathbf{0}: \mathbf{1} \to \mathbf{N}$ and $\mathbf{S}: \mathbf{N} \to \mathbf{N}$ is defined as follows:

$$\begin{aligned} \mathbf{N} &\equiv \{x \in N \mid \top\}, \\ |\mathbf{0}| &\equiv \{\langle *, 0\rangle\} \\ |\mathbf{S}| &\equiv \{\langle x,y\rangle \in N \times N \mid y = Sx\}. \end{aligned}$$

Proposition 12.4. $(\mathbf{N}, \mathbf{0}, \mathbf{S})$ is a natural numbers object in $T(\mathfrak{L})$.

Proof. Let there be given arrows in $T(\mathfrak{L})$:

$$1 \xrightarrow{g} \alpha \xrightarrow{h} \alpha,$$

where α is a closed term of type PA. Then g must be determined by $|g| \cdot = \cdot \{\langle *, a\rangle\}$, where a is a closed term of type A such that $\vdash a \in \alpha$ in $\mathfrak{L}$. We seek a unique arrow $f: \mathbf{N} \to \alpha$ such that the following diagram commutes:

$$\begin{array}{ccccc} \mathbf{1} & \xrightarrow{\mathbf{0}} & \mathbf{N} & \xrightarrow{\mathbf{S}} & \mathbf{N} \\ \Vert & & \downarrow f & & \downarrow f \\ \mathbf{1} & \xrightarrow[g]{} & \alpha & \xrightarrow[h]{} & \alpha \end{array}$$

What this means is that $\vdash |f| \subseteq \mathbf{N} \times \alpha$, that $|f|$ is functional and that

(*) $\qquad \langle 0, a\rangle \in |f|, \quad f\mathbf{S} \cdot = \cdot hf.$

The following well-known construction of $|f|$, presented here as an informal argument, is taken from Jacobson's textbook *Basic Algebra I*. Note that the argument is completely intuitionistic.

Let Γ be the set of all subsets u of $\mathbf{N} \times \alpha$ having the following properties:

(i) $\langle 0, a\rangle \in u$,

(ii) $\forall_{y \in N} \forall_{x \in A}(\langle y, x\rangle \in u \Rightarrow \langle Sy, h(x)\rangle \in u)$,

where $h(x)$ is the unique $z \in A$ such that $\langle x, z\rangle \in |h|$. Since $\mathbf{N} \times \alpha$ has these properties, clearly Γ is nonempty. Let $|f|$ be the intersection of all $u \in \Gamma$. We shall prove that $f = (\mathbf{N}, |f|, \alpha)$ is the required arrow $\mathbf{N} \to \alpha$.

First, it is easily shown by induction that $\forall_{y \in N} \exists_{x \in A} \langle y, x\rangle \in |f|$. To see that f is a function, it remains to show that if $\langle y, x\rangle \in |f|$ and $\langle y, x'\rangle \in |f|$ then $x = x'$.

Let

$$\sigma = \{y \in N \mid \forall_{x \in A} \forall_{x' \in A}((\langle y, x\rangle \in |f| \wedge \langle y, x'\rangle \in |f|) \Rightarrow x = x')\}.$$

We shall prove by induction on y that $\forall_{y \in N} y \in \sigma$, so that $\mathbf{N} = \sigma$.

To show that $0 \in \sigma$, it suffices to prove that $\forall_{x \in A}(\langle 0, x\rangle \in |f| \Rightarrow x = a)$. So suppose $\langle 0, x\rangle \in |f|$; then $\langle 0, x\rangle \in u$ for all $u \in \Gamma$. Let

$$u_1 \equiv \{\langle 0, a\rangle\} \cup \{\langle Sy, x\rangle \in N \times A \mid \langle Sy, x\rangle \in |f|\}.$$

It is easily shown that $u_1 \in \Gamma$ (see Exercise 4 below), hence $\langle 0, x\rangle \in u_1$, and therefore $x = a$.

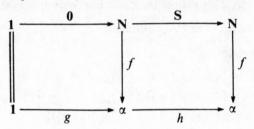

Next, suppose $y \in \sigma$, we claim that $Sy \in \sigma$. Since $y \in N$, we can find $x \in A$ such that $\langle y, x\rangle \in |f|$, hence also $\langle Sy, h(x)\rangle \in |f|$. It will suffice to show that, for any $x' \in A$, $\langle Sy, x'\rangle \in |f|$ implies $x' = h(x)$. So suppose $\langle Sy, x'\rangle \in |f|$, then $\langle Sy, x'\rangle \in u$ for all $u \in \Gamma$. Take

$$u_2 \equiv \{\langle Sy, h(x)\rangle\} \cup \{\langle z, x''\rangle \in N \times A \mid \langle z, x''\rangle \in |f| \wedge z \neq Sy\}.$$

It is easily verified that $u_2 \in \Gamma$ by checking that $\langle 0, a\rangle \in u_2$ and by showing that $\langle y', x''\rangle \in u_2 \Rightarrow \langle Sy', h(x'')\rangle \in u_2$. The latter is shown by relying on the decidability of equality at type N (see Exercise 2 below) and by examining the two separate cases when $y' \neq y$ and $y' = y$. (We leave the first case to the reader (see Exercise 4 below) and present here the second case only. We assume that $\langle y, x''\rangle \in u_2$ and deduce that $\langle Sy, h(x'')\rangle \in u_2$ as follows. Since $y \neq Sy$, clearly $\langle y, x''\rangle \in |f|$. But, since $y \in \sigma$, $x'' = x$, hence $\langle Sy, h(x'')\rangle = \langle Sy, h(x)\rangle \in u_2$.) Therefore $\langle Sy, x'\rangle \in u_2$, and so $x' = h(x)$, as was to be shown.

We have thus completed the proof that f is a function. Moreover, (i) and (ii) hold with u replaced by $|f|$. The reader will have no difficulty in verifying that (ii) does in fact assert the equation $f\mathbf{S} \cdot = \cdot hf$, so that (*) holds.

To prove the uniqueness of f, let g be any arrow $\mathbf{N} \to \alpha$ satisfying (*). Then $|g| \in \Gamma$ and so $|f| \subseteq |g|$. Since f and g are both functions $\mathbf{N} \to \alpha$, it follows that $f \cdot = \cdot g$.

This completes the proof of Proposition 12.4. We summarize Propositions 12.1, 12.2 and 12.4 as follows:

Theorem 12.5. For any type theory $\mathfrak{L}$, $T(\mathfrak{L})$ is a topos (with natural numbers object).

Exercises

1. Complete the details of the proof of Lemma 12.1.
2. For any type theory $\mathfrak{L}$, prove that equality at type N is *decidable*, that is, $\vdash \forall_{x \in N} \forall_{y \in N} (x = y \vee x \neq y)$, where $x \neq y$ is short for $\neg(x = y)$.
3. For any type theory $\mathfrak{L}$, prove that

   $\vdash \forall_{x \in N} (x = 0 \vee \exists_{y \in N} x = Sy)$.
4. Complete the proof of Proposition 12.4.
   (Hint: To prove that $u_1 \in \Gamma$, use Exercise 3 above. To prove that $u_2 \in \Gamma$, consider the subcases $y' = Sy$ and $y' \neq Sy$.)

13 The topos generated by the internal language

In Section 3 we associated with every topos $\mathscr{T}$ a type theory $L(\mathscr{T})$, its internal language, while in Section 12 we associated with every type theory $\mathfrak{L}$ a topos $T(\mathfrak{L})$, the topos generated by $\mathfrak{L}$. In this and the next section we shall study two morphisms

$$\xi_{\mathscr{T}}: \mathscr{T} \to TL(\mathscr{T}), \quad \eta_{\mathfrak{L}}: \mathfrak{L} \to LT(\mathfrak{L})$$

in appropriate categories of toposes and type theories. After having settled on suitable definitions, we shall prove that $\xi_{\mathscr{T}}$ is an equivalence of categories and that $\eta_{\mathfrak{L}}$ is a conservative extension. Finally, we shall extend L and T to functors, making ξ and η into natural transformations.

We had previously (see Definition 3.1) defined a topos as a cartesian closed category with a subobject classifier $(\Omega, \top)$ and a natural numbers object (N, O, S). For our present purposes it is more natural to use an equivalent* definition in which the 'closed' part of the definition is weakened: instead of requiring arbitrary exponents C^A, we shall only require $PA \cong \Omega^A$. More precisely, in place of exponentiation we postulate the *power set structure* $(P, \varepsilon, *)$, meaning that for each object A there is an object PA, an arrow $\varepsilon_A: PA \times A \to \Omega$ and a rule

$$\frac{B \times A \xrightarrow{\;h\;} \Omega}{B \xrightarrow{\;h^*\;} PA}$$

satisfying

$$\varepsilon_A(h^* \times 1_A) \cdot\!=\!\cdot\, h, \quad (\varepsilon_A(k \times 1_A))^* \cdot\!=\!\cdot\, k,$$

for all $h: B \times A \to \Omega$ and $k: B \to PA$. The natural numbers object will be retained.

If $\mathscr{T}$ and $\mathscr{T}'$ are toposes in the present sense, that is, with exponentiation related to power sets, we shall define a *strict logical functor* $F: \mathscr{T} \to \mathscr{T}'$ as a functor which preserves all this structure on the nose, that is, which preserves $1, 0, \times, \pi, \pi', \langle\ \ \rangle, \Omega, \top, P, \varepsilon, *, N, O$ and S, and which in addition preserves δ. Thus, for example,

$$F(1) = 1, \quad F(O_A) \cdot\!=\!\cdot\, O_{F(A)}, \; F(\top) \cdot\!=\!\cdot\, \top,$$
$$F(PA) = P(F(A)), \quad F(h^*) \cdot\!=\!\cdot\, F(h)^*, \quad F(\delta_A) \cdot\!=\!\cdot\, \delta_{F(A)}.$$

Note that a strict logical functor automatically preserves the symbols of the internal language $L(\mathscr{T})$ (see Definition 3.5), hence characteristic morphisms, in view of Proposition 5.4. It follows that a strict logical functor preserves kernels up to isomorphism.

Definition 13.1. The category **Top** has as objects toposes in the weaker

* In the sense that every topos in the new sense is equivalent to one in the old sense (see Proposition 13.3).

sense, that is, with exponentiation relaxed to power sets, and as morphisms strict logical functors.

With each topos $\mathcal{T}$ we shall associate a strict logical functor $\xi_{\mathcal{T}}$: $\mathcal{T} \to TL(\mathcal{T})$. This is defined as follows.

Definition 13.2. If A is an object of $\mathcal{T}$, hence a type of $L(\mathcal{T})$, then

$$\xi_{\mathcal{T}}(A) \equiv \mathbf{A} \equiv \{x \in A \mid \top\}$$

is an object of $TL(\mathcal{T})$. If $f: A \to B$ is an arrow of $\mathcal{T}$, we put

$$\xi_{\mathcal{T}}(f) \equiv \mathbf{f}{:}\mathbf{A} \to \mathbf{B},$$

where $\mathbf{f}$ is a triple $(\mathbf{A}, |\mathbf{f}|, \mathbf{B})$ with

$$|\mathbf{f}| \equiv \mathrm{gph}\, f \equiv \{\langle x, y\rangle \in A \times B \mid fx = y\}$$

being the *graph* of f (actually, in our usual terminology, of $\mathbf{f}$).

It is easy, though tedious, to verify that $\xi_{\mathcal{T}}$ is a strict logical functor. For example,

$$\xi_{\mathcal{T}}(A \times B) \cdot = \cdot \mathbf{A} \times \mathbf{B} \cdot = \cdot \mathbf{A} \times \mathbf{B} \cdot = \cdot \xi_{\mathcal{T}}(A) \times \xi_{\mathcal{B}}(B),$$

according to the definition of $\alpha \times \beta$ in $T(\mathfrak{L})$ in Section 11. Similarly,

$$\xi_{\mathcal{T}}(PA) \cdot = \cdot \mathbf{PA} \cdot = \cdot P\mathbf{A} \cdot = \cdot P(_{\mathcal{T}}(A)),$$

according to the definition of $P\alpha$ in $T(\mathfrak{L})$. However, $\xi_{\mathcal{T}}$ does not preserve exponentiation on the nose, only up to isomorphism, as

$$\xi_{\mathcal{T}}(B^A) \cdot = \cdot \{u \in B^A \mid \top\}, \quad \xi_{\mathcal{T}}(B)^{\xi_{\mathcal{T}}(A)} \cdot = \cdot \{w \in P(A \times B) \mid w{:}\, A \to \mathbf{B}\}.$$

This is why we weakened the definition of a topos to replace exponentiation by power sets. As a final illustration we shall show that $\xi_{\mathcal{T}}$ preserves δ, that is, that $\boldsymbol{\delta}_{\mathbf{A}} \cdot = \cdot \delta_{\mathbf{A}}$.

Indeed, according to the proof of Proposition 12.2, $\delta_{\mathbf{A}}$: $\mathbf{A} \times \mathbf{A} \to \Omega$ is given by

$$|\delta_{\mathbf{A}}| \equiv \{\langle\langle x_1, x_2\rangle, t\rangle \in (A \times A) \times \Omega \mid t = \exists_{x \in A} \langle x, \langle x_1, x_2\rangle\rangle \in |\langle 1_{\mathbf{A}}, 1_{\mathbf{A}}\rangle|\}.$$

In view of Proposition 11.2, the formula following the existential quantifier above may be replaced by

$$\langle x, x_1\rangle \in |1_{\mathbf{A}}| \wedge \langle x, x_2\rangle \in |1_{\mathbf{A}}|,$$

hence by

$$x = x_1 \wedge x = x_2.$$

Therefore

$$|\delta_{\mathbf{A}}| \cdot = \cdot \{\langle\langle x_1, x_2\rangle, t\rangle \in (A \times A) \times \Omega \mid t = (x_1 = x_2)\}$$
$$\cdot = \cdot \mathrm{gph}\, \delta_A \cdot = \cdot |\boldsymbol{\delta}_{\mathbf{A}}|,$$

as was to be proved.

Proposition 13.3. $\xi_{\mathscr{T}}$ is an equivalence of categories.

Proof. We shall show that every object of $TL(\mathscr{T})$ is isomorphic to an object in the image of $\xi_{\mathscr{T}}$ and that $\xi_{\mathscr{T}}$ is full and faithful. (See Part 0, Section 4, Exercise 4.)

An object of $TL(\mathscr{T})$ is a closed term in $L(\mathscr{T})$, say α of type PA, that is, an arrow $\alpha: 1 \to PA$ in $\mathscr{T}$. Let $m: A' \to A$ be a kernel of $\alpha^{\int}: A \to \Omega$, we claim that $\alpha \cong A' \cdot = \cdot \xi_{\mathscr{T}}(A')$. Indeed, consider the arrows $\rho: 1 \to P(A' \times A)$, $\rho': 1 \to P(A \times A')$ given by

$$\rho \equiv \{\langle x', x\rangle \in A' \times A \mid mx' = x\}, \rho' \equiv \{\langle x, x'\rangle \in A \times A' \mid mx' = x\}.$$

Then $(\mathbf{A}', \rho, \alpha)$ and $(\alpha, \rho', \mathbf{A}')$ are seen to be arrows inverse to one another.

To see that $\xi_{\mathscr{T}}$ is full, assume that $(\mathbf{A}, \varphi, \mathbf{B})$ is an arrow $\xi_{\mathscr{T}}(A) \to \xi_{\mathscr{T}}(B)$ in $TL(\mathscr{T})$, where φ is a closed term of type $P(A \times B)$ such that

$$\mathscr{T} \models \forall_{x \in A} \exists !_{y \in B} \langle x, y\rangle \in \varphi.$$

Since description holds in $\mathscr{T}$ (Theorem 5.9), there is a unique arrow $f: A \to B$ such that $fx = y \cdot \underset{\{x,y\}}{=} \cdot \langle x, y\rangle \in \varphi$. It follows that $\varphi \cdot = \cdot \text{gph } f \cdot = \cdot |\mathbf{f}|$, and so $(\mathbf{A}, \varphi, \mathbf{B}) \cdot = \cdot \mathbf{f} \cdot = \cdot \xi_{\mathscr{T}}(f)$.

Thus $\xi_{\mathscr{T}}$ is full. To see that $\xi_{\mathscr{T}}$ is faithful, assume that $f, g: A \rightrightarrows B$ and $\xi_{\mathscr{T}}(f) \cdot = \cdot \xi_{\mathscr{T}}(g)$, then $(\mathbf{A}, |\mathbf{f}|, \mathbf{B}) \cdot = \cdot (\mathbf{A}, |\mathbf{g}|, \mathbf{B})$, hence $\mathscr{T} \models \text{gph } f = \text{gph } g$ and therefore $\mathscr{T} \models \forall_{x \in A}(fx = gx)$. Since internal equality implies external equality (Proposition (5.1), we have $f \cdot = \cdot g$.

Exercises

1. Show how to define arbitrary equalizers in terms of kernels, hence deduce that strict logical functors preserve equalizers up to isomorphism.
2. Complete the verification that $\xi_{\mathscr{T}}$ is a strict logical functor.
3. Complete the proof of Proposition 13.3. (In particular, to show that $(\alpha, \rho', \mathbf{A}')$ is an arrow, note that $\alpha \cdot = \cdot \ulcorner \text{char } m \urcorner \cdot = \cdot \{x \in A \mid \exists_{x' \in A'} mx' = x\}$ by Proposition 5.4, hence that $\mathscr{T} \models \forall_{x \in A}(x \in \alpha \Rightarrow \exists !_{x' \in A'} \langle x, x'\rangle \in \rho')$ by Proposition 6.1.)

14 The internal language of the topos generated

We begin this section by settling on a suitable category of type theories. We then set up a pair of functors between type theories and toposes.

Definition 12.1. The category **Lang** of type theories has as objects type theories based on equality. Its morphisms $\tau: \mathfrak{L} \to \mathfrak{L}'$ are *translations.* These send types to types so as to preserve $1, N, \Omega, P$ and $\times$. They send closed terms to closed terms of corresponding types so as to preserve $*$, 0, $S, \in$, $=, \{\ \}$ and $\langle\ \rangle$ up to provable equality. They also send variables to variables in a prescribed way: with ith variable of type A to ith variable of type $\tau(A)$. Finally, they send theorems to theorems: more precisely, if $\Gamma \vdash_X P$ in $\mathfrak{L}$ then $\tau(\Gamma) \vdash_{\tau(X)} \tau(p)$ in $\mathfrak{L}'$. Two translations τ and τ' between the same languages are said to be *equal* if $\vdash_{\tau(X)} \tau(a) = \tau'(a)$ for all terms $a \equiv a(X)$ of any type A, where X contains all free variables occurring in a.

For example,

$$\tau(\{x \in A \mid x = a\}) = \{x' \in \tau(A) \mid x' = \tau(a)\},$$

where x and x' are the ith variables of types A in $\mathfrak{L}$ and $\tau(A)$ in $\mathfrak{L}'$ respectively. Note that a translation is completely determined by its action on types and closed terms, provided it is known which is the ith variable of any given type in the target language.

Definition 14.2. The translation $\eta_{\mathfrak{L}}: \mathfrak{L} \to LT(\mathfrak{L})$ is defined as follows. For each type A of $\mathfrak{L}$ we put

$$\eta_{\mathfrak{L}}(A) \equiv \mathbf{A} \equiv \{x \in A \mid \top\}.$$

(Note that the notation $\mathbf{A}$ coincides with that given in the special cases $A = 1, \Omega$ and N in Sections 11 and 12.) For each closed term a of type A in $\mathfrak{L}$ we put

$$\lambda_{\mathfrak{L}}(a) \equiv \mathbf{a}: \mathbf{1} \to \mathbf{A}, \quad \text{where } |\mathbf{a}| \equiv \{\langle *, a \rangle\}.$$

(We note that the notation $\mathbf{a}$ coincides with that given in the special cases $a \equiv \top$ or 0 in Section 12.)

A translation $\tau: \mathfrak{L} \to \mathfrak{L}'$ is called a *conservative extension* if, for each closed formula p of $\mathfrak{L}$, $\vdash \tau(p)$ in $\mathfrak{L}'$ implies $\vdash p$ in $\mathfrak{L}$.

Proposition 14.3. For every type A of a type theory $\mathfrak{L}$, $\eta_{\mathfrak{L}}$ induces a biunique correspondence between closed terms of type PA in $\mathfrak{L}$ modulo provable equality and terms of type $\mathbf{PA} \equiv \{u \in PA \mid \top\}$ in $LT(\mathfrak{L})$. In particular, $\eta_{\mathfrak{L}}$ is a conservative extension.

Proof. A closed term of type $\mathbf{PA}$ in $LT(\mathfrak{L})$ is an arrow $f: \mathbf{1} \to \mathbf{PA}$ in $T(\mathfrak{L})$ whose graph $|f|$ is a term of type $P(1 \times PA)$ in $\mathfrak{L}$ such that $\vdash \forall_{z \in 1} \exists!_{u \in PA} \langle z, u \rangle \in |f|$, that is, $\vdash \exists!_{u \in PA} \langle *, u \rangle \in |f|$. Letting

$$\alpha \equiv \{x \in A \mid \exists_{u \in PA}(x \in u \wedge \langle *, u \rangle \in |f|)\},$$

we readily see that $\vdash |f| = \{\langle *, \alpha \rangle\}$, hence $f \cdot = \cdot \boldsymbol{\alpha}$. Thus $\eta_{\mathfrak{L}}$ is surjective. To show that it is one-to-one, assume $\boldsymbol{\alpha} \cdot = \cdot \boldsymbol{\alpha}'$, then $\vdash |\boldsymbol{\alpha}| = |\boldsymbol{\alpha}'|$, that is, $\vdash \{\langle *, \alpha \rangle\} = \{\langle *, \alpha' \rangle\}$, and so $\vdash \alpha = \alpha'$. Since $\boldsymbol{\Omega} \cong P1$ in $T(\mathfrak{L})$, $\eta_{\mathfrak{L}}$ is a conservative extension, which fact may also be shown directly.

A stronger result for pure type theory $\mathfrak{L}_0$ will be shown later (Corollary 20.4).

In general, $\eta_{\mathfrak{L}}: \mathfrak{L} \to LT(\mathfrak{L})$ is far from being an isomorphism of languages. *$LT(\mathfrak{L})$ may have more types than $\mathfrak{L}$*: in addition to types of the form $\mathbf{A}$, corresponding to types A of $\mathfrak{L}$, there are types α coming from objects of the topos $T(\mathfrak{L})$, that is, closed terms of type PA in $\mathfrak{L}$. (On the other hand, it is conceivable that the terms $\mathbf{A} = \mathbf{A}'$, even when $A \neq A'$, although one may then infer that they have the same type $PA = PA'$.) *$LT(\mathfrak{L})$ may have more terms than $\mathfrak{L}$*: a term of type $\mathbf{A}$ in $LT(\mathfrak{L})$ is an arrow $\varphi: \mathbf{1} \to \mathbf{A}$ in $T(\mathfrak{L})$, whose graph $|\varphi|$ is a term of type $P(1 \times A)$ in $\mathfrak{L}$ satisfying $\vdash \exists!_{x \in A} \langle *, x \rangle \in |\varphi|$; only if there is a closed term a of type A such that $\vdash \langle *, a \rangle \in |\varphi|$ can we infer that $\varphi = |\mathbf{a}|$. Different terms of $\mathfrak{L}$ cannot collapse in $LT(\mathfrak{L})$, for, when a and a' are distinct in $\mathfrak{L}$, then not $\vdash a = a'$, so that $\mathbf{a} \cdot \neq \cdot \mathbf{a}'$ in $T(\mathfrak{L})$.

In view of this discussion, the following result becomes evident.

Proposition 14.4

(1) $\eta_{\mathfrak{L}}$ is injective on terms (as long as, for any terms a, a' of the same type in $\mathfrak{L}$, $\vdash a = a'$ implies that a and a' are identified in $\mathfrak{L}$).

(2) $\eta_{\mathfrak{L}}$ is surjective on terms (say, of type $\mathbf{A}$), if and only if $\mathfrak{L}$ has the unique existence property: from $\vdash \exists!_{x \in A} \varphi(x)$ one may infer that there is a closed term a of type A such that $\vdash \varphi(a)$.

We shall say that $\eta_{\mathfrak{L}}$ is *almost* an isomorphism if it is bijective on terms and $\mathfrak{L}$ has enough types, that is, any object α of $T(\mathfrak{L})$ is isomorphic to one of the form $\mathbf{A}'$, which is necessarily a kernel of $\alpha^{\int}: 1 \to PA$ in $T(\mathfrak{L})$, if α is a term of type PA in $\mathfrak{L}$.

For example, $\eta_{L(\mathscr{T})}$ is injective (because internal equality in $\mathscr{T}$ implies external equality) and surjective, because $\mathscr{T}$ has description. In any case, it is almost an isomorphism, because $\eta_{L(\mathscr{T})} = L(\xi_{\mathscr{T}})$ and $\xi_{\mathscr{T}}$ is an equivalence, in view of Lemma 14.6 below and Proposition 13.3.

We shall complete this section by showing how both L and T may be extended to functors.

Definition 14.5. The functor L: **Top** $\to$ **Lang** assigns to each object $\mathscr{T}$ of **Top** its internal language $L(\mathscr{T})$. It assigns to each strict logical functor $F: \mathscr{T} \to \mathscr{T}'$ a translation $L(F)$ as follows. For a type A of $L(\mathscr{T})$, that is, an object A of $\mathscr{T}$, we put

$$L(F)(A) \equiv F(A).$$

For a closed term a of type A in $L(\mathscr{T})$, that is, an arrow $a: 1 \to A$ in $\mathscr{T}$, we put

$$L(F)(a) \equiv F(a).$$

We leave it to the reader to check that $L(F)$ as defined above is indeed a translation and that L is a functor. We also leave it as an exercise to show that L is full and faithful. Instead we shall prove the following useful result.

Lemma 14.6. For any topos $\mathscr{T}$,

$$L(\xi_{\mathscr{T}}) = \eta_{L(\mathscr{T})}.$$

Proof. For types A and closed terms a we have

$$L(\xi_{\mathscr{T}})(A) = \xi_{\mathscr{T}}(A) = \mathbf{A} = \eta_{L(\mathscr{T})}(A),$$
$$L(\xi_{\mathscr{T}})(a) = \xi_{\mathscr{T}}(a) = \mathbf{a} = \eta_{L(\mathscr{T})}(a),$$

using Definitions 14.5, 13.1 and 14.2 in this order. Moreover, if x is the ith variable of type A in $L(\mathscr{T})$ and x' is the ith variable of type $\mathbf{A}$ in $LTL(\mathscr{T})$, we have

$$L(\xi_{\mathscr{T}})(x) = x' = \eta_{L(\mathscr{T})}(x),$$

again by Definitions 14.5 and 14.2.

Definition 14.7. The functor T: **Lang** $\to$ **Top** assigns to each language $\mathfrak{L}$ the topos $T(\mathfrak{L})$, generated by it. For any translation $\tau: \mathfrak{L} \to \mathfrak{L}'$ we define $T(\tau): T(\mathfrak{L}) \to T(\mathfrak{L}')$ as follows. If α is an object of $T(\mathfrak{L})$, say a closed term α of type PA in $\mathfrak{L}$, we put

$$T(\tau)(\alpha) \equiv \tau(\alpha),$$

a closed term of type $P(\tau(A))$ in $\mathfrak{L}'$, hence an object of $T(\mathfrak{L}')$. Now let $f: \alpha \to \beta$ be an arrow in $T(\mathfrak{L})$, that is, a triple $(\alpha, |f|, \beta)$ with $|f|$ a closed term in $\mathfrak{L}$ of type $P(A \times B)$ satisfying

$$\vdash |f| \subseteq \alpha \times \beta, \qquad \vdash \forall_{x \in A}(x \in \alpha \Rightarrow \exists!_{y \in B} \langle x, y \rangle \in |f|).$$

Then we define $T(\tau)(f): \tau(\alpha) \to \tau(\beta)$ by

$$|T(\tau)(f)| \equiv \tau(|f|)$$

and verify that $(\tau(\alpha), \tau(|f|), \tau(\beta))$ is an arrow in $\mathfrak{L}'$, that is, $\vdash \tau(|f|) \subseteq \tau(\alpha) \times \tau(\beta)$, etc. It is a routine matter to check that $T(\tau)$ is a strict logical functor, which even preserves exponentials.

For example,

$$T(\tau)(\alpha \times \beta) \cdot = \cdot \tau(\alpha \times \beta) \cdot = \cdot \tau(\alpha) \times \tau(\beta) \cdot = \cdot T(\tau)(\alpha) \times T(\tau)(\beta)$$

and

$$T(\tau)(\delta_\alpha) \cdot = \cdot \delta_{T(\tau)(\alpha)}$$

because

$$|T(\tau)(\delta_\alpha)| \cdot = \cdot \tau(|\delta_\alpha|)$$
$$\cdot = \cdot \tau(\{\langle\langle x, y\rangle, x = y\rangle \in (A \times A) \times \Omega \mid x \in \alpha \wedge y \in \alpha\})$$
$$\cdot = \cdot \{\langle\langle x', y'\rangle, x' = y'\rangle \in (\tau(A) \times \tau(A)) \times \Omega \mid x' \in \tau(\alpha) \wedge y' \in \tau(\alpha)\}$$
$$\cdot = \cdot |\delta_{\tau(\alpha)}|$$
$$\cdot = \cdot |\delta_{T(\tau)(\alpha)}|.$$

Having defined the functors L and T we may verify that ξ and η as given by Definitions 13.2 and 14.2 are natural transformations. This will be left as an exercise.

We conclude this section by pointing out the possibility of changing types in $LT(\mathfrak{L})$. A type α of $LT(\mathfrak{L})$ is of course a subtype of $\mathbf{A} \equiv \{x \in A \mid \top\}$ and gives rise to the term $\boldsymbol{\alpha}$ of type $P\mathbf{A}$, an arrow $\mathbf{1} \to P\mathbf{A}$ in $T(\mathfrak{L})$ with graph $|\boldsymbol{\alpha}| = \{\langle *, \alpha\rangle\}$.

Remark 14.8. In $T(\mathfrak{L})$ we have the following 'change of type' identities:

$$\{x \in \alpha \mid \varphi(x)\} \cdot = \cdot \{x \in \mathbf{A} \mid x \in \boldsymbol{\alpha} \wedge \varphi(x)\}$$
$$\forall_{x \in \alpha} \varphi(x) \cdot = \cdot \forall_{x \in \mathbf{A}} (x \in \boldsymbol{\alpha} \Rightarrow \varphi(x)),$$
$$\exists_{x \in \alpha} \varphi(x) \cdot = \cdot \exists_{x \in \mathbf{A}} (x \in \boldsymbol{\alpha} \wedge \varphi(x)),$$

and similarly for $\exists!$.

Exercises

1. Prove that the functor L: **Top** → **Lang** is full and faithful.
2. Prove that ξ and η are natural transformations $\xi: \mathrm{id} \to TL$, $\eta: \mathrm{id} \to LT$.
3. Prove Remark 14.8.
4. Show that, if the definition of morphism is sufficiently relaxed, $\eta_{\mathfrak{L}}$ will become an isomorphism.

15 Toposes with canonical subobjects

We would like to show that T is left adjoint to L. To this purpose we need a natural transformation $\varepsilon: TL \to \mathrm{id}$ to accompany the natural transformation $\eta: \mathrm{id} \to LT$ constructed in Section 14. The obvious candidate for $\varepsilon_{\mathcal{T}}$ is an 'inverse' of the equivalence $\xi_{\mathcal{T}}: \mathcal{T} \to TL(\mathcal{T})$ constructed in Section 13. Such a functor $\varepsilon_{\mathcal{T}}$ must exist, since $\xi_{\mathcal{T}}$ is an equivalence of categories. The problem is to ensure that $\varepsilon_{\mathcal{T}}$ is a 'strict' logical functor, that is to say, that it preserves everything on the nose. This can easily be achieved if we insist that from now on all toposes have 'canonical' subobjects. If the reader is not happy with this restriction and is willing to sacrifice strictness, he may skip a page or so and proceed directly to Definition 15.2.

We recall that a *subobject* of A is a monomorphism $B \to A$ (or sometimes, waving hands, the object B itself). An *isomorphism* between subobjects $B \to A$ and $C \to A$ is an isomorphism $B \rightleftarrows C$ such that the following triangle commutes:

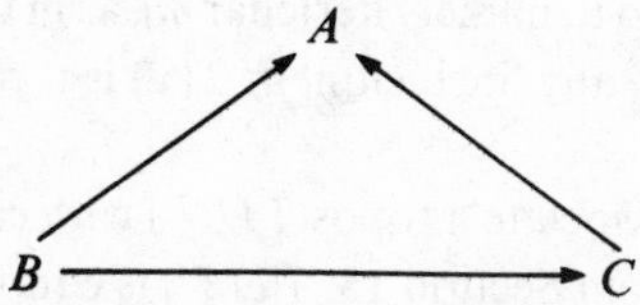

Definition 15.1. A topos *with canonical subobjects* is a topos where to each object A is associated a subset of the set of all subobjects, called 'canonical', such that every subobject of A is isomorphic to a unique canonical one and such that furthermore the following postulates are satisfied:

C1. $1_A : A \to A$ is a canonical subobject of A.

C2. If $f: B \to A$ is a canonical subobject of A and $g: C \to B$ is a canonical subobject of B, then $fg: C \to A$ is a canonical subobject of A.

C3. If $f: B \to A$ and $g: D \to C$ are canonical subobjects of A and C respectively, then $f \times g: B \times D \to A \times C$ is a canonical subobject of $A \times C$.

C4. If $f: B \to A$ is a canonical subobject of A, then $Pf: PB \to PA$ is a canonical subobject of PA, where, for an indeterminate arrow $v: 1 \to PB$,

$$(Pf)v \cdot \underset{v}{=} \cdot \{x \in A \mid \exists_{y \in B}(y \in v \wedge fy = x)\}.$$

A topos with canonical subobjects has canonical kernels. Thus, to each morphism $h: A \to \Omega$ we may associate its *canonical kernel* $\ker h: \operatorname{Ker} h \to A$ namely *the* canonical subobject of A which is isomorphic to *a* kernel of h, that is, which has characteristic morphism h. Then we have

$$\operatorname{char}(\ker h) \cdot = \cdot h, \quad \ker(\operatorname{char} m) \cdot = \cdot m,$$

where $h: A \to \Omega$ is any arrow and $m: B \to A$ is any canonical monomorphism.

Let $\mathbf{Top}_0$ be the category whose objects are toposes (as in Section 13) with canonical subobjects and whose morphisms are strict logical functors (as in Section 13) which preserve canonical subobjects. (A strict logical functor must, in any case, send a subobject onto a subobject.) Since every strict logical functor preserves characteristic morphisms, hence kernels, up to isomorphism, the morphisms of $\mathbf{Top}_0$ will preserve canonical kernels on the nose.

How common are toposes with canonical subobjects? It is clear that the category of sets has canonical subobjects, namely subsets. Consequently, also functor categories have canonical subobjects, namely subfunctors. Therefore also reflective subcategories of functor categories have canonical subobjects, in particular, so does the category of sheaves on a topological space.

Given any type theory $\mathfrak{L}$, we constructed the topos $T(\mathfrak{L})$ generated by it in Section 12. For any monomorphism $m: \beta \to \alpha$ in $T(\mathfrak{L})$ we easily find the canonical subobject corresponding to it, namely ker(char m), as in the proof of Proposition 12.2. More generally, any 'inclusion' in $T(\mathfrak{L})$ is a canonical monomorphism.

With any topos $\mathscr{T}$ we may now associate a topos $TL(\mathscr{T})$ with canonical subobjects. Moreover, as was shown in Section 13, $TL(\mathscr{T})$ is equivalent to $\mathscr{T}$ as a category.

We shall use the same letter L for the functor $L: \mathbf{Top}_0 \to \mathbf{Lang}$ induced by the functor $L: \mathbf{Top} \to \mathbf{Lang}$ discussed in Section 14. As remarked above, for any object $\mathfrak{L}$ of **Lang**, $T(\mathfrak{L})$ is an object of $\mathbf{Top}_0$, so we may also write $T: \mathbf{Lang} \to \mathbf{Top}_0$.

Definition 15.2. Let $\mathscr{T}$ be a topos with canonical subobjects. We construct the strict logical functor $\varepsilon_{\mathscr{T}}: TL(\mathscr{T}) \to \mathscr{T}$ as follows. Given an object α of $TL(\mathscr{T})$, say a closed term of type PA in $L(\mathscr{T})$, that is, an arrow $\alpha: 1 \to PA$ in $\mathscr{T}$, we put

$$\varepsilon_{\mathscr{T}}(\alpha) \equiv \mathrm{Ker}(\alpha^{\int}),$$

the canonical subobject of A whose characteristic morphism is $\alpha^{\int}: A \to \Omega$ in $\mathscr{T}$. Given an arrow $f: \alpha \to \beta$, where $|f|$ is a closed term of type $P(A \times B)$ satisfying the usual two conditions, we take $\varepsilon_{\mathscr{T}}(f): \varepsilon_{\mathscr{T}}(\alpha) \to \varepsilon_{\mathscr{T}}(\beta)$ as the unique arrow $g: \mathrm{Ker}(\alpha^{\int}) \to \mathrm{Ker}(\beta^{\int})$ such that, if we write $k_\alpha \equiv \ker(\alpha^{\int})$, for an indeterminate arrow $x: 1 \to \mathrm{Ker}(\alpha^{\int})$, in $L(\mathscr{T})$,

$$\vdash_x \langle k_\alpha x, k_\beta g x \rangle \in |f|.$$

The existence of the unique arrow g follows by description (Theorem 5.9) once we verify that

$$\vdash \forall_{x \in \mathrm{Ker}(\alpha^{\int})} \exists!_{y \in \mathrm{Ker}(\beta^{\int})} \langle k_\alpha x, k_\beta y \rangle \in |f|.$$

To see this, we argue informally in the internal language, as follows. Let $x \in \mathrm{Ker}(\alpha^{\int})$, then $\alpha^{\int} k_\alpha x \cdot = \cdot \top$, that is, $k_\alpha x \in \alpha$. Since

$$\forall_{x' \in A}(x' \in \alpha \Rightarrow \exists!_{y' \in B} \langle x', y' \rangle \in |f|),$$

there is a unique $y' \in B$ such that $\langle k_\alpha x, y' \rangle \in |f|$. Since

$$\forall_{x' \in A} \forall_{y' \in B}(\langle x', y' \rangle \in |f| \Rightarrow (x' \in \alpha \wedge y' \in \beta)),$$

we have $y' \in \beta$. Now

$$\beta \cdot = \cdot \ulcorner \beta^{\int} \urcorner \cdot = \cdot \ulcorner \text{char}\, k_\beta \urcorner \cdot = \cdot \{y' \in B \mid \exists_{y \in \text{Ker}(\beta^{\int})} k_\beta y = y'\},$$

so we may infer from $y' \in \beta$ that there exists $y \in \text{Ker}(\beta^{\int})$ with $k_\beta y = y'$, and this y is also unique, since k_β is a monomorphism. Thus we have a unique $y \in \text{Ker}(\beta^{\int})$ such that $\langle k_\alpha x, k_\beta y \rangle \in |f|$, as claimed.

It remains to verify that $\varepsilon_{\mathscr{T}}$ is a strict logical functor which preserves canonical subobjects.

For example, let us check that $\varepsilon_{\mathscr{T}}(\alpha \times \beta) = \varepsilon_{\mathscr{T}}(\alpha) \times \varepsilon_{\mathscr{T}}(\beta)$. In view of the definition of $\varepsilon_{\mathscr{T}}$, we must show that

$$\text{Ker}((\alpha \times \beta)^{\int}) = \text{Ker}(\alpha^{\int}) \times \text{Ker}(\beta^{\int}).$$

In fact, we shall prove this equality with Ker replaced by ker. As before, we write $k_\alpha \equiv \ker(\alpha^{\int})$ and also $K_\alpha \equiv \text{Ker}(\alpha^{\int})$. We note that, by C3, $k_\alpha \times k_\beta$ is a canonical subobject, hence it will be the same as $k_{\alpha \times \beta}$ provided it has the same characteristic morphism. Indeed, by Proposition 5.4,

$$\begin{aligned} &\ulcorner \text{char}(k_\alpha \times k_\beta) \urcorner \\ &\cdot = \cdot \{\langle x, y \rangle \in A \times B \mid \exists_{\langle x', y' \rangle \in K_\alpha \times K_\beta} \langle x, y \rangle = (k_\alpha \times k_\beta) \langle x', y' \rangle\} \\ &\cdot = \cdot \{\langle x, y \rangle \in A \times B \mid \exists_{x' \in K_\alpha} x = k_\alpha x' \wedge \exists_{y' \in K_\beta} y = k_\beta y'\} \\ &\cdot = \cdot \{\langle x, y \rangle \in A \times B \mid x \in \ulcorner \text{char}\, k_\alpha \urcorner \wedge y \in \ulcorner \text{char}\, k_\beta \urcorner\} \\ &\cdot = \cdot \ulcorner \text{char}\, k_\alpha \urcorner \times \ulcorner \text{char}\, k_\beta \urcorner, \end{aligned}$$

by Proposition 11.2. Finally, since $k_\alpha \equiv \ker(\alpha^{\int})$, this is $\alpha \times \beta \cdot = \cdot \ulcorner \text{char}\, k_{\alpha \times \beta} \urcorner$, hence $k_\alpha \times k_\beta \cdot = \cdot k_{\alpha \times \beta}$, as was to be shown.

In a similar manner one may show that $\varepsilon_{\mathscr{T}}(P\alpha) = P\varepsilon_{\mathscr{T}}(\alpha)$ using C4. We shall leave the remainder of the proof that $\varepsilon_{\mathscr{T}}$ is a strict logical functor as an exercise to the reader. Instead, we shall now verify that it preserves canonical subobjects.

First, let us check this in the special case

$$\varepsilon_{\mathscr{T}}(m_\alpha) \cdot = \cdot k_\alpha,$$

where $m_\alpha : \alpha \to \mathbf{A}$ is the canonical inclusion given by

$$|m_\alpha| \equiv \{\langle x, x \rangle \in A \times A \mid x \in \alpha\}.$$

Now, by definition, $\varepsilon_{\mathscr{T}}(m_\alpha)$ is the unique arrow $g: k_\alpha \to A$ such that

$$\vdash_x \langle k_\alpha x, k_{\mathbf{A}} g x \rangle \in |m_\alpha|,$$

where $x: 1 \to K_\alpha$ is an indeterminate arrow. Since $k_{\mathbf{A}} \cdot = \cdot 1_{\mathbf{A}}$ by C1, this implies $\vdash_x gx = k_\alpha x$, hence $g \cdot = \cdot k_\alpha$, as was to be checked.

Next, let us turn to the general case of a canonical monomorphism $m: \alpha \to \beta$. This means as above that $\vdash \alpha \subseteq \beta$ and that

$$|m|\cdot=\cdot\{\langle x,x\rangle\in A\times A\,|\,x\in\alpha\}.$$

Then, by C2 or directly, $m_\beta m\cdot=\cdot m_\alpha$. Hence

$$k_\alpha\cdot=\cdot\varepsilon_{\mathcal{T}}(m_\alpha)\cdot=\cdot\varepsilon_{\mathcal{T}}(m_\beta)\varepsilon_{\mathcal{T}}(m)\cdot=\cdot k_\beta\varepsilon_{\mathcal{T}}(m).$$

We shall now prove that $\varepsilon_{\mathcal{T}}(m)$ is a canonical monomorphism.

In any case, $\varepsilon_{\mathcal{T}}(m)\cong k$, where k is canonical, hence $k_\beta\varepsilon_{\mathcal{T}}(m)\cong k_\beta k$. Now $k_\beta k$ is canonical by C5, hence $k_\beta\varepsilon_{\mathcal{T}}(m)\cdot=\cdot k_\beta k$, and so $\varepsilon_{\mathcal{T}}(m)\cdot=\cdot k$, as was to be shown.

We shall leave it as an exercise to the reader to show that ε is a natural transformation.

Lemma 15.3. $\varepsilon_{\mathcal{T}}\xi_{\mathcal{T}}=1_{\mathcal{T}}$.

Proof. By Proposition 5.4.,

$$\ulcorner\mathrm{char}\,1_A\urcorner\cdot=\cdot\{x\in A\,|\,\exists_{y\in A}x=y\}\cdot=\cdot\{x\in A\,|\,\top\}\cdot=\cdot\mathbf{A}.$$

Therefore $\mathbf{A}^{\int}\cdot=\cdot\mathrm{char}\,1_A$ and so $\ker(\mathbf{A}^{\int})\cdot=\cdot 1_A$. From the last equation, by taking the source of each side, we obtain $\mathrm{Ker}(\mathbf{A}^{\int})=A$. Therefore, using definitions 13.2 and 15.2, we calculate

$$\varepsilon_{\mathcal{T}}\xi_{\mathcal{T}}(A)=\varepsilon_{\mathcal{T}}(\mathbf{A})=\mathrm{Ker}(\mathbf{A}^{\int})=A$$

on objects. Similarly, on arrows $f\colon A\to A'$ we have

$$\varepsilon_{\mathcal{T}}\xi_{\mathcal{T}}(f)\cdot=\cdot\varepsilon_{\mathcal{T}}(\mathbf{f})\cdot=\cdot g,$$

where $g\colon A\to A'$ is the unique arrow such that, for an indeterminate arrow $x\colon 1\to A$,

$$\vdash_x\langle x,gx\rangle\in|\mathbf{f}|.$$

Since

$$|\mathbf{f}|\equiv\mathrm{gph}\,f\equiv\{\langle x,fx\rangle\,|\,x\in A\},$$

we see that $g\cdot=\cdot f$.

Theorem 15.4. $L\colon\mathbf{Top}_0\to\mathbf{Lang}$ has $T\colon\mathbf{Lang}\to\mathbf{Top}_0$ as a left adjoint, with adjunctions $\eta\colon\mathrm{id}\to LT$ and $\varepsilon\colon TL\to\mathrm{id}$.

Proof. We verify that

$$L(\varepsilon_{\mathcal{T}})\eta_{L(\mathcal{T})}\cdot=\cdot 1_{L(\mathcal{T})},\quad \varepsilon_{T(\mathfrak{L})}T(\eta_{\mathfrak{L}})\cdot=\cdot 1_{T(\mathfrak{L})}.$$

Indeed, the first equation is obtained by applying the functor L to Lemma 15.3 and recalling that $L(\xi_{\mathcal{T}})\cdot=\cdot\eta_{L(\mathcal{T})}$ by Lemma 14.6. To prove the second equation, let us only carry out the calculation for objects α of $T(\mathfrak{L})$, leaving the calculation for arrows to the reader:

$$\varepsilon_{T(\mathfrak{L})}T(\eta_{\mathfrak{L}})(\alpha)\cdot=\cdot\varepsilon_{T(\mathfrak{L})}\eta_{\mathfrak{L}}(\alpha)\cdot=\cdot\varepsilon_{T(\mathfrak{L})}(\boldsymbol{\alpha})$$

by definitions 14.7 and 14.2. Now $\boldsymbol{\alpha}: \mathbf{1} \to P\mathbf{A} = \mathbf{PA}$ gives rise to the arrow $\varepsilon_{T(\mathfrak{L})}(\boldsymbol{\alpha}): \mathrm{Ker}(\mathbf{1}^{\mathfrak{f}}) \to \mathrm{Ker}((\mathbf{PA})^{\mathfrak{f}})$ in $T(\mathfrak{L})$. Recalling from the proof of Lemma 15.3 that $\mathrm{Ker}(\mathbf{A}^{\mathfrak{f}}) = A$, we may write $\varepsilon_{T(\mathfrak{L})}(\boldsymbol{\alpha}): \mathbf{1} \to \mathbf{PA}$. By Definition 15.2, this is the unique arrow $g: \mathbf{1} \to \mathbf{PA}$ such that, for an indeterminate $x: \mathbf{1} \to \mathbf{1}$, in $LT(\mathfrak{L})$, $\vdash_x \langle k_1 x, k_{\mathbf{A}} g x\rangle \in |\boldsymbol{\alpha}|$.

Now, again from the proof of Lemma 15.3, $k_{\mathbf{A}} \equiv \ker(\mathbf{A}^{\mathfrak{f}}) \cdot = \cdot 1_A$. Moreover, in $\mathfrak{L}$ we have $\vdash_x x = *$. Therefore g is determined by $\vdash \langle *, g\rangle \in |\boldsymbol{\alpha}|$. Since $|\boldsymbol{\alpha}| \equiv \{\langle *, \alpha\rangle\}$, it follows that $g \cdot = \cdot \alpha$, as was to be proved.

In Part 0 we had considered full reflective subcategories; but the same definition applies to reflective subcategories which are not full.

Corollary 15.5. $\mathbf{Top}_0$ is a (non-full) reflective subcategory of $\mathbf{Top}$ with reflection $\xi_{\mathcal{T}}: \mathcal{T} \to TL(\mathcal{T})$.

Proof. Given a morphism $G: \mathcal{T} \to \mathcal{T}'$ of $\mathbf{Top}$ with $\mathcal{T}'$ in $\mathbf{Top}_0$, we shall find a unique morphism G' of $\mathbf{Top}_0$ such that $G'\xi_{\mathcal{T}} = G$. Indeed, let $G' \equiv \varepsilon_{\mathcal{T}'} TL(G)$, then

$$G'\xi_{\mathcal{T}} = \varepsilon_{\mathcal{T}'} TL(G)\xi_{\mathcal{T}} = \varepsilon_{\mathcal{T}'}\xi_{\mathcal{T}'}G = G,$$

using the naturality of ξ and Lemma 15.3. On the other hand, assuming $G'\xi_{\mathcal{T}} = G$, we calculate

$$\begin{aligned} \varepsilon_{\mathcal{T}'} TL(G) &= \varepsilon_{\mathcal{T}'} TL(G')TL(\xi_{\mathcal{T}}) \text{ by functoriality of } TL, \\ &= G'\varepsilon_{TL(\mathcal{T})} TL(\xi_{\mathcal{T}}) \text{ by naturality of } \varepsilon, \\ &= G'\varepsilon_{TL(\mathcal{T})} T(\eta_{L(\mathcal{T})}) \text{ by Lemma 14.6}, \\ &= G' \text{ by Theorem 15.4.} \end{aligned}$$

Exercises

1. Complete the proof in Definition 15.2 that $\varepsilon_{\mathcal{T}}$ is a strict logical functor.
2. Prove that ε is a natural transformation $TL \to \mathrm{id}$.
3. Complete the proof of the second equation in Theorem 15.4.
4. Show that, for each object $\mathcal{T}$ of $\mathbf{Top}_0$, $\xi_{\mathcal{T}}\varepsilon_{\mathcal{T}} \cong 1_{TL(\mathcal{T})}$.

16 Applications of the adjoint functors between toposes and type theories

In Section 15 we established the existence of a pair of adjoint functors

$$\mathbf{Top}_0 \underset{T}{\overset{L}{\rightleftarrows}} \mathbf{Lang}$$

between toposes with canonical subobjects and type theories. In this section we shall see what mileage we can get out of this adjointness, which establishes a biunique correspondence between translations $\mathfrak{L} \to L(\mathscr{T})$ and strict logical morphisms $T(\mathfrak{L}) \to \mathscr{T}$ which preserve canonical subobjects. It is reasonable to call either of these two arrows an *interpretation* of $\mathfrak{L}$ in $\mathscr{T}$. We shall exploit the adjointness to carry out a number of constructions in $\mathbf{Top}_0$.

Example 16.1. The free topos.

Let us consider for a moment *pure type theory* $\mathfrak{L}_0$ which has only those types, terms and theorems that it must have by virtue of being a type theory and which also contains no unnecessary equations between types. Clearly, $\mathfrak{L}_0$ is an initial object in the category **Lang**, hence it admits exactly one interpretation into any given topos with canonical subobjects. It follows that $\mathscr{F} = T(\mathfrak{L}_0)$ is an initial object in the category $\mathbf{Top}_0$, which it has become customary to call the *free topos*.

We shall see later how categorical properties of $\mathscr{F}$ may be translated into metatheorems about $\mathfrak{L}_0$. In a more philosophical vein one may argue that $\mathscr{F}$ is the *ideal universe* a moderate intuitionist or constructivist believes in. It is not so easy to describe an 'ideal universe' for a classical mathematician, since he would insist that $\mathrm{Hom}(1, \Omega)$ has exactly two elements.

Example 16.2. Adjoining an indeterminate to a topos.

We had seen earlier how to adjoin an indeterminate arrow $x: 1 \to A$ to a topos $\mathscr{T}$ to obtain a cartesian closed category (or even dogma) $\mathscr{T}[x]$. This is not usually a topos, just as for a field F the ring $F[x]$ is not usually a field. We shall here construct a topos $\mathscr{T}(x)$ and a strict logical functor $\mathscr{T} \to \mathscr{T}(x)$ which has the expected universal property to some extent. (The analogy with the field $F(x)$ is not quite complete here.)

We begin by adjoining a *parameter* to a type theory $\mathfrak{L}$. As in Section 7, if x is a variable of type A, then $\mathfrak{L}(x)$ is the type theory with the same types as $\mathfrak{L}$, where x is counted as a constant, so that closed terms are open terms of $\mathfrak{L}$ containing no free variables other than x and where deducibility for closed terms means $\vdash_x$ in $\mathfrak{L}$. In describing the ith variable of type A in $\mathfrak{L}(x)$ one must of course skip the variable x, say the kth variable of type A in $\mathfrak{L}$.

The canonical translation $\tau_x: \mathfrak{L} \to \mathfrak{L}(x)$ regards $\mathfrak{L}(x)$ as an extension of $\mathfrak{L}$. This is not necessarily a conservative extension, since from $\vdash_x p$, with x of type A not occurring freely in p, one may infer $\vdash p$ if there is a closed term of type A in $\mathfrak{L}$, but not in general.

The translation τ_x has the obvious universal property: for any translation

$\tau: \mathfrak{L} \to \mathfrak{L}'$ and any term a of type $\tau(A)$ in $\mathfrak{L}'$, there is a unique translation $\tau': \mathfrak{L}(x) \to \mathfrak{L}'$ such that $\tau'\tau_x = \tau$ and $\tau(x) = a$. In fact, τ' is the translation which substitutes a for x in any term $\varphi(x)$ of $\mathfrak{L}(x)$ to yield the term $\varphi(a)$ of $\mathfrak{L}'$,

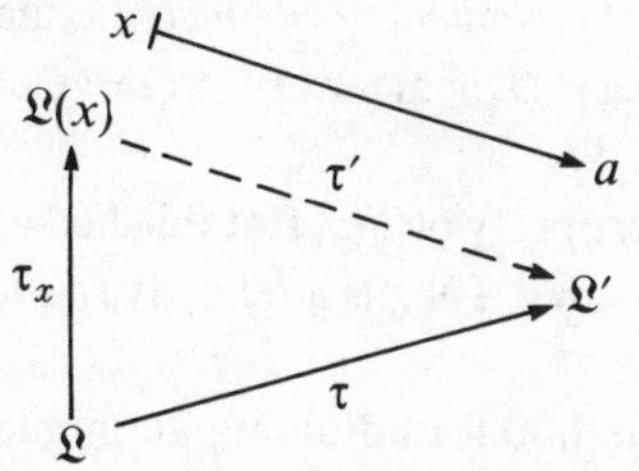

We may use this result to adjoin an indeterminate arrow $\mathbf{1} \to \mathbf{A}$ to the topos $T(\mathfrak{L})$ generated by a type theory $\mathfrak{L}$, namely the arrow $\mathbf{x}: \mathbf{1} \to \mathbf{A}$ in $T(\mathfrak{L}(x))$ with graph $\{\langle *, x \rangle\}$. In fact, the morphism $T(\tau_x): T(\mathfrak{L}) \to T(\mathfrak{L}(x))$ in $\mathbf{Top}_0$ has the universal property: for any morphism $G: T(\mathfrak{L}) \to \mathscr{T}'$ of $\mathbf{Top}_0$ and any arrow $a: 1 \to G(\mathbf{A})$ in $\mathscr{T}'$, there is a unique morphism $G': T(\mathfrak{L}(x)) \to \mathscr{T}'$ of $\mathbf{Top}_0$ such that $G'T(\tau_x) = G$ and $G'(\mathbf{x}) \cdot = \cdot a$.

To prove this, let $\tau: \mathfrak{L} \to \mathfrak{L}' = L(\mathscr{T}')$ correspond to G under the adjointness, to wit

$$\tau = L(G)\eta_{\mathfrak{L}}.$$

Since a is a term of type $\tau(A)$ in $L(\mathscr{T}')$, we may find $\tau': \mathfrak{L}(x) \to L(\mathscr{T}')$ so that

$$(*) \qquad \tau'\tau_x = \tau, \quad \tau'(x) = a.$$

Now let $G': T(\mathfrak{L}(x)) \to \mathscr{T}'$ correspond to τ' under the adjointness, to wit

$$G' = \varepsilon_{\mathscr{T}'} T(\tau').$$

A routine calculation then shows that

$$(**) \qquad G'T(\tau_x) = G, \quad G'(\mathbf{x}) \cdot = \cdot a.$$

Moreover, these two equations imply $(*)$, so that the uniqueness of G' follows from that of τ'.

Next, let us consider the problem of adjoining an indeterminate arrow $1 \to A$ to an *arbitrary* topos $\mathscr{T}$, possibly with canonical subobjects. We may of course just recall that $\mathscr{T} \simeq TL(\mathscr{T})$ and cite the above construction with $\mathfrak{L} = L(\mathscr{T})$. If we wish to be more explicit, we should assert some kind of universal property for the composite functor

$$\mathscr{T} \xrightarrow{\xi_{\mathscr{T}}} TL(\mathscr{T}) \xrightarrow{T(\tau_x)} T(L(\mathscr{T})(x)).$$

Unfortunately $\xi_{\mathscr{T}}$ does not preserve canonical subobjects even if $\mathscr{T}$ has canonical subobjects; it is a morphism of $\mathbf{Top}$ but not of $\mathbf{Top}_0$. We must phrase the universal property rather carefully. Define $\mathscr{T}(\mathbf{x}) \equiv T(L(\mathscr{T})(x))$.

The strict logical functor $T(t_x)\xi_{\mathcal{T}}: \mathcal{T} \to \mathcal{T}(\mathbf{x})$ has this universal property: for any strict logical functor $G: \mathcal{T} \to \mathcal{T}'$ such that $\mathcal{T}'$ has canonical subobjects and any arrow $a: 1 \to G(A)$ in $\mathcal{T}'$, there is a unique strict logical functor $G': \mathcal{T}(\mathbf{x}) \to \mathcal{T}'$ *preserving canonical subobjects* such that $G'T(\tau_x)\xi_{\mathcal{T}} = G$ and $G'(\mathbf{x})\cdot = \cdot a$. Note that G' is meant to preserve canonical subobjects even if G does not.

To deduce this from the universal property of $T(\tau_x)$ established above, we merely cite Corollary 15.5, which says that $\mathbf{Top}_0$ is a reflective subcategory of $\mathbf{Top}$.

There is also a more categorical method for adjoining an indeterminate to an elementary topos due to Joyal, based on a similar construction for Grothendieck toposes in SGA4, which works in fact for arbitrary cartesian categories with equalizers. We shall discuss this in Exercise 2 below.

Example 16.3. Dividing a topos by a filter of propositions.

If $\mathfrak{L}$ is any type theory and F is a set of closed formulas of $\mathfrak{L}$, we may construct a new type theory $\mathfrak{L}/F$ which has the same types and terms as $\mathfrak{L}$ but where provability means $F \vdash$, that is, provability from the *assumptions* F. We recall that F is called a *filter* provided from $p \vdash q$ and $p \in F$ one may infer $q \in F$ and from $p \in F$ and $q \in F$ one may infer $p \wedge q \in F$. When F is a filter, $F \vdash p$ is the same as $p \in F$. Note that $\mathfrak{L} = \mathfrak{L}/\{\top\}$.

Consider now the canonical translation $\tau_F: \mathfrak{L} \to \mathfrak{L}/F$ which introduces new postulates F. Clearly, this has the following universal property: given any translation $\tau: \mathfrak{L} \to \mathfrak{L}'$ such that $\vdash \tau(p)$ in $\mathfrak{L}'$ for all $p \in F$, there is a unique translation $\tau': \mathfrak{L}/F \to \mathfrak{L}'$ such that $\tau'\tau_F = \tau$.

The morphism $T(\tau_F): T(\mathfrak{L}) \to T(\mathfrak{L}/F)$ in $\mathbf{Top}_0$ then has the following universal property: given any morphism $G: T(\mathfrak{L}) \to \mathcal{T}'$ in $\mathbf{Top}_0$ such that $\mathcal{T}' \vDash G(p)$ for all $p \in F$, there is a unique morphism $G': T(\mathfrak{L}/F) \to \mathcal{T}'$ in $\mathbf{Top}_0$ such that $G'T(\tau_F) = G$.

The proof is similar to the proof of the corresponding result in 16.2. As a particular case, if $\mathfrak{L}_0$ is pure type theory, $T(\mathfrak{L}_0/F)$ is an initial object in the full subcategory of $\mathbf{Top}_0$ consisting of those toposes which satisfy the assumptions F.

We may also consider the morphism $T(\tau_F)\xi_{\mathcal{T}}: \mathcal{T} \to \mathcal{T}/F$ in $\mathbf{Top}$, where $\mathcal{T}/F \equiv T(L(\mathcal{T})/F)$. As for the analogous result in 16.2, one may again show that this has the following universal property: for any morphism $G: \mathcal{T} \to \mathcal{T}'$ in $\mathbf{Top}$ such that $\mathcal{T}'$ has canonical subobjects and $\mathcal{T}' \vDash G(p)$ for all $p \in F$, there is a unique morphism $G': \mathcal{T}/F \to \mathcal{T}'$ in $\mathbf{Top}_0$ such that $G'T(\tau_F)\xi_{\mathcal{T}} = G$.

In particular, if $G: \mathcal{T} \to \mathcal{T}'$ is any strict logical functor and if F is the filter of all arrows $p: 1 \to \Omega$ in $\mathscr{T}$ such that $\mathscr{T}' \models G(p)$, then G factors uniquely as $\mathscr{T} \to \mathscr{T}/F \to \mathscr{T}'$. The filter F here is analogous to the kernel of a homomorphism in ring theory.

Example 16.4. Topos of fractions.

Suppose Σ is a set of arrows in a topos $\mathscr{T}$. We want to form a topos $\mathscr{T}\Sigma^{-1}$ with canonical subobjects in which these arrows are invertible and obtain a logical morphism $\mathscr{T} \to \mathscr{T}\Sigma^{-1}$ which is initial among logical morphisms $G: \mathscr{T} \to \mathscr{T}'$ such that $\mathscr{T}'$ has canonical subobjects and $G(\sigma)$ is invertible for all $\sigma \in \Sigma$. The word 'initial' here refers to the existence of a unique strict logical morphism $\mathscr{T}\Sigma^{-1} \to \mathscr{T}'$ which preserves canonical subobjects and such that the following triangle commutes:

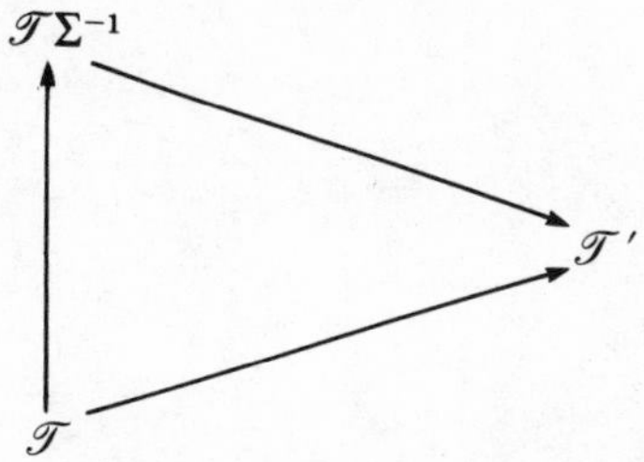

The proposition that an arrow $\sigma: A \to B$ is invertible in the topos $\mathscr{T}$ is a proposition p_σ in the internal language of $\mathscr{T}$. According to Proposition 6.1,

$$p_\sigma \equiv \forall_{y \in B} \exists !_{x \in A}\, \sigma x = y.$$

Letting F_Σ be the filter generated by all p_σ for which $\sigma \in \Sigma$, we may define $\mathscr{T}\Sigma^{-1} \equiv \mathscr{T}/F_\Sigma$ and cite the universal property of Example 16.3.

Example 16.5. The free topos generated by a graph.

We recall (Part 0, Definition 1.2) that a *graph* consists of two classes, the class of arrows and the class of nodes, and two mappings between them, called 'source' and 'target'. One writes $f: A \to B$ for: source$(f) = A$ and target$(f) = B$. A morphism $f: \mathscr{G} \to \mathscr{G}'$ between graphs sends nodes and arrows to nodes and arrows respectively such that $f: A \to B$ in $\mathscr{G}$ implies $F(f): F(A) \to F(B)$ in $\mathscr{G}'$. We thus obtain the category **Grph** of graphs. We shall now consider a functor V: **Lang** $\to$ **Grph.**

With every type theory $\mathfrak{L}$ there is associated its underlying graph $V(\mathfrak{L})$: its nodes are the types of $\mathfrak{L}$ and its arrows $A \to B$ are term forming operations $a \mapsto \varphi(a)$ induced by terms $\varphi(x)$ of type B, x being a free variable of type A. If $\tau: \mathfrak{L} \to \mathfrak{L}'$ is any translation, $V(\tau): V(\mathfrak{L}) \to V(\mathfrak{L}')$ is defined by $V(\tau)(A) \equiv \tau(A)$ and $V(\tau)(\varphi(x)) \equiv \tau(\varphi(x))$.

It is not difficult to construct a left adjoint Λ: **Grph** $\rightarrow$ **Lang** to V. Given a graph $\mathscr{G}$, we define the type theory $\Lambda(\mathscr{G})$ by allowing as basic types all nodes of $\mathscr{G}$, in addition to 1, Ω and N, and by closing the set of terms under the term forming operations $a \mapsto fa$, where $f: A \rightarrow B$ is any arrow of $\mathscr{G}$. (See also Example 1.2.) To show that Λ is the object part of the functor left adjoint to V, we need a morphism $\eta_{\mathscr{G}}: \mathscr{G} \rightarrow V\Lambda(\mathscr{G})$ with the appropriate universal property.

We define $\eta_{\mathscr{G}}(A) \equiv A$, a node A of $\mathscr{G}$ being a type of $\Lambda(\mathscr{G})$ hence a node of $V\Lambda(\mathscr{G})$, and $\eta_{\mathscr{G}}(f) \equiv$ the term forming operation $a \mapsto fa$. Now let $F: \mathscr{G} \rightarrow V(\mathfrak{L})$ be any morphism of graphs, we seek a unique translation τ_F: $\Lambda(\mathscr{G}) \rightarrow \mathfrak{L}$ such that $V(\tau_F)\eta_{\mathscr{G}} = F$.

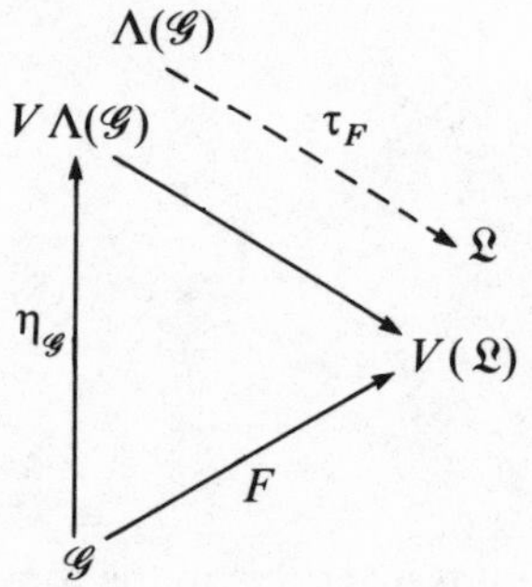

We define τ_F on types by stipulating that for a new basic type A of $\Lambda(\mathscr{G})$, that is, a node of $\mathscr{G}$, $\tau_F(A) \equiv F(A)$. We define τ_F on terms by stipulating that, for any new term forming operation $a \mapsto fa$ of $\Lambda(\mathscr{G})$, f: $A \rightarrow B$ being an arrow of $\mathscr{G}$,

$$\tau_F(fa) \equiv F(f)(\tau_F(a)).$$

We leave it to the reader to check that τ_F thus constructed is the unique translation satisfying $V(\tau_F)\eta_{\mathscr{G}} = F$.

Composing adjoint functors, we see that the functor

$$\mathbf{Top}_0 \xrightarrow{L} \mathbf{Lang} \xrightarrow{V} \mathbf{Grph}$$

has as a left adjoint

$$\mathbf{Grph} \xrightarrow{\Lambda} \mathbf{Lang} \xrightarrow{T} \mathbf{Top}_0.$$

In the special case when $\mathscr{G}$ is the empty graph (no nodes, no arrows), $T\Lambda(\mathscr{G})$ is of course the free topos $\mathscr{F}$ considered in Example 16.1.

We shall discuss briefly two variations of Example 16.5.

A *multigraph* consists of two classes, the class X of arrows and the class Y

of nodes, and two mappings:

$$\text{source:}\quad X \to Y^*,$$
$$\text{target:}\quad X \to Y,$$

where Y^* is the free monoid generated by Y. An arrow of a multigraph may be written in the style of Gentzen as

$$f\colon\ A_1, A_2, \ldots, A_n \to B.$$

The underlying multigraph of a type theory has as nodes types and as arrows term forming operations

$$(a_1, \ldots, a_n) \mapsto \varphi(a_1, \ldots, a_n)$$

given by arbitrary open terms $\varphi(x_1, \ldots, x_n)$ of the type theory.

The reader will have no difficulty constructing the free type theory and the free topos generated by a multigraph.

Classifications had been introduced in Part I, Section 10, Exercise 2. We recapitulate: a *classification* consists of two classes, the class of entities and the class of types, and a mapping between them which assigns to each entity its type. The category of small classifications is the topos $\mathbf{Sets}^{\to}$. The underlying classification of a type theory has as entities of a given type the closed terms of that type. Again, the reader will be able to construct the free type theory and the free topos generated by a classification.

Exercises

1. Prove the equivalence of $(*)$ and $(**)$ in Example 16.2.
2. Given an object A of the topos $\mathscr{T}$, one forms the *slice topos* $\mathscr{T}/A$, whose objects are arrows $f\colon B \to A$ in $\mathscr{T}$ and whose arrows are commutative triangles in $\mathscr{T}$. The logical functor $H\colon \mathscr{T} \to \mathscr{T}/A$ sends the object B of $\mathscr{T}$ onto the object $\pi_{A,B}\colon A \times B \to A$ of $\mathscr{T}/A$. Among the arrows of $\mathscr{T}/A$, the arrow $\xi\colon H(1) \to H(A)$ given by $\langle \pi_{A,1}, \pi_{A,1} \rangle\colon A \times 1 \to A \times A$ behaves like an indeterminate: given a logical functor $G\colon \mathscr{T} \to \mathscr{T}'$ and an arrow $a\colon 1 \to G(A)$ in $\mathscr{T}'$, there is a logical functor $G'\colon \mathscr{T}/A \to \mathscr{T}'$, unique up to natural isomorphism, such that $G'H = G$ and $G'(\xi) \cdot = \cdot a$. G' is constructed at the object $f\colon B \to A$ by the pullback:

$$\begin{array}{ccc} G'(f) & \longrightarrow & G(B) \\ \big\downarrow & & \big\downarrow{\scriptstyle F(f)} \\ 1 & \underset{a}{\longrightarrow} & G(A) \end{array}$$

$\mathscr{T}/A$ is equivalent to $\mathscr{T}(x)$ in Example 16.2.

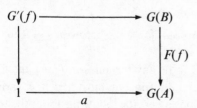

3. If $p: 1 \to \Omega$ is an arrow in $\mathscr{T}$, let (p) be the filter generated by p and $S_p \to 1$ the subobject of 1 which is the kernel of p. Show that the topos $\mathscr{T}/(p)$ of Example 16.3 may also be constructed by adjoining an indeterminate arrow $1 \to S_p$ to $\mathscr{T}$. Deduce that $\mathscr{T}/(p)$ is equivalent to the slice topos $\mathscr{T}/S_p$.

4. Construct the free type theory and hence the free topos generated by (*a*) a multigraph and (*b*) a classification.

17 Completeness of higher order logic with choice rule

We recall that an *interpretation* of a type theory $\mathfrak{L}$ in a topos $\mathscr{T}$ is a translation $\mathfrak{L} \to L(\mathscr{T})$. If $\mathscr{T}$ has canonical subobjects, this corresponds by adjointness to a unique strict logical functor $T(\mathfrak{L}) \to \mathscr{T}$ preserving canonical subobjects. Among the interpretations of $\mathfrak{L}$ there is always the canonical one $\eta_{\mathfrak{L}}: \mathfrak{L} \to LT(\mathfrak{L})$, where $T(\mathfrak{L})$ is the topos generated by $\mathfrak{L}$. Being an adjunction, $\eta_{\mathfrak{L}}$ is initial in the category of all interpretations of $\mathfrak{L}$. Moreover, it is a conservative extension: to know whether a formula p holds in $\mathfrak{L}$ it is necessary and sufficient that its image under $\eta_{\mathfrak{L}}$ holds in $LT(\mathfrak{L})$.

A *standard model* of a type theory $\mathfrak{L}$ is an interpretation $\mathfrak{L} \to L(\mathbf{Sets})$. One wants to introduce a more general notion of 'model' and prove a completeness theorem: every type theory has 'enough' models, that is, for every closed formula p of $\mathfrak{L}$, $\vdash p$ in $\mathfrak{L}$ if and only if $\mathscr{M} \models \tau(p)$ for all models $\tau: \mathfrak{L} \to L(\mathscr{M})$. It is easily seen that, in general, there are not enough standard models. On the other hand, if we admit *all* interpretations as models, the completeness theorem holds trivially, in view of the fact that the initial interpretation $\eta_{\mathfrak{L}}$ is a conservative extension, and so $\vdash p$ in $\mathfrak{L}$ if and only if $T(\mathfrak{L}) \models \mathbf{p}$. We are looking for a definition of 'model' $\tau: \mathfrak{L} \to L(\mathscr{M})$ in which $\mathscr{M}$ resembles the category of sets more closely than an arbitrary topos does.

Definition 17.1. A *model* of a type theory $\mathfrak{L}$ is an interpretation $\tau: \mathfrak{L} \to L(\mathscr{M})$ in which the topos $\mathscr{M}$, also sometimes called a 'model', has the following properties:

M0. Not $\mathscr{M} \models \bot$;

M1. if $\mathscr{M} \models p \vee q$ then $\mathscr{M} \models p$ or $\mathscr{M} \models q$;

M2. if $\mathscr{M} \models \exists_{x \in A} \varphi(x)$ then $\mathscr{M} \models \varphi(a)$ for some arrow $a: 1 \to A$ in $\mathscr{M}$.

Here p, q and $\varphi(x)$ are assumed to be any formulas in $L(\mathscr{M})$, the first two without free variables, the last admitting at most the free variable x. Algebraically the above conditions assert the following about the terminal object 1 of $\mathscr{M}$:

M0′. 1 is not initial;
M1′. 1 is indecomposable, that is, it is not the union of two proper subobjects;
M2′. 1 is projective.

For M2′ see Proposition 6.7. Logically these conditions assert that truth in $\mathscr{M}$ is much like truth in **Sets**. If $\mathscr{M}$ is a Boolean topos, we may also infer the following:

M3. if $\mathscr{M} \models \varphi(a)$ for all arrows $a: 1 \to A$, then $\mathscr{M} \models \forall_{x \in A} \varphi(x)$.

Algebraically M3 asserts:

M3′. 1 is a *generator* of $\mathscr{M}$, that is, the functor $\Gamma = \mathrm{Hom}(1, -): \mathscr{M} \to$ **Sets** is faithful.

Unfortunately this functor Γ is not a logical functor, although it preserves products up to isomorphism. As regards exponents, we have

$$\Gamma(B^A) \cong \mathrm{Hom}(A, B) \subseteq \Gamma(B)^{\Gamma(A)}.$$

Also, being left exact*, Γ preserves monomorphisms; in particular, it sends subobjects of A to subsets of $\Gamma(A)$. However, this mapping $\mathrm{Sub}\, A \to \mathrm{Sub}\, \Gamma(A)$ need not be surjective. We may therefore regard $\mathscr{M}$ essentially as a subcategory of **Sets**, with the same products up to isomorphism but with fewer subobjects, that is, not every ordinary subset need correspond to a subobject in $\mathscr{M}$.

Thus $\mathscr{M}$ is essentially what Henkin calls a 'non-standard' or 'general' model. In a widely quoted article he showed that classical higher order logic has enough models. For expository purposes he confined attention to type theory with the axiom of choice, as in this case the proof becomes much simpler.

Following Henkin, we shall also assume for the time being that $\mathfrak{L}$ satisfies the rule of choice (see Definition 7.1). From this it follows easily (see Proposition 17.4 below) that $LT(\mathfrak{L})$ also satisfies the rule of choice, that is, the topos $T(\mathfrak{L})$ has choice: all epimorphisms split (see Proposition 7.5 and the remark following it). Recall from Corollary 7.4 that now $\mathfrak{L}$ and $LT(\mathfrak{L})$ are classical, hence $T(\mathfrak{L})$ is a Boolean topos. We shall obtain our models of $\mathfrak{L}$ by dividing $T(\mathfrak{L})$ by all possible prime filters.

In what follows we shall consider a pre-ordered set in which the order relation is written $\vdash$. We shall abbreviate '$p \vdash q$ and $q \vdash p$' as '$p \dashv\vdash q$' and regard $\dashv\vdash$ as an equality relation, so that the pre-ordered set becomes a partially ordered set. We shall furthermore assume that it is a lattice: any two elements p and q have a least upper bound $p \vee q$ and a greatest lower bound $p \wedge q$. For good measure, we shall also assume that a lattice has a

* A functor is called *left exact* if it preserves finite limits.

largest element $\top$ and a smallest element $\bot$. A lattice is called *distributive* if, for any elements p, q and r,

$$p \wedge (q \vee r) \dashv\vdash (p \wedge q) \vee (p \wedge r).$$

From this it follows that the opposite lattice, that is, the same set with the reverse order, is also distributive, namely

$$p \vee (q \wedge r) \dashv\vdash (p \vee q) \wedge (p \vee r).$$

Definition 17.2. A subset P of a lattice is called a *filter* provided

F1. if $p \in P$ and $p \vdash q$ then $q \in P$,

F2. if $p \in P$ and $q \in P$ then $p \wedge q \in P$.

A filter P is called *proper* if

F3. $\bot \notin P$

and *prime* if furthermore

F4. if $p \vee q \in P$ then $p \in P$ or $q \in P$.

An *ultrafilter* is a maximal proper filter. In a Heyting algebra this means that, for every element p,

F5. either $p \in P$ or $\neg p \in P$,

where $\neg p \equiv p \Rightarrow \bot$.

Clearly, every ultrafilter is prime and, in a Boolean algebra, every prime filter is an ultrafilter.

We record for later use:

Lemma 17.3. Every filter in a distributive lattice is the intersection of all prime filters containing it.

Proof. Let p be an element not in the filter F; we seek a prime filter containing F but not p. Let P be maximal among filters containing F but not p; it remains to show that P is prime. Suppose $q \notin P$ and $q' \notin P$. Then the filter generated by P and q must contain p, and so must the filter generated by P and q'. Therefore $r \wedge q \vdash p$ and $r' \wedge q' \vdash p$ for some $r, r' \in P$. Hence, by distributivity, $(r \wedge r') \wedge (q \vee q') \vdash p$, and so $q \vee q' \notin P$. This shows that P is prime as required.

We recall that a topos $\mathscr{T}$ *has choice* if its internal language $L(\mathscr{T})$ satisfies the rule of choice.

Proposition 17.4. Suppose $\mathfrak{L}$ is a type theory which satisfies the rule of choice, then so do (*a*) $\mathfrak{L}/P$, for any filter P of closed formulas,

(*b*) $LT(\mathfrak{L})$.

Proof. To show (*a*), suppose

$$p \vdash \forall_{x\in A}(x\in\alpha \Rightarrow \exists_{y\in B}\varphi(x, y))$$

for some $p\in P$. We may write this as

$$\vdash_{x} (p \wedge x\in\alpha) \Rightarrow \exists_{y\in B}\varphi(x, y).$$

Now, since $\mathfrak{L}$ satisfies the rule of choice, so does $\mathfrak{L}(x)$, by Lemma 7.2. Therefore

$$\vdash_{x} (p \wedge x\in\alpha) \Rightarrow \exists!_{y\in B}\psi(x, y)$$

for some formula $\psi(x, y)$ such that $\psi(x, y) \vdash_{\{x,y\}} \varphi(x, y)$. We may write this as

$$p \vdash_{x} x\in\alpha \Rightarrow \exists!_{y\in B}\psi(x, y),$$

so $\mathfrak{L}/P$ satisfies the rule of choice.

To show (*b*), suppose

(1) $\qquad T(\mathfrak{L}) \vDash \forall_{x\in\alpha}\exists_{y\in\beta}\langle x, y\rangle \in \boldsymbol{\rho},$

where α and β are terms in $\mathfrak{L}$ of types PA and PB respectively and $\boldsymbol{\rho}: \mathbf{1} \to P(\alpha \times \beta)$ is an arrow in $T(\mathfrak{L})$ with graph $|\boldsymbol{\rho}| \equiv \{\langle *, \rho\rangle\}$, where ρ is a term of type $P(A \times B)$ in $\mathfrak{L}$ satisfying $\vdash \rho \subseteq \alpha \times \beta$ (see Proposition 14.3).

According to Remark 14.8, (1) may be rewritten in $LT(\mathfrak{L})$ as

(2) $\qquad \vdash \forall_{x\in\mathbf{A}}(x\in \boldsymbol{\alpha} \Rightarrow \exists_{y\in\mathbf{B}}(y\in\boldsymbol{\beta} \wedge \langle x, y\rangle \in\boldsymbol{\rho}))$

As this is the image of a formula in $\mathfrak{L}$ under the conservative extension $\eta_{\mathfrak{L}}$, we have in $\mathfrak{L}$

(3) $\qquad \vdash \forall_{x\in A}(x\in\alpha \Rightarrow \exists_{y\in B}(y\in\beta \wedge \langle x, y\rangle \in\rho)).$

Since $\mathfrak{L}$ satisfies the rule of choice, we may infer

(4) $\qquad \vdash \forall_{x\in A}(x\in\alpha \Rightarrow \exists!_{y\in B}\langle x, y\rangle \in\sigma)$

for some term σ such that $\vdash \sigma \subseteq \rho$.

Applying $\eta_{\mathfrak{L}}$ to (4) and recalling Remark 14.7 once more, we obtain

(5) $\qquad T(\mathfrak{L}) \vDash \forall_{x\in\alpha}\exists!_{y\in\beta}\langle x, y\rangle \in\boldsymbol{\sigma},$

where $T(\mathfrak{L}) \vDash \boldsymbol{\sigma} \subseteq \boldsymbol{\rho}$. By description, there is a unique arrow $f: \alpha \to \beta$ in $T(\mathfrak{L})$ such that $\langle x, y\rangle \in\boldsymbol{\sigma}$ may be rewritten as $fx = y$, hence

(6) $\qquad T(\mathfrak{L}) \vDash \forall_{x\in\alpha}\langle x, fx\rangle \in\boldsymbol{\rho}.$

Thus $T(\mathfrak{L})$ has choice.

Taking $\mathfrak{L} = L(\mathscr{T})$ in Proposition 17.4, we immediately obtain the following:

Corollary 17.5. If $\mathscr{T}$ has choice and P is a filter of arrows $1 \to \Omega$ in $\mathscr{T}$, then $\mathscr{T}/P \equiv T(L(\mathscr{T})/P)$ has choice.

Theorem 17.6. Every type theory which satisfies the rule of choice has enough models.

Proof. We look at interpretations of the form

$$\sigma_p\colon \mathfrak{L} \to \mathfrak{L}/P \xrightarrow{\eta_{\mathfrak{L}/P}} LT(\mathfrak{L}/P),$$

where P is a prime filter of closed formulas in $\mathfrak{L}$. By Proposition 17.4, $\mathscr{M} \equiv T(\mathfrak{L}/P)$ has choice, and so satisfies M2. It satisfies M0 and M1 since P is a prime filter. (For, if $\mathscr{M} \models \mathbf{p} \vee \mathbf{q}$, then $P \vdash p \vee q$, hence $P \vdash p$ or $P \vdash q$, hence $\mathscr{M} \models \mathbf{p}$ or $\mathscr{M} \models \mathbf{q}$.)

Now let p be a closed formula of $\mathfrak{L}$ and suppose that $\vdash \sigma_P(p)$ for all prime filters P. Thus, for all P, $\vdash \eta_{\mathfrak{L}/P}(p)$, hence $\vdash p$ in $\mathfrak{L}/P$, since $\eta_{\mathfrak{L}/P}$ is a conservative extension, and so $p \in P$. Therefore $\vdash p$, by Lemma 17.3.

We shall see later that Theorem 17.6 remains valid even without the assumption that $\mathfrak{L}$ satisfies the rule of choice. At any rate, the rule of choice was only used to show that $T(\mathfrak{L}/P)$ has choice. Without this conclusion we have really proved a more general result, stated here in algebraic form:

Proposition 17.7. Every topos is a subdirect product of toposes whose terminal objects are indecomposable. If the given topos has choice, the terminal objects of the components of the product are also projective generators.

Proof. The proposition asserts that for every topos $\mathscr{T}$ there is a monomorphism $M\colon \mathscr{T} \to \Pi_{i \in I} \mathscr{T}_i$ in **Top**, where the $\mathscr{T}_i$ have the stated property. We construct M by taking $\mathscr{T}_i \equiv \mathscr{T}/P_i$ and stipulating that M composed with the projection $\pi_j\colon \Pi_{i \in I} \mathscr{T}_i \to \mathscr{T}_j$ shall be

$$\mathscr{T} \xrightarrow{\xi_{\mathscr{T}}} TL(\mathscr{T}) \xrightarrow{T(\tau_j)} T(L(\mathscr{T})/P_j),$$

where P_j is any prime filter in the Heyting algebra of arrows $1 \to \Omega$ of $\mathscr{T}$ and $\tau_j\colon L(\mathscr{T}) \to L(\mathscr{T})/P_j$ is the canonical translation.

Let us calculate the effect of M on any object A of $\mathscr{T}$ and any arrow $f\colon A \to B$ of $\mathscr{T}$. In view of definitions 13.2 and 14.7,

$$\begin{aligned} \pi_j(M(A)) &= T(\tau_j)\xi_{\mathscr{T}}(A) \\ &= T(\tau_j)(\mathbf{A}) \\ &= \tau_j(\mathbf{A}) \\ &= \mathbf{A} \bmod P_j, \end{aligned}$$

$$
\begin{aligned}
|\pi_j(M(f))| \cdot = \cdot & |T(t_j(\xi_{\mathscr{T}}(f))| \\
\cdot = \cdot & |T(t_j)(\mathbf{f})| \\
\cdot = \cdot & \tau_j(|\mathbf{f}|) \\
\cdot = \cdot & \{\langle x, y\rangle \in A \times B \mid fx = y\} \bmod P_j.
\end{aligned}
$$

It follows that M is injective on objects as well as faithful. Such a functor, often called an *embedding*, is clearly a monomorphism in **Top**.

See Remark 17.8 in Supplement, page 250.

Exercises

1. Prove that a topos $\mathscr{M}$ has the disjunction property M1 if and only if the terminal object 1 of $\mathscr{M}$ is indecomposable.
2. If $\mathscr{M}$ is a Boolean topos, deduce M3 from M1 and M2.
3. Show that the intersection of all ultrafilters of a Heyting algebra containing a given filter F is the set of all elements p such that $\neg\neg p \in F$.
4. In proving his famous incompleteness theorem, Gödel constructed a sentence γ in pure type theory $\mathfrak{L}_0$ such that **Sets** $\models \gamma$ but $\beta \not\vdash \gamma$, where $\beta \equiv \forall_{t\in\Omega}(t \vee \neg t)$ is the Boolean axiom. Since there is only one interpretation $\mathfrak{L}_0/\beta \to L(\mathbf{Sets})$, show that pure classical type theory $\mathfrak{L}_0/\beta$ does not have enough standard models.

18 Sheaf representation of toposes

Proposition 17.7 may be strengthened considerably. Not only are the toposes $\mathscr{T}/P$, where P ranges over the prime filters of the Heyting algebra $\mathrm{Hom}_{\mathscr{T}}(1, \Omega)$, the components of a subdirect product representation of $\mathscr{T}$, they are the stalks of a sheaf on a compact space from which $\mathscr{T}$ may be recaptured as the topos of global sections, just as a commutative ring may be recaptured as the ring of global sections of a sheaf of local rings on its spectrum, a standard result in algebraic geometry.

To describe the topological space in question, we shall first discuss the spectrum of a distributive lattice L in general. We recall that a filter on the opposite lattice L^{op}, which is also distributive if L is, is called an 'ideal'. For clarity, we repeat:

Definition 18.1. An *ideal* Q of a lattice satisfies:

I1. if $p \in Q$ and $q \vdash p$ then $q \in Q$;

I2. if $p \in Q$ and $q \in Q$ then $p \vee q \in Q$.

An ideal is *proper* provided

I3. $\top \notin Q$

and *prime* provided moreover

I4. if $p \wedge q \in Q$ then $p \in Q$ or $q \in Q$.

For what follows we make the crucial observation that *prime ideals are precisely the complements of prime filters.* We shall write pf(L) for the set of prime filters of L.

The *spectrum* spec(L) of a lattice L is the set of prime ideals of L topologized with basic open sets determined by elements $p \in L$:

$$\triangle(p) = \{Q \in \mathrm{pf}(L^{\mathrm{op}}) \mid p \notin Q\}.$$

If we replace the prime ideals by their complements, spec(L) may be described equivalently as the set of prime filters with basic open sets

$$\nabla(p) = \{P \in \mathrm{pf}(L) \mid p \in P\}.$$

This is to be distinguished from the *cospectrum* of L which also consists of prime filters but has as basic open sets

$$\nabla^*(p) = \{P \in pf(L) \mid p \notin P\}.$$

Lemma 18.2. The basic open sets of the spectrum of a distributive lattice L satisfy:

$$\begin{aligned} \nabla(p) \cup \nabla(q) &= \nabla(p \vee q), \\ \nabla(p) \cap \nabla(q) &= \nabla(p \wedge q), \\ \nabla(\bot) &= \varnothing, \\ \nabla(\top) &= \mathrm{spec}\ L, \end{aligned}$$

$$\nabla(p) \subseteq \nabla(q) \text{ if and only if } p \vdash q.$$

Proof. To check the last statement, suppose $p \nvdash q$, then the filter (p) generated by p does not contain the element q. By Lemma 17.3, (p) is contained in a prime filter P not containing q. Therefore $P \in \nabla(p)$ and $P \notin \nabla(q)$, so $\nabla(p) \nsubseteq \nabla(q)$.

An arbitrary open set is of course a union of basic open sets. An important observation asserts the following.

Proposition 18.3. In the spectrum of a distributive lattice, every basic open set is compact and, conversely, every compact open set is basic. In particular the whole spectrum is compact.

Proof. In the following argument we use ideals, not filters. Suppose

$$\triangle(p) = \bigcup_{i \in I} \triangle(p_i) = \{Q \in \mathrm{pf}(L^{\mathrm{op}}) \mid K \nsubseteq Q\},$$

where K is the ideal generated by the set $\{p_i | i \in I\}$. Then p belongs to the intersection of all prime ideals Q containing K, hence to K, by Lemma 17.3. Therefore

$$p \vdash p_{i_1} \vee \cdots \vee p_{i_n}$$

for a finite subset $\{i_1, \ldots, i_n\}$ of I, and so

$$\triangle(p) = \triangle(p_{i_1}) \cup \cdots \cup \triangle(p_{i_n}).$$

Conversely, let $V = \bigcup_{p \in K} \triangle(p)$ be any open set. If V is compact, then K must contain a finite subset $\{p_1, \ldots, p_n\}$ such that $V = \bigcup_{i=1}^{n} \triangle(p_i)$. In fact, $V = \triangle(p)$ for $p = p_1 \vee \cdots \vee p_n$.

Remark 18.4. If a topological space X has a basis $\mathscr{B}$ which is closed under finite intersections, as does the spectrum of a distributive lattice, then a functor $F: \mathscr{B}^{\mathrm{op}} \to \mathbf{Sets}$ extends uniquely to a sheaf $G: \mathrm{Op}(X)^{\mathrm{op}} \to \mathbf{Sets}$ by the construction

$$G(V) = \varprojlim \{F(B) | B \subseteq V \,\&\, B \in \mathscr{B}\}$$

if and only if F has the *sheaf property* with respect to coverings of elements of $\mathscr{B}$. By this we mean that if $B = \bigcup_{i \in I} B_i$, for B, $B_i \in \mathscr{B}$, and given a family of elements $\{x_i \in F(B_i) | i \in I\}$ which is *compatible* in the sense that

$$F(B_i \cap B_j \to B_i)(x_i) = F(B_i \cap B_j \to B_j)(x_j),$$

for all i, $j \in I$, then there exists a unique $x \in F(B)$ such that

$$F(B_i \to B)(x) = x_i,$$

for all $i \in I$.

Let us now look at a topos $\mathscr{T}$ and define its spectrum Spec $\mathscr{T}$ as the spectrum of the Heyting algebra of arrows $1 \to \Omega$ in $\mathscr{T}$. With any object A of $\mathscr{T}$ we associate the functor $\bar{A}$ defined on the basic open set $\triangledown(p)$ of Spec $\mathscr{T}$ by

$$\bar{A}(\triangledown(p)) = \mathrm{Hom}_{\mathscr{T}/(p)}(1, A),$$

with the obvious restriction mappings

$$\bar{A}(\triangledown(q)) \to \bar{A}(\triangledown(p))$$

when $\triangledown(p) \subseteq \triangledown(q)$, that is, $p \vdash q$ in $L(\mathscr{T})$.

Theorem 18.5. For any object A of a topos $\mathscr{T}$, the contravariant functor $\bar{A}$ defined on basic open sets by $\bar{A}(\triangledown(p)) = \mathrm{Hom}_{\mathscr{T}/(p)}(1, A)$ may be extended to a sheaf on Spec $\mathscr{T}$. Note, in particular, that $\bar{A}(\mathrm{Spec}\,\mathscr{T}) = \mathrm{Hom}_{\mathscr{T}}(1, A)$. Moreover, a stalk of this sheaf at a point P of Spec $\mathscr{T}$ is given by $\mathrm{Hom}_{\mathscr{T}/P}(1, A)$.

Proof. In view of Remark 18.4 and Proposition 18.3, it suffices to check the sheaf property for finite covers of basic open subsets of Spec $\mathscr{T}$. So suppose

$$\nabla(p) = \bigcup_{i=1}^{n} \nabla(q_i),$$

so that $p \vdash q_1 \vee \cdots \vee q_n$ in $L(\mathscr{T})$, and let $a_i \in \mathrm{Hom}_{\mathscr{T}/(q_i)}(1, A)$ be such that, for all $i, j \in I$, the restrictions of a_i and a_j to $\mathscr{T}/(q_i \wedge q_j)$ coincide. Then we claim that there is a unique $a \in \mathrm{Hom}_{\mathscr{T}/(p)}(1, A)$ such that, for all $i \in I$, the restriction of a to $\mathscr{T}/(q_i)$ is a_i.

According to the construction of $\mathscr{T}/(q_i)$, a_i is given by its graph $|a_i| \equiv \alpha_i$, which is a term of type $P(1 \times A)$ in $L(\mathscr{T})$ such that the proposition

$$\mathrm{Fcn}\,\alpha_i \equiv \exists!_{x \in A} \langle *, x \rangle \in \alpha_i$$

belongs to the principal filter (q_i), that is,

$$q_i \vdash \mathrm{Fcn}\,\alpha_i$$

in $L(\mathscr{T})$. We are given that, for all $i, j \in I$,

$$q_i \wedge q_j \vdash \alpha_i = \alpha_j$$

in $L(\mathscr{T})$. Now the following is a theorem of intuitionistic type theory:

$$\bigwedge_{i,j=1}^{n} ((q_i \wedge q_j) \Rightarrow \alpha_i = \alpha_j), \quad \bigvee_{i=1}^{n} q_i \vdash \exists!_{u \in P(1 \times A)} \bigwedge_{i=1}^{n} (q_i \Rightarrow u = \alpha_i).$$

Indeed, this result holds more generally if $P(1 \times A)$ is replaced by any type B and may be called *definition by cases.*

Using the fact that $\mathscr{T}/(p) \models q_1 \vee \cdots \vee q_n$, we may therefore infer that

$$\mathscr{T}/(p) \models \exists!_{u \in P(1 \times A)} \bigwedge_{i=1}^{n} (q_i \Rightarrow u = \alpha_i).$$

By description, there is a unique term α of type $P(1 \times A)$ in $L(\mathscr{T})$ such that

$$\mathscr{T}/(p) \models \bigwedge_{i=1}^{n} (q_i \Rightarrow \alpha = \alpha_i),$$

hence, for all $i = 1, \ldots, n$.

$$\mathscr{T}/(q_i) \models \alpha = \alpha_i.$$

We claim that α is the graph of a function $a\colon 1 \to A$ in $\mathscr{T}/(p)$. To prove this, we must show that $\mathscr{T}/(p) \models \mathrm{Fcn}\,\alpha$. Now we have

$$\mathscr{T}/(q_i) \models \mathrm{Fcn}\,\alpha_i, \quad \mathscr{T}/(q_i) \models \alpha = \alpha_i,$$

hence

$$\mathscr{T}/(q_i) \models \mathrm{Fcn}\,\alpha.$$

Therefore

$$\mathscr{T} \vDash (q_1 \vee \cdots \vee q_n) \Rightarrow \mathrm{Fcn}\,\alpha,$$

whence $\mathscr{T} \vDash p \Rightarrow \mathrm{Fcn}\,\alpha$, as remained to be shown. The sheaf property has now been established, hence $\bar{A}$ may be extended to a sheaf on Spec $\mathscr{T}$.

What are the stalks of this sheaf? In general, the stalk of a sheaf G on a topological space X at a point $x \in X$ is given by

$$G_x = \varinjlim \{G(V) \mid x \in V\},$$

where V ranges over all open sets containing x or only over all basic open sets containing x. In the present case, we have the stalk

$$\begin{aligned} \bar{A}_P &= \varinjlim \{\bar{A}(\nabla(p)) \mid P \in \nabla(p)\} \\ &= \varinjlim \{\mathrm{Hom}_{\mathscr{T}/(p)}(1, A) \mid p \in P\} \\ &= \mathrm{Hom}_{\mathscr{T}/P}(1, A). \end{aligned}$$

We leave it to the reader to check that $\mathrm{Hom}_{\mathscr{T}/P}(1, A)$ is indeed a colimit of the $\mathrm{Hom}_{\mathscr{T}/(p)}(1, A)$.

John Gray has defined a sheaf G with values in a category $\mathscr{A}$ that need not be the category of sets. It assigns to each open set V an object $G(V)$ of $\mathscr{A}$ such that, if

$$V = V_1 \cup \cdots \cup V_n,$$

then the following is an equalizer diagram in $\mathscr{A}$:

$$G(V) \to \prod_{i=1}^{n} G(V_i) \rightrightarrows \prod_{i,j=1}^{n} G(V_i \cap V_j).$$

Another way of expressing essentially the same fact as Theorem 18.5 is the following.

Proposition 18.6. Assigning to each basic open set $\nabla(p)$ of Spec $\mathscr{T}$ the topos $\mathscr{T}/(p)$, one obtains a sheaf $S(\mathscr{T})$ on Spec $\mathscr{T}$ with values in **Top**. Its stalks are the toposes $\mathscr{T}/P$, where P ranges over the prime filters of $\mathrm{Hom}_{\mathscr{T}}(1, \Omega)$.

Proof. According to the general definition of category-valued sheaves above, it suffices to show that, if

$$p \dashv\vdash q_1 \vee \cdots \vee q_n$$

in $L(\mathscr{T})$, then the following is an equalizer diagram in **Top**:

$$\mathscr{T}/(p) \to \prod_{i=1}^{n} \mathscr{T}/(q_i) \rightrightarrows \prod_{i,j=1}^{n} \mathscr{T}/(q_i \wedge q_j).$$

This may be shown by a straightforward application of 'definition by cases'. We let the reader check that $\mathcal{T}/P$ is the colimit of the $\mathcal{T}/(p)$. The relation between the sheaf $S(\mathcal{T})$ and the set-valued sheaves $\bar{A}$ is given by

$$\bar{A}(\nabla(p)) = \mathrm{Hom}_{S(\mathcal{T})(\nabla(p))}(1, A).$$

In analogy with ring theory, one may call a topos $\mathcal{T}$ *local* if it is non-trivial and has the disjunction property:

$$\text{if } \mathcal{T} \vDash p \vee q \text{ then } \mathcal{T} \vDash p \text{ or } \mathcal{T} \vDash q.$$

Equivalently, this says that $\mathcal{T}(1, \Omega)$ has exactly one maximal proper ideal. For a filter P on $\mathrm{Hom}_{\mathcal{T}}(1, \Omega)$, clearly $\mathcal{T}/P$ is local if and only if P is prime.

Corollary 18.7. Every topos is equivalent to the topos of global sections of a sheaf of local toposes.

Exercises

1. Complete the proof of Theorem 18.5 and write out a proof of Proposition 18.6.

2. Call a non-trivial topos *cosimple* if the Heyting algebra $\mathrm{Hom}(1, \Omega)$ has exactly two elements and *cosemisimple* if $\mathrm{Hom}(1, \Omega)$ is Boolean. Show that the stalks of the sheaf $S(\mathcal{T})$ are all cosimple if and only if $\mathcal{T}$ is cosemisimple.

3. Define the *cospectrum* of a topos $\mathcal{T}$ as the cospectrum of the Heyting algebra $\mathrm{Hom}_{\mathcal{T}}(1, \Omega)$. Show how to obtain a presheaf $S^*(\mathcal{T})$ on this cospectrum defined by $S^*(\mathcal{T})(\nabla^*(p)) = \mathcal{T}/\mathrm{Alt}(p)$, where $\mathrm{Alt}(p) \equiv \{q \in \mathrm{Hom}_{\mathcal{T}}(1, \Omega) \mid \mathcal{T} \vDash p \vee q\}$ is the filter of *alternatives* of p.

4. Prove the theorem on definition by cases in the text.

5. Prove the following second version of definition by cases (which is intuitionistically distinct from the first version in the text): if $p_1, \ldots, p_n$ are closed formulas and $a_1, \ldots, a_n$ are terms of type A, then

$$\bigwedge_{i,j=1}^{n} (p_i \vee p_j \vee a_i = a_j), \neg \bigwedge_{i=1}^{n} p_i \vdash \exists!_{x \in A} \bigwedge_{i=1}^{n} (x = a_i \vee p_i).$$

6. Prove that the topos of global sections of the sheaf associated with the presheaf $S^*(\mathcal{T})$ of Exercise 3 is the topos $\mathcal{T}$ and that its stalks are the toposes $\mathcal{T}/T_P$, where with each prime filter P one associates the filter $T_P \equiv \{p \in \mathrm{Hom}_{\mathcal{T}}(1, \Omega) \mid \exists_{q \notin P} \mathcal{T} \vDash p \vee q\}$.

7. Prove that the intersection of all maximal proper ideals of a distributive lattice is the set of all elements p such that, for all elements q, $p \vee q = \top$ implies $q = \top$.

8. Prove that a distributive lattice has exactly one maximal proper ideal if and only if, for all elements p and q, $p \vee q = \top$ implies $p = \top$ or $q = \top$. Hence deduce that the two characterizations of a local topos in the text are equivalent.

19 Completeness without assuming the rule of choice

In this section we shall generalize the completeness theorem for higher order logic discussed in Section 17 by dropping the condition that the rule of choice holds. The proof, already suggested by Henkin for classical logic, will involve adjoining infinitely many so-called 'Henkin constants'. From our point of view, these are just parameters. Models for intuitionistic first order logic had been discussed by Beth and Kripke. We shall here follow the proof of the completeness theorem for intuitionistic first order logic by Aczel.

First, let us explain what is meant by adjoining an infinite set X of variables to a type theory $\mathfrak{L}$. Each term of $\mathfrak{L}(X)$ is a term of $\mathfrak{L}(X')$ for some finite subset X' of X. We define $\Gamma \vdash p$ in $\mathfrak{L}(X)$ to mean $\Gamma \vdash_X p$ in $\mathfrak{L}$, it being understood that X contains all free variables in Γ and in p. By this we mean that $\Gamma' \vdash_{X'} p$ for some finite subset Γ' of Γ and some finite subset X' of X containing all free variables in Γ' and in p.

In what follows, we shall need a large supply of variables; a countable number will not do in general. As usual, we assume that the class of terms of a language is a set in a sufficiently large universe. Let λ be the cardinality of the set of closed terms of $\mathfrak{L}$ or $\aleph_0$, whichever is larger. We shall assume that Y is a new set of variables containing λ variables of each type, given once and for all, in addition to the countably many variables which already occur in formulas of $\mathfrak{L}$

The general completeness theorem for higher order logic asserts:

Theorem 19.1. Given any type theory $\mathfrak{L}$, $\mathcal{M} \models \tau(p)$ for all models $\tau: \mathfrak{L} \to L(\mathcal{M})$ if and only if $\vdash p$ in $\mathfrak{L}$.

Proof. Clearly, when $\vdash p$ in $\mathfrak{L}$ then $\vdash \tau(p)$ in $L(\mathcal{M})$, hence $\mathcal{M} \models \tau(p)$. Conversely, suppose p_0 is a closed formula of $\mathfrak{L}$ such that not $\vdash p_0$ in $\mathfrak{L}$. We shall construct a model

$$\mathfrak{L} \xrightarrow{\subseteq} \mathfrak{L}_\infty \xrightarrow{\eta_{\mathfrak{L}_\infty}} LT(\mathfrak{L}_\infty)$$

of $\mathfrak{L}$ such that not $\vdash \eta_{\mathfrak{L}_\infty}(p_0)$ in $LT(\mathfrak{L}_\infty)$, that is, not $\vdash p_0$ in $\mathfrak{L}_\infty$. Here $\mathfrak{L}_\infty$ will be constructed from $\mathfrak{L}$ by adjoining an infinite set of parameters and imposing an infinite set of conditions.

We form a sequence of type theories

$$\mathfrak{L} = \mathfrak{L}_0 \xrightarrow{\subseteq} \mathfrak{L}_1 \xrightarrow{\subseteq} \mathfrak{L}_2 \xrightarrow{\subseteq} \cdots,$$

where $\mathfrak{L}_n = \mathfrak{L}(X_n)/P_n$ is defined as follows, with $X_n \subseteq Y$.

(*a*) $X_0 = \emptyset$; for $n \geqslant 1$, $X_n \supseteq X_{n-1}$ contains exactly one variable $x_\alpha \in Y$ for each term α of type PA in $\mathfrak{L}(X_{n-1})$ such that $\exists_{x\in A}\, x\in\alpha$ belongs to P_{n-1}; but we insist that $Y - X_n$ still contains λ variables to each type.

(*b*) $P_0 = (\top)$ is the smallest filter of closed formulas of $\mathfrak{L}$; for $n \geqslant 1$, $P_n \supseteq P_{n-1}$ is a prime filter of closed formulas of $\mathfrak{L}(X_n)$ containing all formulas of the form $x_\alpha \in \alpha$, but not containing p_0.

We shall verify by mathematical induction that these parallel constructions can be carried out.

(*a*) The construction of X_n as a subset of Y presupposes that X_n has at most λ variables of each type, which will follow from the fact that $\mathfrak{L}(X_{n-1})$ has at most λ closed terms. Indeed, this is so for $\mathfrak{L}(X_0) = \mathfrak{L}$. Assuming that $\mathfrak{L}(X_{n-1})$ has at most λ closed terms, consider the closed terms of $\mathfrak{L}(X_n)$. Each of these is obtained from an open term $\psi(x_1, \ldots, x_m)$ of $\mathfrak{L}(X_{n-1})$ by substituting x_{α_i} for x_i. The number of such ψ is at most λ, e.g. since the universal closure of each formula $\psi = \psi$ is a closed term of $\mathfrak{L}(X_{n-1})$. Replacing the x_i we obtain at most $\lambda^m = \lambda$ substitution instances for each ψ, hence at most $\lambda^2 = \lambda$ closed terms in $\mathfrak{L}(X_n)$.

(*b*) The construction of P_n presupposes that P_{n-1} together with the formulas $x_\alpha \in \alpha$ does not entail p_0 in $\mathfrak{L}(X_n)$. Indeed, suppose p_0 could be derived from these assumptions, then this derivation would only involve a finite number of formulas of P_{n-1} and a finite number of formulas $x_\alpha \in \alpha$, as well as only a finite number of elements of X_n, so

$$F, x_{\alpha_1} \in \alpha_1, \ldots, x_{\alpha_m} \in \alpha_m \vdash_{X'} p_0,$$

where F is a finite subset of P_{n-1} and X' is a finite subset of X_n. But then

$$F, \exists_{x\in A_1} x \in \alpha_1, \ldots, \exists_{x\in A_m} x \in \alpha_m \vdash_{X'} p_0,$$

and so $p_0 \in P_{n-1}$, which contradicts the construction of P_{n-1}.

Now let $\mathfrak{L}_\infty = \mathfrak{L}(X_\infty)/P_\infty$, where

$$X_\infty = \bigcup_{n=0}^{\infty} X_n, \quad P_\infty = \bigcup_{n=0}^{\infty} P_n.$$

Moreover, $\Gamma \vdash p$ in $\mathfrak{L}_\infty$ shall mean that $\Gamma \cup P_\infty \vdash_{X_\infty} p$ in $\mathfrak{L}(X_\infty)$, that is, $\Gamma' \cup P_n \vdash_{X_n} p$ for some finite subset $\Gamma' \subseteq \Gamma$ and some natural number n. In particular, $\vdash p$ in $\mathfrak{L}_\infty$ if and only if $p \in P_\infty$, that is, $p \in P_n$ for some n.

We claim that the interpretation

$$\mathfrak{L} \xrightarrow{\subseteq} \mathfrak{L}_\infty \xrightarrow{\eta_{\mathfrak{L}_\infty}} LT(\mathfrak{L}_\infty)$$

is a model, that is, $T(\mathfrak{L}_\infty)$ satisfies M0, M1 and M2. As is easily seen, it suffices to prove:

(0) $\perp \notin P_\infty$; in fact $p_0 \notin P_\infty$;

(1) if $p \vee q \in P_\infty$ then $p \in P_\infty$ or $q \in P_\infty$;

(2) if $\exists_{x\in A}\varphi(x) \in P_\infty$ then $\varphi(a) \in P_\infty$ for some term a of type A in $\mathfrak{L}(X_\infty)$.

For example, to see why (1) implies M1, suppose $p'\colon \mathbf{1} \to \mathbf{\Omega}$ and $q'\colon \mathbf{1} \to \mathbf{\Omega}$ are arrows in $T(\mathfrak{L}_\infty)$ such that $T(\mathfrak{L}_\infty) \models p' \vee q'$. As already observed earlier, we may write $p' \equiv \mathbf{p}$ and $q' \equiv \mathbf{q}$, where p and q are closed formulas of $\mathfrak{L}_\infty$, hence we know that $\vdash \mathbf{p} \vee \mathbf{q}$ in $LT(\mathfrak{L}_\infty)$. Since $\eta_{\mathfrak{L}_\infty}$ is conservative, it follows that $\vdash p \vee q$ in $\mathfrak{L}_\infty$, that is, $p \vee q \in P_\infty$. Then by (1), $p \in P_\infty$ or $q \in P_\infty$, so that $T(\mathfrak{L}_\infty) \models p'$ or $T(\mathfrak{L}_\infty) \models q'$, which establishes M2 for $T(\mathfrak{L}_\infty)$.

We shall now prove (0) to (2).

(0) Suppose $p_0 \in P_\infty$, then $p_0 \in P_n$ for some n, contradicting the construction of P_n.

(1) Suppose $p \vee q \in P_\infty$, then $p \vee q \in P_n$, for some n, hence $p \in P_n \subseteq P_\infty$ or $q \in P_n \subseteq P_\infty$.

(2) Suppose $\exists_{x\in A}\varphi(x)$ belongs to P_∞, then it belongs to P_{n-1} for some $n \geqslant 1$. Let $\alpha \equiv \{x \in A \mid \varphi(x)\}$, then $\exists_{x\in A}\, x \in \alpha$ belongs to P_{n-1}, hence $x_\alpha \in \alpha$ belongs to P_n, therefore $\varphi(x_\alpha) \in P_n \subseteq P_\infty$.

Finally, if $T(\mathfrak{L}_\infty) \models p_0$, then $p_0 \in P_\infty$, since $\eta_{\mathfrak{L}_\infty}$ is conservative, contradicting (0). This completes the proof of the completeness theorem.

What do models of intuitionistic type theories look like? Models of first order intuitionistic logic had been constructed by Beth and Kripke. In one such construction, associated with the name 'Kripke', the rôle of the topos of sets is taken by the functor category $\mathbf{Sets}^{\mathscr{P}}$, where $\mathscr{P}$ is a nonempty preordered set. Strictly speaking, for this to be called a model, $\mathscr{P}$ should have an initial element. The topos semantics of Section 9 will then assure that $\mathbf{Sets}^{\mathscr{P}}$ satisfies M1 in addition to M0 and M2, which hold in any case. General models of intuitionistic type theory will bear the same relation to such Kripke models as Henkin models of classical type theory do to **Sets**.

Theorem 19.2. For any type theory $\mathfrak{L}$ there is a partially ordered set $\mathscr{P}$ and a faithful left exact functor $\Gamma\colon T(\mathfrak{L}) \to \mathbf{Sets}^{\mathscr{P}}$. $\mathscr{P}$ will have an initial element if $\mathfrak{L}$ is consistent and has the disjunction and existence properties, e.g , if $\mathfrak{L}$ is the internal language of a model. In that case, Γ is near exact.

Proof. In what follows, Y will be a given set containing λ variables of each type, as in the proof of Proposition 19.1. We shall construct a preordered set

$\mathscr{P}$ as follows. Its elements are pairs (X, P), where P is a 'saturated' prime filter of closed formulas in $\mathfrak{L}(X)$, X being any subset of Y such that Y-X has λ variables of each type. By P being *saturated* we mean that, if $\exists_{x\in A}\varphi(x)$ belongs to P, then $\varphi(a)$ belongs to P for some closed term a of type A in $\mathfrak{L}$. The condition on Y-X assures that there are enough additional free variables left. The set $\mathscr{P}$ is partially ordered by stipulating that $(X, P) \leqslant (X', P')$ if and only if $X \subseteq X'$ and $P \subseteq P'$.

Let us now abbreviate $T(\mathfrak{L})$ as $\mathscr{T}$. In fact, we may as well assume that $\mathscr{T}$ is any topos and that $\mathfrak{L} = L(\mathscr{T})$. We form $\Gamma: \mathscr{T} \to \mathbf{Sets}^{\mathscr{P}}$ as follows: for each object A of $\mathscr{T}$ and each arrow $f: A \to B$ in $\mathscr{T}, \Gamma(A): \mathscr{P} \to \mathbf{Sets}$ and $\Gamma(f)$: $\Gamma(A) \to \Gamma(B)$ are defined by

$$\Gamma(A)(X, P) \equiv \mathrm{Hom}_{\mathscr{T}(X)/P}(1, A), \quad \Gamma(f)(X, P) \equiv \mathrm{Hom}_{\mathscr{T}(X)/P}(1, f).$$

Why is $\Gamma(A)$ a functor? If $(X, P) \leqslant (X', P')$, there is a canonical translation $\tau: \mathfrak{L}(X)/P \to \mathfrak{L}(X')/P'$, hence a logical functor $T(\tau): \mathscr{T}(X)/P \to \mathscr{T}(X')/P'$, and this induces a mapping $\mathrm{Hom}_{\mathscr{T}(X)/P}(1, A) \to \mathrm{Hom}_{\mathscr{T}(X')/P'}(1, A)$ which sends the arrow $a: 1 \to A$ of $\mathscr{T}(X)/P$ onto the arrow $T(\tau)(a)$ of $\mathscr{T}(X')/P'$. The reader will easily verify that $\Gamma(f)$ is a natural transformation. That Γ is left exact follows from the fact that limits in $\mathbf{Sets}^{\mathscr{P}}$ are defined componentwise.

Finally we note that, if $\mathfrak{L}$ is consistent and has the disjunction and existence properties, $(\emptyset, (\top))$ is an initial element of $\mathscr{P}$. Again, that Γ is near exact (see Remark 17.8 in Supplement) follows since colimits in $\mathbf{Sets}^{\mathscr{P}}$ are defined componentwise.

20 Some basic intuitionistic principles

In this and the following sections, we show that various constructivist or intuitionistic principles hold in pure type theory.

In as much as intuitionists are willing to believe in a formal language, they do believe in the *disjunction property*:

if $\vdash p \vee q$ then $\vdash p$ or $\vdash q$.

We shall prove that this is indeed valid in pure intuitionistic type theory $\mathfrak{L}_0$. It is not valid in pure classical type theory $\mathfrak{L}_0/(\beta)$, where $\beta \equiv \forall_{t\in\Omega}(t \vee \neg t)$ is the Boolean axiom. For example, Gödel constructed a sentence γ for which neither $\beta \vdash \gamma$ nor $\beta \vdash \neg\gamma$, yet, of course, $\beta \vdash \gamma \vee \neg\gamma$.

It is often claimed that classical mathematics is non-constructive. Here is an example of a non-constructive existence proof based on the Boolean axiom, which we learned from van Dalen:

Theorem. There exist irrational numbers a and b such that a^b is rational.

Proof. Either $\sqrt{2}^{\sqrt{2}}$ is rational or not. In the first case take $a = b = \sqrt{2}$, in the second case take $a = \sqrt{2}^{\sqrt{2}}$ and $b = \sqrt{2}$.

It so happens that for this theorem a constructive proof is easily available: take $a = \sqrt{2}$ and $b = 2\log_2 3$. Intuitionists insist that all existence proofs be constructive. In particular, they ought to believe in the *existence property*:

$$\text{if } \vdash \exists_{x \in A} \varphi(x) \text{ then there is a term } a \text{ of type } A \text{ such that } \vdash \varphi(a).$$

Again we shall prove this for pure intuitionistic type theory $\mathfrak{L}_0$. This too fails in pure classical type theory $\mathfrak{L}_0/(\beta)$. For example, take

$$\varphi(x) \equiv (x = 0 \wedge \gamma) \vee (x \neq 0 \wedge \neg\gamma),$$

where γ is Gödel's undecidable sentence.

Clearly, on the assumption $\gamma \vee \neg\gamma$ such an x exists, thus $\beta \vdash \exists_{x \in N} \varphi(x)$. On the other hand, if we could prove $\beta \vdash \varphi(n)$ for some *standard numeral* $n \equiv S^k 0$, we would also be able to prove γ (in case $k = 0$) or $\neg\gamma$ (in case $k \neq 0$).

This counterexample breaks down if we admit terms of type N other than standard numerals. For example, if we adjoin to $\mathfrak{L}_0/(\beta)$ a minimization operator $\mu_{x \in N}$ (the smallest $x \in N$ such that), then, whenever classically we can prove $\exists_{x \in N} \varphi(x)$, we can also prove $\varphi(n)$ for $n \equiv \mu_{x \in N} \varphi(x)$. Presumably, an intuitionist would reject all non-standard numerals, that is, he would be committed to the belief that all terms of type N are provably equal to terms of the form $S^k 0$, where k is a natural number. Yet he should accept the minimization operator applied to $\varphi(x)$ whenever $\vdash \exists_{x \in N} \varphi(x)$, because in pure type theory $\mathfrak{L}_0$ it will always yield a standard numeral, as will follow from the discussion below and Lemma 20.3.

The minimization operator is a special case of the *unique existence property*:

$$\text{if } \vdash \exists!_{x \in A} \psi(x) \text{ then there is a term } a \text{ of type } A \text{ such that } \vdash \psi(a).$$

Indeed, if $A = N$ and $\vdash \exists_{x \in N} \varphi(x)$, let

$$\psi(x) \equiv \varphi(x) \wedge \forall_{y \in N}(y < x \Rightarrow \neg\varphi(y)),$$

then the smallest x such that $\varphi(x)$ is the unique x such that $\psi(x)$. The unique existence property is acceptable to both intuitionists and classical mathematicians, but for different reasons; it holds in $\mathfrak{L}_0$ automatically (see Lemma 20.3 below), but in $\mathfrak{L}_0/(\beta)$ only if a description operator or a minimization operator is adjoined.

It is more difficult to find an example in classical mathematics where the existence property breaks down even if a description operator is allowed. Here is a candidate with $A = P(PN)$: let $\chi(x)$ be the formal sentence which asserts that x is a non-principal ultrafilter on the set N and put

$$\varphi(x) \equiv (\exists_{x \in A} \chi(x) \Rightarrow \chi(x)).$$

Then clearly $\beta \vdash \exists_{x \in A} \varphi(x)$, yet we conjecture that not $\beta \vdash \varphi(a)$ for any closed term α of type $A = P(PN)$.

In fact, Feferman has shown that, if $\chi(x)$ means that x is a well-ordering of the reals and if $\varphi(x)$ is defined as above, then $\vdash \varphi(a)$ for no closed term a of Zermolo–Fraenkel set theory, hence surely for no term of classical type theory.

There might be an even simpler example where the existence property breaks down at type $A = PN$, if one could find a formula $\varphi(x)$ which asserts that the subset x of N is not definable in classical type theory.

As already mentioned, both disjunction and existence properties hold in pure intuitionistic type theory $\mathfrak{L}_0$. There are several ways of proving these metatheorems. We shall here adopt a method which resembles the realizability arguments known to logicians, but which depends on a categorical construction due to Peter Freyd. We shall postpone the details of this construction to a later section; for the moment we are content to state its most relevant properties.

Theorem 20.1. Every topos $\mathscr{T}$ possesses a 'Freyd cover', a topos $\hat{\mathscr{T}}$ and a strict logical functor $G: \hat{\mathscr{T}} \to \mathscr{T}$ with the following properties:

(1) All arrows $1 \to N$ in $\hat{\mathscr{T}}$ have the form $S^k 0$ for some $k \in \mathbb{N}$.

Moreover, the internal logic of $\hat{\mathscr{T}}$ is described as follows, where we have written $\check{A} \equiv G(A)$, $\check{p} \equiv G(p)$ and $\check{\varphi}(x) \equiv G'(\varphi(x))$, $G': \hat{\mathscr{T}}[x] \to \mathscr{T}[x]$ being the canonical extension of G such that $G'(x) \cdot \underset{x}{=} \cdot x$:

(2) $\hat{\mathscr{T}} \models \top$ always;

(3) $\hat{\mathscr{T}} \models \bot$ never;

(4) $\hat{\mathscr{T}} \models p \wedge q$ iff $\hat{\mathscr{T}} \models p$ and $\hat{\mathscr{T}} \models q$;

(5) $\hat{\mathscr{T}} \models p \vee q$ iff $\hat{\mathscr{T}} \models p$ or $\hat{\mathscr{T}} \models q$;

(6) $\hat{\mathscr{T}} \models p \Rightarrow q$ iff (i) $\mathscr{T} \models \check{p} \Rightarrow \check{q}$
and (ii) $\hat{\mathscr{T}} \models p$ implies $\hat{\mathscr{T}} \models q$;

(7) $\hat{\mathscr{T}} \models \forall_{x \in A} \varphi(x)$ iff (i) $\mathscr{T} \models \forall_{x \in \check{A}} \check{\varphi}(x)$
and (ii) $\hat{\mathscr{T}} \models \varphi(a)$ for all a: $1 \to A$ in $\hat{\mathscr{T}}$;

(8) $\hat{\mathscr{T}} \models \exists_{x \in A} \varphi(x)$ iff $\hat{\mathscr{T}} \models \varphi(a)$ for some a: $1 \to A$ in $\hat{\mathscr{T}}$.

Finally, if $\mathscr{T} = T(\mathfrak{L})$ is the topos generated by a type theory, $\hat{\mathscr{T}}$ has canonical subobjects and G preserves them.

We note that if A is a type of $\mathfrak{L}_0$, it is automatically an object in every topos, hence we need not distinguish notationally between the object A of $\hat{\mathscr{T}}$ and the object $\check{A}$ of $\mathscr{T}$. Similarly, if p is a closed formula of $\mathfrak{L}_0$, we need not distinguish between the arrow $p: 1 \to \Omega$ in $\hat{\mathscr{T}}$ and the arrow $\check{p}: 1 \to \Omega$ in $\mathscr{T}$.

While the proof of Theorem 20.1 will be left to a later section, we shall use the theorem now to prove some of the more basic intuitionistic principles.

Proposition 20.2. The disjunction property holds in $\mathfrak{L}_0$: if $\vdash p \vee q$ then $\vdash p$ or $\vdash q$.

Proof. Let p and q be closed formulas of $\mathfrak{L}_0$ and suppose $\vdash p \vee q$. Then $p \vee q$ holds in any topos, in particular, $\hat{\mathscr{F}} \vDash p \vee q$, where $\hat{\mathscr{F}}$ is the Freyd cover of the free topos $\mathscr{F}$. By (5) of Theorem 20.1, $\hat{\mathscr{F}} \vDash p$ or $\hat{\mathscr{F}} \vDash q$. Now apply the logical functor $G: \hat{\mathscr{F}} \to \mathscr{F}$ and remember that $G(p) \cdot = \cdot p$. Therefore, $\mathscr{F} \vDash p$ or $\mathscr{F} \vDash q$. This shows that $\vdash p$ or $\vdash q$ in $L(\mathscr{F}) \equiv LT(\mathfrak{L}_0)$. Since $\mathfrak{L}_0 \to LT(\mathfrak{L}_0)$ is a conservative extension (see Proposition 14.3), $\vdash p$ or $\vdash q$ in $\mathfrak{L}_0$.

The last part of this argument can also be done as follows. If the arrow $p \cdot = \cdot \top$ in $\mathscr{F}$, then its graph $\{\langle *, p \rangle\}$ is provably equal to $\{\langle *, \top \rangle\}$, from which it easily follows that $\vdash p$ in $\mathfrak{L}_0$.

Lemma 20.3. The unique existence property holds in $\mathfrak{L}_0$: if $\vdash \exists!_{x \in A} \psi(x)$ then $\vdash \psi(a)$ for some closed term a of type A. In particular, every arrow $1 \to N$ in the free topos $\mathscr{F}$ has the form $S^k 0$ for some $k \in \mathbb{N}$.

Proof. We are given that $\vdash \exists!_{x \in A} \psi(x)$ and wish to show that $\vdash \psi(a)$ for some closed term a of type A. We proceed by induction on the construction of A.

If $A = 1$, take $a \equiv *$.

If $A = \Omega$, take $a \equiv \psi(\top)$. (See the proof of Lemma 12.3.)

If $A = PB$, take $a \equiv \{y \in B \mid \exists_{v \in PB}(\psi(v) \wedge y \in v)\}$.

If $A = B \times C$, we first find a term b of type B such that $\vdash \exists!_{z \in C} \psi(\langle b, z \rangle)$, then a term c of type C such that $\vdash \psi(\langle b, c \rangle)$.

If $A = N$, we proceed as follows. By assumption, ψ determines an arrow $n: 1 \to N$ in the free topos $\mathscr{F} \equiv T(\mathfrak{L}_0)$ with graph

$$|n| \equiv \{\langle *, x \rangle \in 1 \times N \mid \psi(x)\}.$$

Now let $F: \mathscr{F} \to \hat{\mathscr{F}}$ be the unique arrow in $\mathbf{Top}_0$ from the initial object $\mathscr{F}$, then $F(n): 1 \to N$ in $\hat{\mathscr{F}}$. By (1) of Theorem 20.1, $F(n) \cdot = \cdot S^k 0$ for some $k \in \mathbb{N}$. Therefore, $n \cdot = \cdot GF(n) \cdot = \cdot S^k 0$ in $\mathscr{F}$. From the way 0 and S are defined in $\mathscr{F}$ (see Section 12, just before Proposition 12.4), it follows that $|n| \equiv \{\langle *, S^k 0 \rangle\}$. Take $a \equiv S^k 0$ in $\mathfrak{L}_0$, then $\vdash \langle *, a \rangle \in |n|$, hence $\vdash \psi(a)$ in $\mathfrak{L}_0$.

The proof by induction is now complete.

The reader may have noticed that the discussion of the case $A = PB$ above is reminiscent of the proof of Proposition 14.3. It is therefore not surprising that, for the special type theory $\mathfrak{L}_0$, we have the following strengthened form of Proposition 14.3.

Corollary 20.4. The canonical translation $\mathfrak{L}_0 \to LT(\mathfrak{L}_0)$ induces a biunique correspondence between terms of type A in $\mathfrak{L}_0$ modulo provable equality and terms of type $\mathbf{A} \equiv \{x \in A \mid \top\}$ in $LT(\mathfrak{L}_0)$, that is, arrows $\mathbf{1} \to \mathbf{A}$ in the free topos $T(\mathfrak{L}_0)$.

Proof. For the present purpose, it is instructive to distinguish notationally between A and $\mathbf{A}$. A term of type $\mathbf{A}$ in $LT(\mathfrak{L}_0)$ is an arrow $f: \mathbf{1} \to \mathbf{A}$ in $T(\mathfrak{L}_0)$, hence its graph $|f|$ is a term of type $P(1 \times A)$ in $\mathfrak{L}_0$ such that $\vdash \exists!_{x \in A} \langle *, x \rangle \in |f|$. By Lemma 20.3, there is a closed term a of type A in $\mathfrak{L}_0$ such that $\vdash \langle *, a \rangle \in |f|$, hence $|f| \cdot = \cdot \{\langle *, a \rangle\}$. Therefore, $f \cdot = \cdot \mathbf{a}$ in $T(\mathfrak{L}_0)$, where $\mathbf{a}$ is the image of a under the canonical translation.

Proposition 20.5. The existence property holds in $\mathfrak{L}_0$: if $\exists_{x \in A} \varphi(x)$ then $\vdash \varphi(a)$ for some closed term a of type A.

Proof. Suppose $\vdash \exists_{x \in A} \varphi(x)$. Then this holds in any topos, in particular, $\hat{\mathscr{F}} \models \exists_{x \in A} \varphi(x)$. By (8) of Theorem 20.1, there is an arrow $a: 1 \to A$ in $\hat{\mathscr{F}}$ such that $\hat{\mathscr{F}} \models \varphi(a)$. Now apply the logical functor $G: \hat{\mathscr{F}} \to \mathscr{F}$, then we obtain $\mathscr{F} \models \varphi(\breve{a})$, where $\breve{a} \equiv G(a)$ is an arrow $1 \to A \equiv G(A)$ in $\mathscr{F}$. Therefore, $\vdash \varphi(\breve{a})$ in $L(\mathscr{F}) \equiv LT(\mathfrak{L}_0)$. By Corollary 20.4, $\breve{a} \equiv \mathbf{a}'$ is the image of a term a' of type A in $\mathfrak{L}_0$ under the canonical translation $\mathfrak{L}_0 \to LT(\mathfrak{L}_0)$. It follows that $\vdash \varphi(a')$ in $\mathfrak{L}_0$.

Remark 20.6. With the exception of the unique existence property, the results in this section can also be obtained by algebraic methods, which is in fact the way Freyd proved them.

Let us recall from Proposition 6.7 the following characterization of a *projective* object C in a topos $\mathscr{T}$: if $\mathscr{T} \models \forall_{z \in C} \exists_{x \in A} \varphi(z, x)$ then $\mathscr{T} \models \forall_{z \in C} \varphi(z, fz)$ for some arrow $f: C \to A$ in $\mathscr{T}$.

Condition (8) of Theorem 20.1 can now be read as saying that the terminal object 1 is projective in $\hat{\mathscr{T}}$. Now the free topos $\mathscr{F}$ is a retract of $\hat{\mathscr{F}}$: we have $F: \mathscr{F} \to \hat{\mathscr{F}}$ and $G: \hat{\mathscr{F}} \to \mathscr{F}$ such that $GF = \mathrm{id}$. It then easily follows that the terminal object 1 of $\mathscr{F}$ is also projective. (One must use the fact that the logical functor F preserves epimorphisms.) Therefore, for any type A in $\mathfrak{L}_0$, if $\mathscr{F} \models \exists_{x \in A} \varphi(x)$ then $\mathscr{F} \models \varphi(a)$ for some arrow $a: 1 \to A$ in $\mathscr{F}$.

We are close to a proof of Proposition 20.5: we have established the existence property for $L(\mathscr{F})$ but not yet for $\mathfrak{L}_0$. To obtain the result for $\mathfrak{L}_0$ one must still have recourse to Lemma 20.3 or Corollary 20.4.

It is even easier to give an algebraic proof of Proposition 20.2. Recall from Section 8 that an object C of a topos is *indecomposable* provided, whenever $k: D \to C$ and $l: E \to C$ are two arrows which are *jointly epimorphic*, in the sense that $[k,l]: D + E \to C$ is an epimorphism, then either k or l is an epimorphism. The analogue of Proposition 6.7, with projectivity replaced by indecomposability, is easily stated.

Proposition 20.7. The object C of the topos $\mathscr{T}$ is indecomposable if and only if, whenever $\mathscr{T} \vDash \forall_{z \in C}(\varphi(z) \vee \psi(z))$, then either $\mathscr{T} \vDash \forall_{z \in C}\varphi(z)$ or $\mathscr{T} \vDash \forall_{z \in C}\psi(z)$.

The proof is left to the reader. (See Proposition 8.6.)

Condition (5) of Theorem 20.1 can now be interpreted as saying that the terminal object 1 of $\hat{\mathscr{T}}$ is indecomposable. Since $\mathscr{F}$ is a retract of $\hat{\mathscr{F}}$, it easily follows that the terminal object 1 of $\mathscr{F}$ is indecomposable. This establishes the disjunction property for $L(\mathscr{F})$, from which one easily deduces it for $\mathfrak{L}_0$.

Exercises

1. Prove Proposition 20.7.
2. Carry out the details of the algebraic proof that 1 is an indecomposable projective in the free topos $\mathscr{F}$.
3. Show that C is indecomposable (projective, injective) in $\mathscr{T}$ if and only if 1 is indecomposable (projective, injective) in $\mathscr{T}(z)$, where z is an indeterminate of type C, that is, in $\mathscr{T}/C$ (see Exercise 2 of Section 16).
4. Prove that if C is indecomposable, then $\mathrm{Hom}(C,-)$ preserves binary coproducts. (See Remark 17.8 in the supplement for the case $C = 1$.) Show the converse if C is projective.

21 Further intuitionistic principles

While the disjunction and existence properties appear to be basic to the philosophy of intuitionism, some other principles are less intuitive. Here are some that will be established as metatheorems about pure type theory $\mathfrak{L}_0$ in this section.

Disjunction property with parameters: if $\vdash \forall_{x \in PC}(\varphi(x) \vee \psi(x))$ then either $\vdash \forall_{x \in PC}\varphi(x)$ or $\vdash \forall_{x \in PC}\psi(x)$, and similarly with PC replaced by Ω.

Troelstra's uniformity rule: if $\vdash \forall_{x \in PC}\exists_{y \in N}\varphi(x, y)$ then $\vdash \exists_{y \in N}\forall_{x \in PC}\varphi(x, y)$.

Independence of premisses: if $\vdash \neg q \Rightarrow \exists_{x \in A}\varphi(x)$ then $\vdash \exists_{x \in A}(\neg q \Rightarrow \varphi(x))$.

Markov's rule: if $\vdash \forall_{x\in A}(\varphi(x) \vee \neg\varphi(x))$ and $\vdash \neg\forall_{x\in A}\neg\varphi(x)$ then $\vdash \exists_{x\in A}\varphi(x)$.

The proofs of the first two of these rules depend on an addition to Theorem 20.1, which will be displayed as a lemma.

Lemma 21.1. (*a*) If $x: 1 \to \Omega$ is an indeterminate over $\mathcal{T}$ then there is an arrow $\hat{x}: 1 \to \Omega$ in $\mathcal{T}(x)^\wedge$ such that $G: \mathcal{T}(x)^\wedge \to \mathcal{T}(x)$ sends $\hat{x}$ to x.

(*b*) If $x: 1 \to PC$ is an indeterminate over $\mathcal{T}$ and $\hat{C}$ is an object of $\mathcal{T}(x)^\wedge$ such that $G(\hat{C}) = C$, then there is an arrow $\hat{x}: 1 \to P\hat{C}$ in $\mathcal{T}(x)^\wedge$ such that $G(\hat{x}) \cdot = \cdot x$.

The proof of this lemma, like the proof of Theorem 20.1, will be postponed to the next section.

Proposition 21.2. The disjunction and existence properties hold with a parameter x of type $A = PC$ or $A = \Omega$, that is,

(*a*) if $\vdash \forall_{x\in A}(\varphi(x) \vee \psi(x))$ then $\vdash \forall_{x\in A}\varphi(x)$ or $\vdash \forall_{x\in A}\psi(x)$;

(*b*) if $\vdash \forall_{x\in A}\exists_{y\in B}\varphi(x, y)$ then $\vdash \forall_{x\in A}\varphi(x, \beta(x))$, where $\beta(x)$ is a term of type B.

Proof. (*a*) Suppose $\vdash \forall_{x\in A}(\varphi(x) \vee \psi(x))$, where $A = PC$ or $A = \Omega$, that is, $\vdash_x \varphi(x) \vee \psi(x)$, where $\vdash_x$ is provability in $\mathfrak{L}_0(x)$. Now recall (see Example 16.2) that we may define $\mathcal{F}(x) \equiv T(\mathfrak{L}_0(x))$, since $\mathcal{F} \equiv T(\mathfrak{L}_0)$, and that $\mathcal{F}(x)$ has the usual universal property in $\mathbf{Top}_0$. In particular, the unique arrow $\mathcal{F} \to \mathcal{F}(x)^\wedge$ in $\mathbf{Top}_0$ may be extended uniquely to an arrow $F': \mathcal{F}(x) \to \mathcal{F}(x)^\wedge$ such that $F'(x) \cdot = \cdot \hat{x}$, where $\hat{x}$ is as in Lemma 21.1. Now $\mathcal{F}(x) \models \varphi(x) \vee \psi(x)$, hence also $\mathcal{F}(x)^\wedge \models \varphi(\hat{x}) \vee \psi(\hat{x})$, as is seen by applying the logical functor F'. Therefore, by (5) of Theorem 20.1, $\mathcal{F}(x)^\wedge \models \varphi(\hat{x})$ or $\mathcal{F}(x)^\wedge \models \psi(\hat{x})$. Applying the logical functor $G: \mathcal{F}(x)^\wedge \to \mathcal{F}(x)$, we then obtain $\mathcal{F}(x) \models \varphi(x)$ or $\mathcal{F}(x) \models \psi(x)$, hence $\vdash \varphi(x)$ or $\vdash \psi(x)$. (See the last part of the proof of Proposition 20.2.)

(*b*) First let us prove the unique existence property with a parameter of type A. Suppose $\vdash_x \exists!_{y\in B}\psi(x, y)$. We then prove that there is a term $\beta(x)$ of type B such that $\vdash_x \psi(x, \beta(x))$, by induction on the construction of B. See the proof of Lemma 20.3, but replace A by B. The argument is exactly the same as that given there; in particular, when $B = N$, we find that $\beta(x) \equiv S^k 0$ in $\mathfrak{L}_0(x)$ for some $k \in \mathbb{N}$.

Now let us turn to the general existence property with parameters x of type $A = PC$ or Ω. Suppose $\vdash_x \exists_{y\in B}\varphi(x, y)$, that is, $\mathcal{F}(x) \models \exists_{y\in B}\varphi(x, y)$. Applying the functor $F': \mathcal{F}(x) \to \mathcal{F}(x)^\wedge$ as in (*a*), we obtain $\mathcal{F}(x)^\wedge \models \exists_{y\in B}\varphi(\hat{x}, y)$. By (8) of Theorem 20.1, there is an arrow $b: 1 \to B$ in

$\mathscr{F}(x)^\wedge$ such that $\mathscr{F}(x)^\wedge \vDash \varphi(\hat{x}, b)$. Now apply the functor $G: \mathscr{F}(x)^\wedge \to \mathscr{F}(x)$ and put $G(b) \equiv \beta'(x)$, then $\mathscr{F}(x) \vDash \varphi(x, \beta'(x))$.

Unfortunately, $\beta'(x)$ here is a term in $L(\mathscr{F}(x)) \equiv LT(\mathfrak{L}_0(x))$ and not in $\mathfrak{L}_0(x)$. Nonetheless, by an easy extension of Corollary 20.4, the canonical translation $\mathfrak{L}_0(x) \to LT(\mathfrak{L}_0(x))$ induces a biunique correspondence $\beta(x) \mapsto \beta'(x)$ between terms of type B in $\mathfrak{L}_0(x)$ and terms of type $\mathbf{B} \equiv \{x \in B \mid \top\}$ in $LT(\mathfrak{L}_0(x))$. It now easily follows that $\vdash_{\bar{x}} \varphi(x, \beta(x))$.

Corollary 21.3. Troelstra's uniformity rule holds for $\mathfrak{L}_0$: for $A = PC$ or Ω, if $\vdash \forall_{x \in A} \exists_{y \in N} \varphi(x, y)$ then $\vdash \exists_{y \in N} \forall_{x \in A} \varphi(x, y)$.

Proof. Suppose $\vdash \forall_{x \in A} \exists_{y \in N} \varphi(x, y)$. For $B = N$, the above proof yields $\beta'(x) \equiv S^k 0$ in $L(\mathscr{F}(x))$, hence also $\beta(x) \equiv S^k 0$ in $\mathfrak{L}_0(x)$. Thus $\vdash \forall_{x \in A} \varphi(x, n)$ with $n \equiv S^k 0$, and so $\vdash \exists_{y \in N} \forall_{x \in A} \varphi(x, y)$.

Corollary 21.4. In the free topos $\mathscr{F}$, every arrow $A \to N$, with $A = PC$ or Ω, factors through the terminal object 1, hence has the form $A \to 1 \xrightarrow{S^k 0} N$.

Proof. Let $f: A \to N$ be an arrow in $\mathscr{F}$. Now f is determined by its graph $|f| \equiv \{\langle x, y \rangle \in A \times N \mid \varphi(x, y)\}$ such that $\vdash \forall_{x \in A} \exists!_{y \in N} \varphi(x, y)$. As we saw above (in the proof of the unique existence property with parameter x), it follows that $\vdash \forall_{x \in A} \varphi(x, n)$ for some $n \equiv S^k 0$. It is then easily verified that

$f \cdot = \cdot n \bigcirc_A : A \to 1 \xrightarrow{n} N$.

Note that the disjunction property fails for a parameter x of type N, as is seen by taking

$$\varphi(x) \equiv x \text{ is even}, \quad \psi(x) \equiv x \text{ is odd}.$$

Also the unique existence property fails for a parameter of type N. For example, one can prove that

$$\vdash \forall_{x \in N} \exists!_{y \in N} ((x = 0 \wedge y = 0) \vee (x \neq 0 \wedge Sy = x)).$$

Yet there does not exist in $\mathfrak{L}_0$ a name for the predecessor function, that is, a term $\beta(x)$ of type N such that

$$\vdash \beta(0) = 0 \wedge \forall_{x \in N} (\beta(Sx) = x).$$

Of course, the unique existence property holds in $L(\mathscr{F}(x))$, as in the internal language of any topos (see Theorem 5.9).

The question remains: does the existence property hold in $L(\mathscr{F}(x))$ when x is of type N? This is equivalent to asking whether 1 is projective in $\mathscr{F}(x)$, that is, N is projective in $\mathscr{F}$. (See Exercise 3 of Section 20.) The answer to this question is 'yes', but a proof of this fact would take us beyond the scope of this book. It requires reflection principles and other techniques from proof theory, on which the categorical viewpoint does not yet appear to shed any light.

We recall from Example 16.3 that $\mathscr{F}/(p) \equiv T(\mathfrak{L}_0/(p))$ is initial in the category of all toposes $\mathscr{T}$ in $\mathbf{Top}_0$ such that $\mathscr{T} \models p$.

Lemma 21.5. In $\mathfrak{L}_0$, if $p \equiv \neg q$ then either $\vdash \neg p$ or $(\mathscr{F}(p))\hat{} \models p$, so that there is a unique arrow $F: \mathscr{F}/(p) \to (\mathscr{F}(p))\hat{}$ in $\mathbf{Top}_0$.

Proof. By (6) of Theorem 20.1, we have: $\hat{\mathscr{T}} \models \neg q$ iff (i) $\mathscr{T} \models \neg q$ and (ii) not $\hat{\mathscr{T}} \models q$. In particular,

$$(\mathscr{F}/(\neg q))\hat{} \models \neg q \quad \text{iff not}(\mathscr{F}/(\neg q))\hat{} \models q.$$

Suppose at worst that $(\mathscr{F}/(\neg q))\hat{} \models q$. Applying the logical functor $G: (\mathscr{F}/(\neg q))\hat{} \to \mathscr{F}/(\neg q)$, we obtain that $\mathscr{F}/(\neg q) \models q$, hence that $\mathscr{F}/(\neg q) \models \bot$, therefore $\mathscr{F} \models \neg\neg q$, and so $\vdash \neg\neg q$.

Proposition 21.6. If $p \equiv \neg q$, then the disjunction and existence properties hold in $\mathfrak{L}_0/(p)$:

(*a*) if $p \vdash r \vee s$ then $p \vdash r$ or $p \vdash s$;

(*b*) if $p \vdash \exists_{x \in A} \varphi(x)$ then $p \vdash \varphi(a)$ for some term a in $\mathfrak{L}_0$.

Here (*b*) is equivalent to *independence of premisses* for p:

(*b'*) if $\vdash p \Rightarrow \exists_{x \in A} \varphi(x)$ then $\vdash \exists_{x \in A}(p \Rightarrow \varphi(x))$.

We shall leave the proof of this as an exercise, as it is quite similar to that of Proposition 21.2.

For the proof of Markov's principle, we need the following generalization of Proposition 21.2(*a*). A similar generalization holds for Proposition 21.2(*b*).

Lemma 21.7. If Q is a type of $\mathfrak{L}_0$ such that $\mathbf{Q} \equiv \{x \in Q \mid \top\}$ is injective in $\mathscr{F}$, then from $\vdash \forall_{x \in Q}(\varphi(x) \vee \psi(x))$ in $\mathfrak{L}_0$ one may infer that $\vdash \forall_{x \in Q} \varphi(x)$ or $\vdash \forall_{x \in Q} \psi(x)$.

Proof. Since $\mathbf{Q}$ is injective in $\mathscr{F}$, the singleton arrow $\iota_{\mathbf{Q}}: \mathbf{Q} \to P\mathbf{Q}$ splits, so we can find an arrow e: $P\mathbf{Q} \to \mathbf{Q}$ such that $e\iota_{\mathbf{Q}} \cdot = \cdot 1_{\mathbf{Q}}$. Since $\mathscr{F} \equiv T(\mathfrak{L}_0)$, we have $T(\mathfrak{L}_0(y)) \equiv \mathscr{F}(y)$, with indeterminate arrow $\mathbf{y}: \mathbf{1} \to P\mathbf{Q}$ (see Example 16.2). Now let us assume that $\vdash \forall_{x \in Q}(\varphi(x) \vee \psi(x))$, then $\mathscr{F}(y) \models \forall_{x \in Q}(\varphi(x) \vee \psi(x))$, hence $\mathscr{F}(y) \models \varphi(e\mathbf{y}) \vee \psi(e\mathbf{y})$. By Proposition 21.2(*a*), $\mathscr{F}(y) \models \varphi(e\mathbf{y})$ or $\mathscr{F}(y) \models \psi(e\mathbf{y})$. Now apply the unique logical functor $\mathscr{F}(y) \to \mathscr{F}(x)$ which extends the identity functor on $\mathscr{F}$ and which sends $\mathbf{y}$ onto $\iota_{\mathbf{Q}}\mathbf{x}$. Then we obtain $\mathscr{F}(x) \models \varphi(x)$ or $\mathscr{F}(x) \models \psi(x)$. Finally, since the canonical translation $\mathfrak{L}_0(x) \to LT(\mathfrak{L}_0(x))$ is a conservative extension, we have $\vdash_{\bar{x}} \varphi(x)$ or $\vdash_{\bar{x}} \psi(x)$, hence $\vdash \forall_{x\in Q}\varphi(x)$ or $\vdash \forall_{x\in Q}\psi(x)$.

Proposition 21.8. Markov's rule holds in $\mathfrak{L}_0$: if $\vdash \forall_{x\in A}(\varphi(x) \vee \neg\varphi(x))$, then from $\vdash \neg\forall_{x\in A}\neg\varphi(x)$ one may infer that $\vdash \exists_{x\in A}\varphi(x)$.

Proof. To warm up, we first give the usual proof when $A = N$. Assume that $\varphi(x)$ is *decidable*, that is, $\vdash \forall_{x\in N}(\varphi(x) \vee \neg\varphi(x))$, and also that $\vdash \neg\forall_{x\in N}\neg\varphi(x)$. In view of the second assumption, **Sets** $\vDash \neg\forall_{x\in N}\neg\varphi(x)$. Since **Sets** is Boolean, we may infer that **Sets** $\vDash \exists_{x\in N}\varphi(x)$. Therefore, **Sets** $\vDash \varphi(n)$ for some $n \equiv S^k0$, and so not $\vdash \neg\varphi(n)$. Since $\varphi(x)$ is decidable, $\vdash \varphi(n)$ and therefore $\vdash \exists_{x\in N}\varphi(x)$.

Next, let us prove the result for an arbitrary type A in $\mathfrak{L}_0$. In $\mathscr{F}$, A give rise to an object $\mathbf{A} \equiv \{x\in A|\top\}$, and it follows by induction on the construction of A that

$$\mathbf{A} \cong \mathbf{N}^k \times P\mathbf{A}_1 \times \cdots \times P\mathbf{A}_n$$

for some k, $n\in\mathbb{N}$. (Recall that $\mathbf{\Omega} \cong P\mathbf{1}$.) Now $\mathbf{N}^k \cong \mathbf{1}$ (if $k = 0$) or $\mathbf{N}^k \cong \mathbf{N}$ (if $k > 0$). Moreover, $P\mathbf{B}$ in injective in any topos (see Proposition 6.4) and any product of injectives is injective. Therefore, $\mathbf{A} \cong \mathbf{Q}$ or $\mathbf{N} \times \mathbf{Q}$, where Q is a type of $\mathfrak{L}_0$ such that $\mathbf{Q}$ is injective. It then clearly suffices to prove Markov's rule in the two special cases $A = Q$ and $A = N \times Q$, where $\mathbf{Q}$ is injective.

Case 1: $A = Q$. Suppose $\vdash \forall_{x\in Q}(\varphi(x) \vee \neg\varphi(x))$, then $\vdash \forall_{x\in Q}\varphi(x)$ or $\vdash \forall_{x\in Q}\neg\varphi(x)$, by Lemma 21.7. Now suppose further that $\vdash \neg\forall_{x\in Q}\neg\varphi(x)$, then the second alternative is ruled out, and so $\vdash \forall_{x\in Q}\varphi(x)$. Let t be a closed term of type Q, then $\vdash \varphi(t)$, hence $\vdash \exists_{x\in Q}(\varphi(x)$.

Case 2: $A = N \times Q$. Suppose that $\vdash \forall_{x\in N}\forall_{y\in Q}(\varphi(x, y) \vee \neg\varphi(x, y))$ and that $\vdash \neg\forall_{x\in N}\forall_{y\in Q}\neg\varphi(x, y)$. In view of the second assumption, **Sets** $\vDash \exists_{x\in N}\exists_{y\in Q}\varphi(x, y)$, and therefore there is a natural number S^k0 in **Sets** such that **Sets** $\vDash \exists_{y\in Q}\varphi(S^k0, y)$, hence *not* $\vdash \forall_{y\in Q}\neg\varphi(S^k0, y)$. However, by the first assumption $\vdash \forall_{y\in Q}(\varphi(S^k0, y) \vee \neg\varphi(S^k0, y))$, and so, by Lemma 21.7, $\vdash \forall_{y\in Q}(S^k0, y)$. Again, let t be any closed term of type Q, then $\vdash \varphi(S^k0, t)$, hence $\vdash \exists_{x\in N}\exists_{y\in Q}\varphi(x, y)$, as was to be shown.

Remark 21.9. Along the lines of Remark 20.6, we shall discuss the algebraic interpretations of some of the results of this section.

Lemma 21.1 allows us to infer that, for x of type PC or Ω, $\mathscr{F}(x)$ is a retract of its Freyd cover.

Underlying Proposition 21.2 is the algebraic statement that $P\mathbf{C}$ and $\mathbf{\Omega}$ are indecomposable projectives in $\mathscr{F}$. This establishes the disjunction and existence properties for $L(\mathscr{F}(x))$. The corresponding properties for $\mathfrak{L}_0(x)$ follow easily, but the unique existence property requires an extra argument as above.

Lemma 21.5 allows one to infer that, for $p \equiv \neg q$, either $\vdash \neg p$ or else $\mathscr{F}/(p)$ is a retract of its Freyd cover.

Underlying Proposition 21.6 is the algebraic statement that, for any arrow $p: 1 \to \Omega$ in $\mathscr{F}$, the corresponding subobject of 1, Ker p, is an indecomposable projective in $\mathscr{F}$. This is equivalent to saying that 1 is an indecomposable projective in $\mathscr{F}/(p)$. Thus we have the disjunction and existence properties for $L(\mathscr{F}/(p))$, from which the corresponding properties for $\mathfrak{L}_0/(p)$ then follow easily, provided some attention is paid to the unique existence property.

Lemma 21.7 really depends on the following algebraic fact: if **C** is an injective object in $\mathscr{F}$ corresponding to a type C of $\mathfrak{L}_0$, then **C** is indecomposable, in view of the observations that the singleton arrow $\mathbf{C} \to P\mathbf{C}$ is a monomorphism (Proposition 5.8) and that $P\mathbf{C}$ is indecomposable (see Remark 20.6). For the same reason, we can also assert that **C** is projective.

Exercises

1. Give a logician's proof of Proposition 21.6 analogous to that of Proposition 21.2.

2. Give an algebraic proof that $A = PC$ and $A = \Omega$ are indecomposable projectives in the free topos $\mathscr{F}$ or, equivalently, that 1 is an indecomposable projective in $\mathscr{F}(x)$ when x is of type $A = PC$ or Ω.

3. Do the same if $A = \mathrm{Ker}(\neg q)$, assuming that not $\vdash \neg\neg q$. Note that $\mathscr{F}(x)$, with x of type A, is then the same as $\mathscr{F}/(\neg q)$.

4. Show that any injective object in the free topos $\mathscr{F}$ corresponding to a type of $\mathfrak{L}_0$ is an indecomposable projective.

5. Show that $(a) \Rightarrow (b) \Rightarrow (c)$ for any closed formula p of $\mathfrak{L}_0$:
   (a) p is *hereditary*, that is, for any nondegenerate topos $\mathscr{T}$, if $\mathscr{T} \models p$ then $\hat{\mathscr{T}} \models p$;
   (b) p is refutable or p is *Freydian*, that is, $(\mathscr{F}/(p))^\wedge \models p$;
   (c) p *satisfies independence of premisses*, that is, if $\vdash p \Rightarrow \exists_{x\in A}\varphi(x)$ then $\vdash \exists_{x\in A}(p \Rightarrow \varphi(x))$.

6. Show that $\perp$ is hereditary, that $p \wedge q$ is hereditary if p and q are, and that $p \Rightarrow q$ is hereditary if q is.

7. Show that $p \Rightarrow q$ is Freydian if not $(p \Rightarrow q) \vdash p$.

8. If $\beta \equiv \forall_{t\in\Omega}(t \vee \neg t)$, show that **Sets**$\hat{}$ satisfies either β nor $\neg\beta$, but that it does satisfy $\neg\neg\beta$.

9. Assuming that $\neg\neg\beta$ is not a theorem of pure intuitionistic type theory, show that $\neg\neg\beta \Rightarrow \beta$ is Freydian but not hereditary. The assumption follows from the existence of a non-Boolean topos in which $\mathrm{Hom}(1, \Omega)$ is Boolean, e.g. **Sets**$^{\mathscr{M}}$ when $\mathscr{M}$ is any monoid which is not a group. (See Section 9, exercises 3 and 6, and also Exercise 9.)

22 The Freyd cover of a topos

In this section we shall construct the Freyd cover of a topos and finally prove Theorem 20.1 and Lemma 21.1, on which Sections 20 and 21 depend.

Definition 22.1. If $\mathscr{T}$ is a category with a terminal object 1, its *Freyd cover* consists of a category $\hat{\mathscr{T}}$ with terminal object and a functor $G: \hat{\mathscr{T}} \to \mathscr{T}$ preserving the terminal object constructed as follows.

Write $\Gamma \equiv \Gamma_{\mathscr{T}} \equiv \mathrm{Hom}_{\mathscr{T}}(1, -): \mathscr{T} \to$ **Sets**. Then the category $\hat{\mathscr{T}} \equiv$ (**Sets**, Γ) has as objects triples (X, ξ, U), where X is a set, U is an object of $\mathscr{T}$ and ξ: $X \to \Gamma(U)$ is a mapping, and as arrows commutative squares

$$\begin{array}{ccc} X & \xrightarrow{\ \varphi\ } & Y \\ {\scriptstyle \xi}\downarrow & & \downarrow{\scriptstyle \eta} \\ \Gamma(U) & \xrightarrow[\Gamma(t)]{} & \Gamma(V) \end{array}$$

that is, pairs (φ, t) such that $\varphi: X \to Y$ in **Sets** and $t: U \to V$ in $\mathscr{T}$ subject to the equation $\eta\varphi = \Gamma(t)\xi$. Composition is defined componentwise: $(\psi, s)(\varphi, t) \equiv (\psi\varphi, st)$.

It is clear that $\hat{\mathscr{T}}$ is a category with terminal object $\{*\} \to \Gamma(1)$ and that it comes equipped with a functor $G: \hat{\mathscr{T}} \to \mathscr{T}$ preserving the terminal object, where

$$G(X, \xi, U) \equiv U. \quad G(\varphi, t) \equiv t.$$

It is also clear that there is a functor $\Sigma: \hat{\mathscr{T}} \to$ **Sets**, where

$$\Sigma(X, \xi, U) \equiv X, \quad \Sigma(\varphi, t) \equiv \varphi.$$

Thus each object A of $\hat{\mathscr{T}}$ has the form $A = (\Sigma(A), \lambda_A, \breve{A})$ with

$$\lambda_A : \Sigma(A) \to \Gamma(\breve{A}),$$

where we have written $\breve{A} \equiv G(A)$ for the underlying object in $\mathscr{T}$. Moreover, each arrow $f: A \to B$ in $\hat{\mathscr{T}}$ must be such that the following square commutes:

$$\begin{array}{ccc}
\Sigma(A) & \xrightarrow{\Sigma(f)} & \Sigma(B) \\
{\scriptstyle \lambda_A}\big\downarrow & & \big\downarrow{\scriptstyle \lambda_B} \\
\Gamma(\breve{A}) & \xrightarrow[\Gamma(\breve{f})]{} & \Gamma(\breve{B})
\end{array}$$

where $\breve{f} \equiv G(f)$ is the underlying arrow in $\mathscr{T}$. This means that there are given arrows $\Sigma(f)$ in **Sets** and $\breve{f} \equiv G(f)$ in $\mathscr{T}$ satisfying

$$\lambda_B \Sigma(f) = \Gamma(\breve{f})\lambda_A,$$

in other words, for all $\alpha \in \Sigma(A)$,

(#) $\qquad \lambda_B(\Sigma(f)(\alpha)) = \breve{f}\lambda_A(\alpha).$

We mention, in particular, that

$$\Sigma(1) \equiv \{*\}, \quad \breve{1} \equiv 1, \quad \lambda_1(*) \equiv 1_1,$$

so that $\bigcirc_A$: $A \to 1$ is given by

$$\Sigma(\bigcirc_A): \Sigma(A) \to \{*\}, \quad \bigcirc_{\breve{A}}: \breve{A} \to 1.$$

Let us now look at an arrow $a: 1 \to A$ in $\hat{\mathscr{T}}$. In view of (#) above and the definitions of $\Sigma(1)$ and λ_1, a is determined by arrows $\Sigma(a)$ in **Sets** and $\breve{a} \equiv G(a)$ in $\mathscr{T}$ satisfying

$$\lambda_A(\Sigma(a)(*)) = \breve{a}\lambda_1(*) = \breve{a}.$$

We have thus proved the following:

***Lemma* 22.2.** An arrow $a: 1 \to A$ in the Freyd cover $\hat{\mathscr{T}}$ of a category $\mathscr{T}$ with terminal object is completely determined by its image under Σ.

This result is easily strengthened to obtain a natural isomorphism $\mathrm{Hom}_{\hat{\mathscr{T}}}(1, -) \cong \Sigma$. (See Exercise 1 below.)

We are now in a position to discuss the Freyd cover of a topos.

***Proposition* 22.3.** If $\mathscr{T}$ is a topos, then so is $\hat{\mathscr{T}}$ and G is a strict logical functor. Moreover, if $\mathscr{T} = T(\mathfrak{L})$ is the topos generated by a type theory, then $\hat{\mathscr{T}}$ has canonical subobjects and G preserves them.

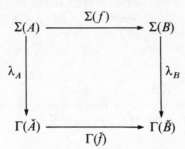

Proof. To assure that $\hat{\mathscr{T}}$ is a topos, we must first of all specify objects N and Ω and operations on objects yielding new objects $A \times B$ and PA. Since G is supposed to be a strict logical functor, these take the following form:

$$\lambda_N: \Sigma(N) \to \Gamma(N),$$
$$\lambda_\Omega: \Sigma(\Omega) \to \Gamma(\Omega),$$
$$\lambda_{A\times B}: \Sigma(A \times B) \to \Gamma(\check{A} \times \check{B}),$$
$$\lambda_{PA}: \Sigma(PA) \to \Gamma(P\check{A}),$$

where we now specify:

$$\Sigma(N) \equiv \mathbb{N}, \quad \lambda_N(k) \equiv S^k 0 \quad (k \in \mathbb{N});$$
$$\Sigma(\Omega) \equiv \Gamma(\Omega) \dot{\cup} \{\top\}, \quad \lambda_\Omega(p, i) \equiv p \quad (p \in \Gamma(\Omega), \quad i = 0, 1);$$
$$\Sigma(A \times B) \equiv \Sigma(A) \times \Sigma(B),$$
$$\lambda_{A\times B}(\alpha, \beta) \equiv \langle \lambda_A(\alpha), \lambda_B(\beta) \rangle \quad (\alpha \in \Sigma(A), \beta \in \Sigma(B));$$
$$\Sigma(PA) \equiv \mathrm{Hom}_{\mathscr{T}}(A, \Omega), \quad \lambda_{PA}(f) \equiv \ulcorner \check{f} \urcorner \quad (f \in \mathrm{Hom}_{\mathscr{T}}(A, \Omega)).$$

The reader will have noticed that we take the disjoint union $\Gamma(\Omega) \dot{\cup} \{\top\}$ to mean $\Gamma(\Omega) \times \{0\} \cup \{(\top, 1)\}$.

To assume that $\hat{\mathscr{T}}$ is a topos, we must also specify the arrows $0: 1 \to N$, $S: N \to N, \top: 1 \to \Omega$, $\pi_{A,B}: A \times B \to A$, $\pi'_{A,B}: A \times B \to B$ and $\varepsilon_A: PA \times A \to A$. Since G is supposed to be a strict logical functor, each of these arrows will be determined by its image under Σ, and so we specify:

$$\Sigma(0)(*) \equiv 0,$$
$$\Sigma(S)(k) \equiv k + 1 \quad (k \in \mathbb{N}),$$
$$\Sigma(\top)(*) \equiv (\top, 1),$$
$$\Sigma(\pi_{A,B})(\alpha, \beta) \equiv \alpha \quad (\alpha \in \Sigma(A), \quad \beta \in \Sigma(B)),$$
$$\Sigma(\pi'_{A,B})(\alpha, \beta) \equiv \beta \quad (\alpha \in \Sigma(A), \quad \beta \in \Sigma(B)),$$
$$\Sigma(\varepsilon_A)(f, \alpha) \equiv \Sigma(f)(\alpha) \quad (f \in \mathrm{Hom}_{\mathscr{T}}(A, \Omega), \quad \alpha \in \Sigma(A)).$$

The reader may have to do a little work to check that condition (#) is satisfied in all cases, for example, that

$$\Gamma(\varepsilon_{\check{A}})\lambda_{PA\times A}(f, \alpha) = \lambda_\Omega(\Sigma(f)(\alpha)).$$

To assure that $\hat{\mathscr{T}}$ is a topos, we must finally specify operations which, from arrows $f: A \times B \to \Omega$, $h: A \to \Omega$ and monomorphisms $m: B \to A$, will produce new arrows and objects $f^*: A \to PB$, $\ker h: \mathrm{Ker}\, h \to A$ and $\mathrm{char}\, m: A \to \Omega$.

To construct f^* it suffices to specify $\Sigma(f^*): \Sigma(A) \to \Sigma(PB) \equiv \mathrm{Hom}_{\mathscr{T}}(B, \Omega)$, which we define by its action on any $\alpha \in \Sigma(A)$ as follows:

$$\Sigma(f^*)(\alpha) \equiv f_\alpha,$$

where

$$\Sigma(f_\alpha)(\beta) \equiv \Sigma(f)(\alpha, \beta) \quad (\beta \in \Sigma(B))$$

and

$$\check{f}_\alpha \equiv \check{f}\langle \lambda_A(\alpha)\bigcirc_{\check{B}}, 1_{\check{B}}\rangle.$$

Of course, $G(f^*) \equiv (G(f))^*$.

The reader is invited to check that the equations of a cartesian category with power set structure are now satisfied.

Next, we wish to construct the object Ker h and the arrow ker h: Ker $h \to A$ for a given arrow $h: A \to \Omega$ in $\hat{\mathscr{T}}$.

We let

$$\Sigma(\text{Ker } h) \equiv \{\alpha \in \Sigma(A) \mid \lambda_A(\alpha) \text{ factors through ker } \check{h}\},$$

that is, an element α of $\Sigma(A)$ belongs to $\Sigma(\text{Ker } h)$ if and only if there is an arrow g_α: $1 \to \text{Ker } \check{h}$ in $\mathscr{T}$ such that $\lambda_A(\alpha) = (\ker \check{h})g_\alpha$. Since ker $\check{h}$ is a monomorphism, this arrow g_α is uniquely determined by α, so we define $\lambda_{\text{Ker } h}$ by

$$\lambda_{\text{Ker } h}(\alpha) \equiv g_\alpha.$$

We also define $\Sigma(\ker h)$ to be the inclusion $\Sigma(\text{Ker } h) \subseteq \Sigma(A)$.

Finally, we wish to construct the arrow char m: $A \to \Omega$ for a given monomorphism $m: B \to A$. It is easily checked that m is a monomorphism if and only if both $\Sigma(m)$ and $\check{m} \equiv G(m)$ are monomorphisms in **Sets** and $\mathscr{T}$ respectively. To construct char m, it suffices to define $\Sigma(\text{char } m)$. For any $\alpha \in \Sigma(A)$, we put

$$\Sigma(\text{char } m)(\alpha) \equiv (p_\alpha, 1),$$

where

$$p_\alpha \equiv (\text{char } \check{m})\lambda_A(\alpha).$$

It is now a routine exercise to show that Ω in $\hat{\mathscr{T}}$ is a subobject classifier, that is,

$$\text{char}(\ker h) \cdot = \cdot h, \quad \ker(\text{char } m) \cong m.$$

Thus $\hat{\mathscr{T}}$ is a topos and, if we look at the above definitions carefully, we see that G is a strict logical functor.

To complete the proof of Proposition 22.3, we need to consider canonical subobjects in $\hat{\mathscr{T}}$. We shall say that a monomorphism $m: B \to A$ in $\hat{\mathscr{T}}$ is *canonical* if both $\Sigma(m)$ and $\check{m} \equiv G(m)$ are canonical. For $\Sigma(m)$ this means that $\Sigma(B) \subseteq \Sigma(A)$ and for $\check{m}$ this presupposes that $\mathscr{T}$ has canonical subobjects. It is easy to verify conditions C1 to C3 of Definition 15.1. For example, condition C3 would require us to check that if $\Sigma(B) \subseteq \Sigma(A)$ and $\Sigma(D) \subseteq \Sigma(C)$ then $\Sigma(B \times D) \subseteq \Sigma(A \times C)$. This is evident, since $\Sigma(B \times D) = \Sigma(B) \times \Sigma(D)$.

To verify C4 for $\hat{\mathscr{T}}$ is a little more difficult. Given that $m: B \to A$ is a canonical monomorphism in $\hat{\mathscr{T}}$, we wish to show that $P(m): PB \to PA$ is a canonical monomorphism in $\hat{\mathscr{T}}$, where P is regarded here as a covariant functor. Since m is canonical, we know that $\check{m}$ is canonical and that $\Sigma(m)$ is an inclusion $\Sigma(B) \subseteq \Sigma(A)$. Since $\check{m}$ is canonical, so is $P(\check{m})$, since $\mathscr{T}$ satisfies C4. Thus, it only remains to show that $\Sigma(PB) \subseteq \Sigma(PA)$. Unfortunately, this is not strictly correct, if we persist in defining $\Sigma(PA) \equiv \mathrm{Hom}_{\mathscr{T}}(A, \Omega)$. For our present purpose, we should redefine $\Sigma(PA)$ in a more suitable manner.

Let us assume that $\mathscr{T}$ satisfies the following additional property for canonical subobjects:

C5. if A and B are objects of $\mathscr{T}$, then there is at most one canonical monomorphism $B \to A$.

When this condition holds, an arrow $f: A \to \Omega$ is completely determined by the object $\mathrm{Ker}\, f$, never mind the arrow $\ker f$.

We have previously used the word 'subobject' loosely, sometimes for a monomorphism and sometimes for its source. If A is an object of $\mathscr{T}$, which is assumed to satisfy C5, we may write $\mathrm{Sub}_{\mathscr{T}} A$ for the set of canonical subobjects of A in the latter sense. Now suppose that $\mathscr{T}$ satisfies C5, then clearly so does $\hat{\mathscr{T}}$. Therefore, we may change the definition of $\Sigma(PA)$ to the following:

$$\Sigma(PA) \equiv \mathrm{Sub}_{\mathscr{T}}(A),$$

with the result that, when $B \in \mathrm{Sub}_{\mathscr{T}}(A)$, then $\Sigma(PB) \subseteq \Sigma(PA)$, as required.

How common is the condition C5? It is clearly satisfied by **Sets** and functor categories. What concerns us now are 'linguistic' toposes of the form $\mathscr{T} = T(\mathfrak{L})$. An object of $T(\mathfrak{L})$ is a closed term α of type PA and a canonical subobject of α is another closed term β of the same type PA such that $\vdash \beta \subseteq \alpha$. The canonical monomorphism $\beta \to \alpha$ is given by its graph $\{\langle y, x\rangle \in A \times A \mid y \in \beta \wedge y = x\}$, there being no other canonical monomorphism from β to α. Thus $T(\mathfrak{L})$ clearly satisfies C5.

The proof of Proposition 22.3 is now complete.

Remark. In Proposition 22.3, we could have replaced '$\mathscr{T} = T(\mathfrak{L})$' by '$\mathscr{T}$ has canonical subobjects and satisfies C5'. We do not know whether the result holds without this condition.

We are finally in a position to prove Theorem 20.1 and Lemma 21.5.

Proof of Theorem 20.1.

In view of Proposition 22.3, we need only verify the clauses (1) to (8), except for (2) and (4), which hold in any topos.

(1) An arrow $n\colon 1 \to N$ in $\hat{\mathcal{T}}$ is determined by $\Sigma(n)\colon \{*\} \to \mathbb{N}$, according to Lemma 22.2. Putting $\Sigma(n)(*) = k$, one easily calculates $\check{n}\cdot = \cdot\,\lambda_N(k)\cdot = \cdot\, S^k0$ in $\mathcal{T}$, hence also $n\cdot = \cdot\, S^k0$ in $\hat{\mathcal{T}}$.

(3) If $\hat{\mathcal{T}} \vDash \bot$, then surely $\hat{\mathcal{T}} \vDash p$ for any arrow $p\colon 1 \to \Omega$ in $\hat{\mathcal{T}}$. Using Lemma 22.2, let p be defined by $\Sigma(p)(*) \equiv (\top, 0)$; which differs from $\Sigma(\top)(*) \equiv (\top, 1)$. Therefore $p\cdot \neq \cdot \top$, which contradicts the assertion that $\hat{\mathcal{T}} \vDash p$.

(5) will be left as an exercise, to be based on (7), in a manner similar to (8) below.

(6) The necessity of conditions (i) and (ii) is obvious, so let us assume these conditions and show that $\hat{\mathcal{T}} \vDash p \Rightarrow q$. We distinguish two cases.

In case $p\cdot = \cdot \top$, (ii) implies that $q\cdot = \cdot \top$, hence surely that $\hat{\mathcal{T}} \vDash p \Rightarrow q$.

In case $p\cdot \neq \cdot \top$, $\Sigma(p)(*) \neq (\top, 1)$, by Lemma 21.2. Hence, in view of (i),

$$\Sigma(p)(*) = (\check{p}, 0) = (\check{p} \wedge \check{q}, 0) = \Sigma(p \wedge q)(*),$$

since also $p \wedge q\cdot \neq \cdot \top$. Therefore, $p \wedge q\cdot = \cdot\, p$, by Lemma 22.2, that is, $\hat{\mathcal{T}} \vDash p \Rightarrow q$.

(7) Let $\varphi(x)$ be a formula of $L(\hat{\mathcal{T}})$ with a free variable x of type A. By functional completeness, we have $\vdash_x \varphi(x) = fx$, where $f\colon A \to \Omega$ in $\hat{\mathcal{T}}$. This gives rise to the arrow $\ulcorner f \urcorner\cdot = \cdot \{x \in A \mid \varphi(x)\}\colon 1 \to PA$ in $\hat{\mathcal{T}}$. By Lemma 22.2, $\ulcorner f \urcorner$ is determined by $\Sigma(\ulcorner f \urcorner)$; in fact, since $\ulcorner G(f) \urcorner\cdot = \cdot G(\ulcorner f \urcorner)$,

(†) $\qquad \ulcorner \check{f} \urcorner\cdot = \cdot\, \lambda_{PA}(\Sigma(\ulcorner f \urcorner)(*)).$

Therefore,

$$\hat{\mathcal{T}} \vDash \forall_{x \in A}\varphi(x) \text{ iff } \{x \in A \mid \varphi(x)\}\cdot = \cdot \{x \in A \mid \top\} \text{ iff } \ulcorner f \urcorner\cdot = \cdot \ulcorner \top \bigcirc_A \urcorner$$
$$\text{iff } \Sigma(\ulcorner f \urcorner)(*) = \Sigma(\ulcorner \top \bigcirc_A \urcorner)(*).$$

Now $g \equiv \Sigma(\ulcorner f \urcorner)(*) \in \Sigma(PA) \equiv \mathrm{Hom}_{\hat{\mathcal{T}}}(A, \Omega)$, hence it is determined by arrows $\Sigma(g)\colon \Sigma(A) \to \Sigma(\Omega)$ in **Sets** and $\check{g}\colon \check{A} \to \Omega$ in $\mathcal{T}$. By definition of λ_{PA}, we have

$$\lambda_{PA}(g) \equiv \lambda_{PA}(\Sigma(\ulcorner f \urcorner)(*))\cdot = \cdot \ulcorner \check{g} \urcorner.$$

So, comparing this with (†) above, we see that $\check{g}\cdot = \cdot \check{f}$. Also $h \equiv \Sigma(\ulcorner \top \bigcirc_A \urcorner)(*)$ is determined by $\Sigma(h)$, which sends $\alpha \in \Sigma(A)$ onto $(\top, 1)$, and by $\check{h} \equiv \ulcorner \top \bigcirc_A \urcorner$. Therefore $\hat{\mathcal{T}} \vDash \forall_{x \in A}\varphi(x)$ if and only if

(i′) $\qquad \check{f}\cdot = \cdot \top \bigcirc_A$ in $\mathcal{T}$,

(ii′) $\qquad \Sigma(g)(\alpha) = (\top, 1)$ for all $\alpha \in \Sigma(A)$.

Clearly (i′) asserts that $\{x\in A\,|\,\check{\varphi}(x)\}\cdot\!=\!\cdot\{x\in A\,|\,\top\}$ in $\mathscr{T}$, that is, (i). We claim that (ii′) holds if and only if $\hat{\mathscr{T}}\vDash a\in\ulcorner f\urcorner$ for all $a\colon 1\to A$, that is, (ii) holds.

Indeed, a is determined by $\alpha\equiv\Sigma(a)(*)\in\Sigma(A)$, and so quantification over $a\colon 1\to A$ is equivalent to quantification over $\alpha\in\Sigma(A)$. Moreover, (ii) holds if and only if $\varepsilon_A\langle\ulcorner f\urcorner, a\rangle\cdot\!=\!\cdot\top$, that is, $\Sigma(\varepsilon_A)(\Sigma(\ulcorner f\urcorner)(*),\alpha)=(\top,1)$, that is, $\Sigma(g)(\alpha)=(\top,1)$, which is (ii′).

(8) We recall that

$$\exists_{x\in A}\varphi(x)\equiv\forall_{t\in\Omega}(\forall_{x\in A}(\varphi(x)\Rightarrow t)\Rightarrow t).$$

In view of (7), $\hat{\mathscr{T}}\vDash\exists_{x\in A}\varphi(x)$ is then equivalent to the conjunction of the following:

(i″) $\mathscr{T}\vDash\exists_{x\in\check{A}}\check{\varphi}(x)$,

(ii″) $\hat{\mathscr{T}}\vDash\forall_{x\in A}(\varphi(x)\Rightarrow p)\Rightarrow p$, for every $p\colon 1\to\Omega$.

We claim that the conjunction of (i″) and (ii″) is equivalent to

(iii) $\hat{\mathscr{T}}\vDash\varphi(a)$ for some $a\colon 1\to A$ in $\hat{\mathscr{T}}$.

Clearly, (iii) implies (i″), as is seen by applying the functor G, and (ii″), by elementary logic. Conversely, assume (i″) and (ii″), we shall prove (iii).

As in the proof of (3), let $p\colon 1\to\Omega$ be determined by $\Sigma(p)(*)\equiv(\top,0)$, hence $\check{p}\cdot\!=\!\cdot\top$. Then surely not $\hat{\mathscr{T}}\vDash p$, hence, by (ii″), also not $\hat{\mathscr{T}}\vDash\forall_{x\in A}(\varphi(x)\Rightarrow p)$. Therefore, by (7), either not $\mathscr{T}\vDash\forall_{x\in\check{A}}(\check{\varphi}(x)\Rightarrow\check{p})$ or, for some $a\colon 1\to A$ in $\hat{\mathscr{T}}$, not $\hat{\mathscr{T}}\vDash\varphi(a)\Rightarrow p$. Since $\check{p}\cdot\!=\!\cdot\top$, the first alternative is absurd, so the second alternative must hold. By (6), this means either not $\mathscr{T}\vDash\varphi(a)\Rightarrow\check{p}$, which again is absurd, or $\hat{\mathscr{T}}\vDash\varphi(a)$ but not $\hat{\mathscr{T}}\vDash p$. Thus (iii) holds, as remained to be shown.

The proof of Theorem 20.1 is now complete.

Proof of Lemma 21.1. In case $x\colon 1\to\Omega$ is an indeterminate over $\mathscr{T}$, let $\hat{x}\colon 1\to\Omega$ in $\mathscr{T}(x)^\wedge$ be given by $G(\hat{x})\cdot\!=\!\cdot x$ and $\Sigma(\hat{x})(*)\equiv(x,0)$.

In case $x\colon 1\to PC$ over $\mathscr{T}$, let $\hat{x}\colon 1\to P\hat{C}$ in $\mathscr{T}(x)^\wedge$ be given by $G(\hat{x})\cdot\!=\!\cdot x$ and $\Sigma(\hat{x})(*)\equiv f\colon\hat{C}\to\Omega$, where $\check{f}\cdot\!=\!\cdot x^{\int}$ and $\Sigma(f)(\gamma)\equiv(x^{\int}\lambda_C(\gamma),0)$ for all $\gamma\in\Sigma(\hat{C})$.

Exercises

1. If $\mathscr{T}$ is a category with terminal object, prove that there is a natural isomorphism $\mathrm{Hom}_{\mathscr{T}}(1,-)\cong\Sigma$.

2. If $\mathscr{T}$ is a topos, show that the construction of products, equalizers, exponentials and Ω in $\hat{\mathscr{T}}$ follows from the assumption that G is a strict logical functor.

3. Fill in the missing details in the proof of Proposition 22.3.

4. Write out a proof of (5) of Theorem 20.1 based on (8).

5. (Lawvere and Tierney). If (C, ε, δ) is a cotriple on the topos $\mathscr{A}$ (that is, a triple on $\mathscr{A}^{op}$) and if $C: \mathscr{A} \to \mathscr{A}$ is left exact, show that the category of coalgebras of C (that is, algebras of C^{op}) is a topos.

6. If $\mathscr{T}$ and $\mathscr{S}$ are toposes and $\Gamma: \mathscr{T} \to \mathscr{S}$ is a left exact functor, consider the functor $C: \mathscr{T} \times \mathscr{S} \to \mathscr{T} \times \mathscr{S}$ defined on objects (A, X) by

$$C(A, X) \equiv (A, \Gamma(A) \times X)$$

and on arrows similarly. Show that the category of coalgebras of C is isomorphic to the category $(\mathscr{S}, \Gamma)$, called the topos obtained by *gluing along* Γ. (The Freyd cover of $\mathscr{T}$ is a special case of this, with $\Gamma \equiv \mathrm{Hom}_{\mathscr{T}}(1, -): \mathscr{T} \to$ **Sets**.) Since the forgetful functor from coalgebras to $\mathscr{T} \times \mathscr{S}$ has a right adjoint, show how colimits are constructed in $(\mathscr{S}, \Gamma)$.

7. (*a*) Prove that an arrow (φ, t) in $\hat{\mathscr{T}}$ is an epimorphism if and only if φ is surjective and t is an epimorphism in $\mathscr{T}$.
(*b*) Conclude that 1 is an indecomposable projective in $\hat{\mathscr{T}}$.
(*c*) Let $M \equiv (\varnothing, \lambda_M, 1)$ be the object of $\hat{\mathscr{T}}$ given by the unique mapping $\lambda_M: \varnothing \to \Gamma(1)$. Show that M is the largest proper subobject of 1 in $\hat{\mathscr{T}}$ and that $\hat{\mathscr{T}}/M \cong \mathscr{T}$. Verify that $\hat{\mathscr{T}} \to \hat{\mathscr{T}}/M \simeq \mathscr{T}$ is the functor G of the text.
(*d*) Conclude that 1 is an indecomposable projective in the free topos. (This is Freyd's original argument.)

8. Show that the isomorphism of Exercise 6 together with the dual of Proposition 6.5 in Part 0 allows one to obtain a right adjoint $H: \mathscr{T} \to \hat{\mathscr{T}}$ to the strict logical functor G such that $H(A) = 1_{\Gamma(A)}$. Prove that 1 is projective in $\mathscr{T}$ if and only if H preserves epimorphisms.

9. Check that the Heyting algebra $\mathrm{Hom}_{\hat{\mathscr{T}}}(1, \Omega)$ is obtained from the Heyting algebra $\mathrm{Hom}_{\mathscr{T}}(1, \Omega)$ by adjoining as new largest element $(\top, 1)$.

Historical comments on Part II

Section 1

A theory of types was introduced by Russell (1908) as a safeguard against paradoxes in set theory. In *Principia Mathematica* (1910–13), he and Whitehead used type theory as a foundation of mathematics. Later, Church (1940) developed a system of type theory based on the λ-calculus (see Section 1, Exercise 4). For a more detailed history and a comparison with other systems of set theory, see Hatcher (1982).

While the systems of type theory mentioned so far are based on classical logic, intuitionistic type theories are a relatively recent phenomenon.

Interest in such theories accompanied the trend towards proof-theoretical investigations in intuitionistic set theories (e.g. Girard 1971, 1972, Friedman 1973 and Myhill 1973).

As in Part I, our formal languages have a deduction symbol $\vdash_X$, where the subscript $X \equiv \{x_1,\ldots,x_n\}$ in $\varphi \vdash_X \psi$ contains all variables occurring freely in φ and ψ. This notion appears to be due to Mostowski (1951). Here it takes into account possibly empty types in the internal language of a topos. Also, as in Part I, we allow our languages to be large, so that we can discuss, for example, the internal language of the category of sets.

Section 2

The presentation of intuitionist type theories based on equality follows (Lambek and Scott 1981*b*). A somewhat similar system, partially based on equality, was given by Boileau (1975) and Boileau and Joyal (1981). Already Henkin (1963) discussed a classical type theory of propositional types (Ω being the only basic type) based on equality, and his footnotes acknowledge some even earlier sources. (See also Tarski 1923). However, a detailed comparison of type theories with and without equality probably appears here for the first time.

Section 3

The observation that one can do logic inside a topos was of course known to Lawvere (see also Freyd 1972). The first to publish a formal description of the internal language was W. Mitchell (1972), who exploited it successfully to replace arguments about arrows and diagrams by the familiar set-theoretic reasoning. The internal language was also discovered, apparently independently, by several people, including Bénabou and Joyal. It is called the 'Mitchell–Bénabou language', in the book by Johnstone (1977). For other expositions see (Osius 1975a, Coste 1972, 1973, 1974 and Schlomiuk 1977). The language of Fourman (1974, 1977) involves a description operator and an existence predicate (see the discussion in Boileau and Joyal 1981). In the spirit of Lawvere's algebraic theories, type theories may themselves be viewed as categories, as was done in (Volger 1975*b*) and (Lambek 1980*a*, preprint in 1974). A topos then does not just contain, but actually coincides with its internal language.

The key natural isomorphism in toposes

$$\mathrm{Sub} \cong \mathrm{Hom}(-,\Omega)$$

allows two equivalent ways of interpreting closed formulas in a topos, either as subobjects of 1 or as arrow $1 \to \Omega$. Most authors have chosen the first interpretation; we prefer the second, perhaps influenced by Church (1940), but mainly because subobjects are really equivalence classes of monomorphisms, hence more complicated than simple arrows.

While this choice may be a matter of taste, there are two divergent generalizations to topics outside the scope of this book. On the one hand, categories with a distinguished Heyting algebra object Ω (and suitable machinery for handling quantifiers) permit interpretations of type theory, thus the 'semantical' categories of Volger (1975*b*) and the 'dogmas' of Lambek (1980*a*). On the other hand, in many categories the subobject lattice carries a suitable logical structure, for example, in the 'regular' categories of Reyes (1974) and the 'logical' categories of Makkai and Reyes (1977). The last mentioned work also introduces 'coherent' and 'geometric' logics, often infinitary multi-sorted first order theories, which are particularly suited for the study of Grothendieck toposes with geometric morphisms.

Section 4

The relation between Lawvere's natural numbers object and Peano's axioms in the internal language has been studied elsewhere (e.g. Osius 1975*a*), but is here worked out from first principles. Other interesting observations about the natural numbers object are found in (Freyd 1972).

Section 5

Treatments of equality, characteristic morphisms, singletons and description in the internal language may be found, with obvious variations, in several of the above references. The proof that description holds in a topos follows (Lambek 1980*a*).

Finite coproducts, as well as equalizers, were originally part of the definition of a topos. Mikkelsen (1976) was the first to show that these were dispensible, and Exercise 2 contains a linguistic proof of this result. A neat categorical proof of the same result based on the tripleability of the contravariant power set functor was discovered independently by Paré (1974) and Rattray (Lambek and Rattray 1975*b*).

Section 6

The treatment of monomorphisms and epimorphisms is standard (e.g. Osius 1975*a*). The injectivity of *PC*, though not its linguistic proof, was known since the beginning of the subject. The internal characterization of projectives is due to Freyd (1978).

Section 7

The fact that choice implies the Boolean axiom was discovered by Diaconescu (1975). The linguistic proof given here was inspired by that of Beason (1982).

Section 8

The use of generalized elements seems to be due to Joyal (see Kock *et al.* 1975) and the inductive truth definition of Theorem 8.4 as well. The latter first appeared in print in (Osius 1975*b*) under the name 'Kripke–Joyal semantics' (see also Johnstone 1977), as it generalizes Kripke's original semantics for first order intuitionistic logic (Kripke 1965). However, the clauses for $\vee$ and $\exists$ involves the notion of 'covering' and are closer to the semantics of Beth (see Dummett 1977). We regret that time limitations did not allow us to rewrite this section in the spirit of Exercise 4.

As stressed by Lawvere (1971) and also Joyal, topos semantics provides a common, natural generalization of many proof-theoretic and semantical notions (see also Ščedrov and Scott 1982, Lambek and Scott 1983, Ščedrov 1984).

Section 9

The topos semantics for functor categories contains, as a special case, the higher order analogue of Kripke's semantics for first order intuitionistic logic (see Kripke 1965).

Section 10

For a thorough treatment of sheaf semantics the reader is referred to the pioneering paper by Fourman and Scott (1979) and also to (Ščedrov 1984*a*). An interesting generalization of sheaves are Ω-valued sets due to Higgs (1973, 1984) and discussed in the above articles.

Section 11

Dogmas (= semantical categories) were discussed by Volger (1975*b*) and Lambek (1980*a*). Freyd's theory of allegories is unpublished.

Section 12

The topos generated by a type theory was discovered independently by many people. The first to publish it were Volger (1975*b*) and Fourman (1977), but it also appeared in the theses of Fourman (1974) and Boileau (1975) and in preprints by Coste (1974) and Lambek (1974).

Section 13

That every topos is equivalent to the topos generated by its internal language appears in Volger (1975*b*), Fourman (1977) and Boileau and Joyal (1981). That the power set structure suffices in the definition of a topos appears in (Mikkelson 1976).

Section 14

Unfortunately, not every type theory is the internal language of a topos, unless we make some additional demands on a type theory (see however Exercise 4). This was known to Volger (1975*b*) and is discussed by Fourman (1977).

Section 15

That there is a pair of adjoint functors between toposes and type theories was first shown by Volger (1975*b*); but some logical morphisms appearing in his argument were only shown to be pseudo-functors. To make this argument more precise, one apparently needs to confine attention to toposes with canonical subobjects, not a serious restriction, since every topos is equivalent to one such. This step was taken by Lambek (1980*a*). Coste (1974) discusses a pair of adjoint functors between toposes and what he calls 'formal toposes'.

Section 16

Several of the constructions presented here were first carried out by more categorical methods. Thus, the construction of $\mathscr{T}(x)$ mentioned in Example 2 is due to Joyal in this generality, having originally been done for Grothendieck toposes in (Artin, Grothendieck and Verdier 1972). The result of dividing a topos by a filter is called a 'filter power' in (Johnstone 1977) and is considered there as a special case of a 'topos of fractions', whereas here the latter is treated as a special case of the former. (Of course, our construction does not work for arbitrary categories of fractions.)

Section 17

The completeness theorem for classical type theories with choice is due to Henkin (1950). The categorical treatment presented here follows (Lambek and Moerdijk 1982*a*).

Section 18

This entire section is based on the last mentioned reference. The proof is quite analogous to that of the well-known theorem that every commutative ring is isomorphic to the ring of continuous sections of a sheaf of local rings. The first author had noticed this analogy, but was using what is here called the 'cospectrum' of a topos, and it was the second author who realized that it was more natural to use the 'spectrum' instead.

Section 19

The use of Henkin constants comes, of course, from Henkin (1949, 1950). A categorical proof was first contained in an unpublished manuscript by Lambek and Moerdijk (1982*b*). The present argument follows the proof of the completeness of first order intuitionistic logic by Aczel (1969), which result had also been proved by Beth, Kripke and others (see Smorynski 1973). A related categorical construction appears in (Freyd 1972).

Section 20

For an excellent introduction to intuitionism we refer the reader to (Dummett 1977). The treatment of the internal logic of Freyd covers follows (Lambek and Scott 1983). The inductive clauses in Theorem 20.1 are a higher order version of the 'Aczel slash' (see Aczel 1969, Smorynski 1973). Note that the Freyd cover proofs of the existence property for $\mathfrak{L}_0$ (Proposition 20.5 or Exercise 2) merely witness existential formulas by arrows in the free topos. To replace these arrows by actual terms of the appropriate type in the language requires a syntactical argument, the unique existence property (Lemma 20.3).

Section 21

Most of the intuitionistic principles discussed here are well-known from the literature (e.g. Troelstra (ed.) 1973 and Beeson 1982). The disjunction property with parameters (that is, the statement that objects corresponding to a type of the form *PC* are indecomposable in the free topos) was introduced in (Lambek and Scott 1983). In (1981*c*) we had also studied the existence and disjunction properties modulo *p* (Proposition 21.6). (See also Troelstra 1973.) Underlying Proposition 21.6 is the open problem how to characterize projective subobjects of 1 in the free topos (see Remark 21.9). Moerdijk (1982) discusses existence and disjunction properties for more general theories using a construction somewhat similar to the Freyd cover.

The question in the text concerning the projectivity of N in the free topos, equivalently, the *countable rule of choice* in pure type theory, has of course been solved by proof-theoretical techniques (see Friedman and Ščedrov 1983). Makkai in unpublished notes (1980) cast the logician's proof into a categorical (actually 2-categorical) mould.

Section 22

We had originally proved the existence and disjunction properties of pure type theory in 1978 (Lambek and Scott 1980) using an extension of Kleene–Friedman realizability (Friedman 1973). When we presented these results at a New York/Montreal topos meeting in October 1978, Peter Freyd immediately realized that they showed that 1 is an indecomposable projective in the free topos. Soon after (Freyd 1978), he gave the elegant categorical proof sketched in the exercises. (In fact, this entire section consists of detailed verifications of his calculations.) Ščedrov and Scott (1982) showed that the proofs by realizability and by Freyd's method were essentially the same. The version of the argument using the Aczel slash presented in Section 20 is a bit more direct.

Supplement to Part II, Section 17.

Remark 17.8. On second thought, we can say a little more about the functor $\Gamma \equiv \mathrm{Hom}_{\mathscr{M}}(1,-)\colon \mathscr{M} \to \mathbf{Sets}$ discussed earlier. It is not only left exact but also almost right exact: it preserves finite coproducts (essentially because 1 is indecomposable, see below) and (regular) epimorphisms (because 1 is projective). Such a functor is called *near exact* (see Barr and Wells 1985). Thus, every model $\mathscr{M}$ comes equipped with a near exact functor into Sets. In case $\mathscr{M}$ is Boolean, this functor is also faithful. Why does Γ preserve finite coproducts? It preserves nullary coproducts, because by M0′ there are no arrows $1 \to 0$, and it preserves binary coproducts, because every arrow $c\colon 1 \to A + B$ factors through A or B. Indeed, it follows from Part I, Section 5, Exercise 2 that

$$\mathscr{M} \models \exists_{x \in A}\, c = \kappa_{A,B}x \vee \exists_{y \in B}\, c = \kappa'_{A,B}y.$$

Hence, by M1, $\mathscr{M} \models \exists_{x \in A} c = \kappa_{A,B}x$ or $\mathscr{M} \models \exists_{y \in B} c = \kappa'_{A,B}y$. Since $\kappa_{A,B}$ and $\kappa'_{A,B}$ are easily seen to be monomorphisms, we may apply Part II, Lemma 5.2 to infer that c factors through A or B.

III

Representing numerical functions in various categories

Introduction to Part III

After a brief review of recursive functions, we discuss which of them can be represented by arrows in cartesian closed categories with weak natural numbers object, by arrows in toposes and by elements of C-monoids. These problems are translated into related questions about typed λ-calculi, type theories and extended λ-calculi respectively.

1 Recursive functions

In this section we consider number-theoretic functions which conform to our intuitive notions of being effectively calculable or, to put it in different terms, which can be computed by some kind of algorithm or program. This notion was made precise in the 1930s by the logicians Church, Gödel, Kleene and Turing. We recapitulate some relevant definitions.

Let $\mathscr{C}$ be a set of functions $\mathbb{N}^k \to \mathbb{N}$ (in the category of sets), where $\mathscr{N} = \{0, 1, 2, \ldots\}$ is the set of natural numbers. Such functions will be called *numerical*. Not to be too fussy, let us write $h(a) \equiv h(a_1, \ldots, a_k)$ for such a function instead of $a \mapsto h(a)$, where $a \equiv (a_1, \ldots, a_k) \in \mathbb{N}^k$. We say that $\mathscr{C}$ is *closed under*

substitution if, whenever $g_1(a), \ldots, g_m(a)$ and $h(b_1, \ldots, b_m) \in \mathscr{C}$, then $h(g_1(a), \ldots, g_m(a)) \in \mathscr{C}$,

primitive recursion if, whenever $g(a)$ and $h(n, m, a) \in \mathscr{C}$, then also $f(n, a) \in \mathscr{C}$ where $f(0, a) = g(a)$ and $f(S(n), a) = h(n, f(n, a), a)$;

(*restricted*) minimization if, whenever $g(a, n) \in \mathscr{C}$ and, for all $a \in \mathbb{N}^k$, there exists $n \in \mathbb{N}$ such that $g(a, n) = 0$, then $f(a) = \mu_n(g(a, n) = 0) \in \mathscr{C}$, where $\mu_n(\cdots n \cdots)$ means 'the least n such that $\cdots n \cdots$.'

By a *basic* numerical function we shall understand one of the following:

the *successor function*	$S(n) \equiv n + 1$,
the *zero function*	$0(n) \equiv 0$,
the *projections*	$p_1^k(n_1, \ldots, n_k) \equiv n_i (i = 1, \ldots, k)$.

The set of *primitive recursive functions* is the smallest set $\mathscr{C}$ containing the basic functions and closed under substitution and primitive recursion.

The set of (*total*) *recursive functions* is the smallest set $\mathscr{C}$ containing the basic functions and closed under substitution, primitive recursion and (restricted) minimization.

Soon after Dedekind had introduced the primitive recursive functions in 1888, it was realized that they failed to capture completely the notion of 'computable numerical function'. There are numerical functions computable by some algorithm which are not primitive recursive. One can show this either by Cantor's diagonal argument or by exhibiting a concrete example of such a function. The first such example is due to Ackermann (1928), and we present here a simplified form due to Rózsa Péter.

Let $\alpha: \mathbb{N} \times \mathbb{N} \to \mathbb{N}$ be defined by

$$\alpha(0, n) = n + 1,$$

$$\alpha(m + 1, 0) = \alpha(m, 1),$$

$$\alpha(m + 1, n + 1) = \alpha(m, \alpha(m + 1, n)).$$

Although $\alpha(m, n)$ is easily seen to be computable, it can be shown that $\alpha(m, m)$ grows faster than any primitive recursive function. Thus $\alpha(m, m)$ is not primitive recursive, hence neither is $\alpha(m, n)$.

Recursive functions however did turn out to capture our intuitive notion of what computable numerical functions should be. In a pioneering paper, Turing (1936–7) showed that these functions are precisely the ones computable on an abstract machine described by him. At about the same time, Church (1936) showed that the recursive functions were precisely those definable in untyped λ-calculus. These observations led to the so-called *Church–Turing Thesis*:

> (CT) The (total) recursive functions exactly capture the intuitive notion of 'computable numerical function'.

Thus (CT) identifies a vague intuitive notion (computable) with a precise mathematical one (recursive). Although not amenable to direct proof, its validity is not seriously doubted. Indeed, various other attempts to define computability led to the same result. Thus the recursive functions also turn out to be the numerical functions calculable by Markov algorithms, by a finite program on an abacus or register machine or, for that matter, on a modern computer (provided we allow unlimited storage).

One may also consider the computability of *partial numerical functions* from $\mathbb{N}^k$ to $\mathbb{N}$, that is, functions $f: D \to \mathbb{N}$ whose domain D is a subset of $\mathbb{N}^k$. A set $\mathscr{C}$ of partial recursive functions is *closed under unrestricted minimization* if, whenever $g(a, n) \in \mathscr{C}$, also $f(a) \equiv \mu_n(g(a, n) = 0)$ is in $\mathscr{C}$. If, for some particular k-tuple a, there is no n such that $g(a, n) = 0$, then $f(a)$ is undefined.

The set of *partial recursive functions* is the smallest set of partial numerical functions containing the basic functions and closed under substitution, primitive recursion and unrestricted minimization. The Church–Turing thesis, extended to partial functions, identifies computable partial numerical functions with partial recursive functions.

We end this section with a technical result (see Kleene 1952, Mendelson 1974, Shoenfield 1967). It uses the fact that, for given k, all partial recursive functions from $\mathbb{N}^k$ to $\mathbb{N}$ have been effectively enumerated in a sequence $f_1, f_2, \ldots$, for example, by enumerating the programs which calculate them. We say that a k-ary relation $R(a_1, \ldots, a_k) \equiv R(a)$ is a *primitive recursive relation* if there is a primitive recursive function $g(a)$ such that, for all $a \in \mathbb{N}^k$, $R(a)$ if and only if $g(a) = 0$.

Kleene Normal Form Theorem. There exists a primitive recursive function $U(m)$ and a primitive recursive $(k + 2)$-ary relation $T_k(e, a, n)$ such that, for any partial recursive function f from $\mathbb{N}^k$ to $\mathbb{N}$ and all $a \in \mathbb{N}^k$, there is a number $e \in \mathbb{N}$ such that $f(a) \equiv f_e(a) = U(\mu_n T_k(e, a, n))$.

The idea of the proof of this theorem is roughly as follows. Consider a program with code number e calculating the partial recursive function $f_e(a)$ for the input $a \in \mathbb{N}^k$. For each $n = 0, 1, 2, \ldots$, let σ_n encode the *stage* of a calculation, that is, the contents of all the registers (all but a finite number of which will be empty) and the node of the program at time n (a program is a labelled graph). Thus, the passage from σ_n to σ_{n+1} represents one step in the calculation. Let $U(n)$ be the content of the output register at time n and let $T_k(e, a, n)$ be the statement which asserts that the node of the eth program with input a at time n is 'stop'. Then $U(\mu_n T_k(e, a, n))$ is the content of the output register when the calculation stops, that is to say, $f_e(a)$.

Exercises

1. Show that the following numerical functions are primitive recursive:

   $m + n$, $m \cdot n$, m^k, $n!$, $\mathrm{pre}(n)$, $m \dot{-} n$, $\min(m, n)$, $\max(m, n)$, $\delta(n)$, $|m - n|$, where

$$\text{pre}(n) = \begin{cases} n-1 & \text{if } n > 0 \\ 0 & \text{if } n = 0, \end{cases}$$

$$m \dot{-} n = \begin{cases} m-n & \text{if } m \geqslant n \\ 0 & \text{if } m < n, \end{cases}$$

$$\delta(n) = \begin{cases} 1 & \text{if } n = 0 \\ 0 & \text{if } n > 0. \end{cases}$$

2. Show that the following numerical functions are recursive:

   $[n/2], [\sqrt{n}], [\log_{10}(n+1)]$,

   where $[\xi]$ denotes the greatest integer not exceeding ξ.

3. Show that for any $k \in \mathbb{N}$ the set of recursive (primitive recursive) numerical k-ary relations is closed under the Boolean operations: conjunction, disjunction and negation.

4. Show that the following numerical unary and binary relations are recursive:

   n is odd, n is a perfect square, $m < n$, $m \neq n$, m divides n, n is a prime number.

5. Show that the following numerical functions are recursive:

   $p(n)$, $\pi(n)$, $\exp_m(n)$,

   where

   $p(0) = 2$, $p(n) = n$th odd prime (if $n \geqslant 1$),

   $\pi(n) =$ number of primes $< n$,

   $\exp_m(n) =$ largest k such that m^k divides n, unless $m = 1$ or $n = 0$, in which case it $= 0$.

6. Show that the following numerical function due to Cantor is recursive:

   $$\mathscr{J}(m, n) = m + \sum_{k=0}^{m+n} k = m + [\tfrac{1}{2}(m+n)(m+n+1)],$$

   and prove that $\mathscr{J}: \mathbb{N} \times \mathbb{N} \to \mathbb{N}$ is a bijection. Letting

   $(k)_0 \equiv \mu_{m \leqslant k} \exists_{n \leqslant k} \mathscr{J}(m, n) = k$,

   $(k)_1 \equiv \mu_{n \leqslant k} \exists_{m \leqslant k} \mathscr{J}(m, n) = k$,

   show that $(k)_0$ and $(k)_1$ are recursive and that $(\mathscr{J}(m, n))_0 = m$, $(\mathscr{J}(m, n))_1 = n$, $\mathscr{J}((k)_0, (k)_1) = k$.

7. If $R(a, n)$ is a primitive recursive relation with $a \in \mathbb{N}^k$ and $n \in \mathbb{N}$, show that the relations

$\exists_{n<m}R(a,n), \forall_{n<m}R(a,n)$

are primitive recursive. If, moreover $\forall_{a\in\mathbb{N}^k}\exists_{n\in\mathbb{N}}R(a,n)$, show that $\mu_{n<m}R(a,n)$ is a primitive recursive function. Hence re-examine the recursive functions, properties and relations appearing in earlier exercises and show that they are all primitive recursive.

2 Representing numerical functions in cartesian closed categories

We are interested in representing numerical functions $\mathbb{N}^k \to \mathbb{N}$ by arrows in cartesian closed categories with weak natural numbers object. We shall write $\S n \equiv S^n 0$ for the nth arrow $1 \to N$.

Definition 2.1. A function $f: \mathbb{N}^k \to \mathbb{N}$ is *representable* in a cartesian closed category $\mathscr{C}$ with a weak natural numbers object N if there is an arrow $f^\dagger$: $N^k \to N$ in $\mathscr{C}$ such that, for every k-tuple $(n_1, \ldots, n_k)$ of natural numbers,

$$f^\dagger\langle \S n_1, \ldots, \S n_k\rangle = \S f(n_1, \ldots n_k),$$

that is, the following diagram commutes:

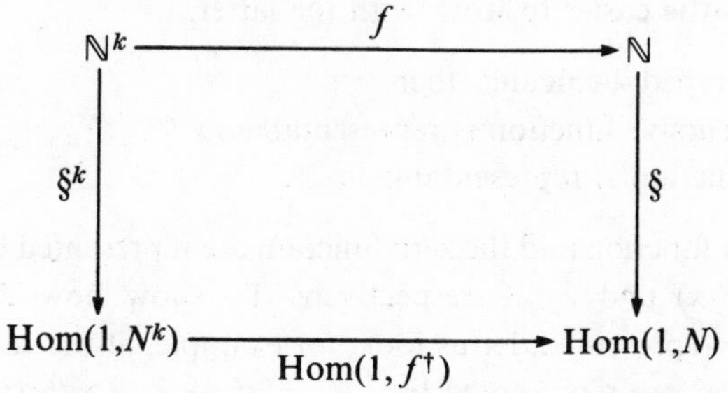

(We have written $\S^k(n_1, \ldots, n_k) \equiv \langle \S n_1, \ldots, \S n_k\rangle$.)

We may also represent numerical functions by closed terms in typed λ-calculi. We write $\#n \equiv S^n 0$ for the nth numeral.

Definition 2.2. A function $f: \mathbb{N}^k \to \mathbb{N}$ is *representable* in a typed λ-calculus $\mathscr{L}$ if there is a closed term F of type N^{N^k} such that

$$F^{\smallint}\langle \#n_1, \ldots, \#n_k\rangle = \#f(n_1, \ldots, n_k)$$

holds in $\mathscr{L}$ for every k-tuple $(n_1, \ldots n_k)$.

In view of the equivalence between typed λ-calculi and cartesian closed categories with weak natural numbers object established in Part I (Theorem 11·3), it is not surprising that definitions 2.1 and 2.2 are also equivalent. More precisely, we have the following result.

Proposition 2.3. A numerical function $f: \mathbb{N}^k \to \mathbb{N}$ is representable in a typed λ-calculus $\mathscr{L}$ if and only if it is representable in the cartesian closed category $\mathbf{C}(\mathscr{L})$ generated by it.

Proof (sketched). By construction of $\mathbf{C}(\mathscr{L})$ (see Part I, Section 11), an arrow $f^{\dagger}: N^k \to N$ has the form $(x \in N^k, \varphi(x))$, where $\varphi(x)$ is a term of type N. In the same vein one easily calculates that $\S n: 1 \to N$ has the form $(z \in 1, \#n)$. Put $F \equiv \lambda_{x \in A} \varphi(x)$, then F has type N^A, where $A \equiv N^k$, and

$$F^{\int} \langle \#n_1, \ldots, \#n_k \rangle = \varphi(\langle \#n_1, \ldots, \#n_k \rangle).$$

A routine calculation, which we leave to the reader, now shows that the equation appearing in Definition 1.1 translates into

$$(z \in 1, F^{\int} \langle \#n_1, \ldots, \#n_k \rangle) = (z \in 1, \#f(n_1, \ldots, n_k)),$$

which is clearly equivalent to the equation in Definition 2.2. Similarly we pass from Definition 2.2 to Definition 2.1.

The following result is due to Marie-France Thibault, who established it for cartesian closed categories with weak natural numbers object, when it had not yet been shown that these were equivalent to typed λ-calculi. In retrospect, we find it a little easier to work with the latter.

Theorem 2.4. If $\mathscr{L}$ is a typed λ-calculus, then
(1) every primitive recursive function is representable in $\mathscr{L}$,
(2) the Ackermann function is representable in $\mathscr{L}$.

Proof. (1) The successor function and the zero function are represented by the closed terms $\lambda_{x \in N} S(x)$ and $\lambda_{x \in N} 0$ respectively. To show how the projection functions are represented, let us look, for example, at the case $n = 3$. Then p_1^3, p_2^3, and p_3^3 are represented by $\lambda_{x \in N^3}\, \pi(\pi(x))$, $\lambda_{x \in N^3}\, \pi'(\pi(x))$ and $\lambda_{x \in N^3} \pi'(x)$ respectively, where $N^3 \equiv (N \times N) \times N$.

Suppose the functions $h(b_1, \ldots, b_m)$ and $f_i(a_1, \ldots, a_k)$ $(i = 1, \ldots, m)$ are represented by the terms $H \in N^{N^m}$ and $F_i \in N^{N^k}$ respectively, then their composite $h(f_1(a_1, \ldots, a_k), \ldots, f_m(a_1, \ldots, a_k))$ is represented by the term $\lambda_{x \in N^k} H^{\int} \langle F_1{}^{\int} x, \ldots, F_m{}^{\int} x \rangle$.

Finally, suppose $g(a_1, \ldots, a_k)$ and $h(n, m, a_1, \ldots, a_k)$ are represented by the terms $G \in N^{N^k}$ and $H \in N^{(N \times N) \times N^k}$ respectively. We wish to represent $f(n, a_1, \ldots, a_k)$ by a term $F \in N^{N \times N^k}$, where $f(0, a_1, \ldots, a_k) = g(a_1, \ldots, a_k)$ and $f(S(n), a_1, \ldots, a_k) = h(n, f(n, a_1, \ldots, a_k), a_1, \ldots, a_k)$.

Let x be a variable of type N^k and put

$$a_x \equiv G^{\int} x, \quad g_x{}^{\int} \langle u, v \rangle \equiv H^{\int} \langle \langle u, v \rangle, x \rangle.$$

Applying Lemma 2.5 below to the language $\mathscr{L}(x)$ with parameter x, we obtain a term $\psi_x(u) \in N$ such that

$$\psi_x(0) \underset{x}{=} a_x \underset{x}{=} G^{\int} x,$$

$$\psi_x(\#(n+1)) \underset{x}{=} g_x{}^{\int} \langle \#n, \psi_x(\#n) \rangle \underset{x}{=} H^{\int} \langle \langle \#n, \psi_x(\#n) \rangle, x \rangle.$$

Replacing x by $\#a \equiv \langle \#a_1, \ldots, \#a_k \rangle$, we obtain

$$\psi_{\#a}(\#0) = G^{\int}(\#a) = \#g(a),$$

$$\psi_{\#a}(\#Sn) = H^{\int} \langle \langle \#n, \psi_{\#a}(\#n) \rangle, \#a \rangle.$$

It follows by induction on n that $\psi_{\#a}(\#n) = \#f(n,a)$. (For $\#f(0,a) = \#g(a)$, $\#f(Sn,a) = \#h(n, f(n,a), a) = H^{\int} \langle \langle \#n, \#f(n,a) \rangle, \#a \rangle$.) Now introduce F of type $N^{N \times N^k}$ by

$$F^{\int} \langle u, x \rangle \underset{\{u,x\}}{=} \psi_x(u),$$

then

$$F^{\int} \langle \#n, \#a \rangle = \#f(n,a)$$

and therefore F represents $f(n,a)$.

This completes the proof of (1), provided we establish the following lemma.

Lemma 2.5. If $\mathscr{L}$ is a typed λ-calculus and $a \in A$ and $g \in A^{N \times A}$ are given terms, we can define a term $\Psi(u) \equiv \Psi_A(a, g, u) \in A$, where u is a variable of type N, such that $\Psi(0) = a$ and $\Psi(S^{n+1}(0)) = g^{\int} \langle S^n(0), \Psi(S^n(0)) \rangle$ for all $n \in \mathbb{N}$.

Proof. The proof is rather similar to the above argument showing that primitive recursion does not lead outside the set of representable numerical functions.

Given terms $c \in N \times A, k \in (N \times A)^{N \times A}$ and a variable $u \in N$, we have $I_{N \times A}(c, k, u) \in N \times A$ such that the following equations hold in $\mathscr{L}$:

(i) $\quad I_{N \times A}(c, k, 0) = c, \quad I_{N \times A}(c, k, S(u)) \underset{u}{=} k^{\int} I_{N \times A}(c, k, u).$

Now let $c \equiv \langle 0, a \rangle$ and suppose k is defined by the equation

$$k^{\int} \langle u, x \rangle \underset{\{u,x\}}{=} \langle S(u), g^{\int} \langle u, x \rangle \rangle.$$

Write $I_{N \times A}(c, k, u) \equiv \langle \Phi(u), \Psi(u) \rangle$, then (i) becomes:

(ii) $\quad \Phi(0) = 0, \quad \Phi(S(u)) \underset{u}{=} S(\Phi(u)),$

(iii) $\quad \Psi(0) = a, \quad \Psi(S(u)) \underset{u}{=} g^{\int} \langle \Phi(u), \Psi(u) \rangle.$

Unfortunately, as long as only a weak natural numbers object is postulated, we cannot infer from (ii) that $\Phi(u) \underset{u}{=} u$ holds in $\mathscr{L}$. However, we can use induction on n to prove the equation $\Phi(S^n(0)) = S^n(0)$, hence (iii) gives rise to

(iv) $\quad \Psi(0) = a, \quad \Psi(S^{n+1}(0)) = g^{\int}\langle S^n(0), \Psi(S^n(0))\rangle,$

as was to be proved.

Now let us return to the proof of Theorem 2.4.

(2) To represent the Ackermann function $\alpha(m, n)$ of Section 1 we seek a term $F \in N^{N \times N}$ such that $F^{\int}\langle \#m, \#n\rangle = \#\alpha(m, n)$ holds in $\mathscr{L}$, where $\#m \equiv S^m(0)$ as before. Define F by the equation

$$F^{\int}\langle u, v\rangle \underset{\{u,v\}}{=} \Psi_A(a, g, u)^{\int} v,$$

where $A \equiv N^N$, $a \equiv \lambda_{x \in N} S(x) \in A$ and $g \in A^{N \times A}$ is given by

$$(g^{\int}\langle u, z\rangle)^{\int} v \underset{\{u,z,v\}}{=} \Psi_N(z^{\int} 1, \lambda_{w \in N \times N}(z^{\int} \pi'_{N,N}(w)), v).$$

Note that g does not really depend on its first argument.

A routine calculation now shows the following equations in $\mathscr{L}$:

(i) $\quad F^{\int}\langle 0, \#n\rangle = \#S(n),$

(ii) $\quad F^{\int}\langle \#(m+1), 0\rangle = F^{\int}\langle \#m, \#1\rangle,$

(iii) $\quad F^{\int}\langle \#(m+1), \#(n+1)\rangle = F^{\int}\langle \#m, F^{\int}\langle \#(m+1), \#n\rangle\rangle.$

The reader may have to do a bit of work to establish (iii). Here is a hint: first reduce the LHS to

$$\Psi_A(a, g, \#m)^{\int} \Psi_N(b, h, \#n),$$

where

$$b \equiv \Psi_A(a, g, \#m)^{\int} \#1,$$

$$h \equiv \lambda_{w \in N \times N}(\Psi_A(a, g, \#m)^{\int} \pi'_{N,N}(w)),$$

then show that this is equal to

$$\Psi_A(a, g, \#m)^{\int}((g^{\int}\langle \#m, \Psi_A(a, g, \#m)\rangle)^{\int} \#n),$$

since g does not depend on its first argument, which is finally reduced to the RHS.

It follows by multiple mathematical induction from (i), (ii) and (iii) that

$$F^{\int}\langle \#m, \#n\rangle = \#\alpha(m, n)$$

holds in $\mathscr{L}$, as was to be shown.

Corollary 2.6. The set of numerical functions representable in a cartesian closed category with weak natural numbers object properly includes the set of primitive recursive functions.

To state a converse of this, we have to specify which cartesian closed category we are talking about. For example, all numerical functions are representable in the category of sets. On the other hand, the set of numerical functions representable in $\mathbf{C}(\mathscr{L}_0)$, the cartesian closed category with weak natural numbers object generated by the pure typed λ-calculus $\mathscr{L}_0$, is a proper subset of the set of recursive functions. This result is also due to Marie-France Thibault.

Theorem 2.7. If a numerical function $\mathbb{N}^k \to \mathbb{N}$ is representable in pure typed λ-calculus $\mathscr{L}_0$, then it is recursive; but not every recursive function is representable in $\mathscr{L}_0$.

Proof (sketched). The proof uses familiar techniques of arithmetization. We shall present the main line of the argument, but omit some of the more gory details.

We begin by enumerating all terms of type N^{N^k}, say $F_0, F_1, F_2, \ldots$. Given $a \equiv (a_1, \ldots, a_k) \in \mathbb{N}^k$ and $e, n \in \mathbb{N}$, let $S(e, a, n)$ be the $(k+2)$-ary relation which asserts that, for some $b \in \mathbb{N}$, n is the Gödel number of a proof that the equation

$$F_e{}^{\mathsf{f}}(\#a_1, \ldots, \#a_k) = \#b$$

holds in $\mathscr{L}_0$.

Since b is uniquely determined by n, $S(e, a, n)$ is seen to be a recursive relation (even a primitive recursive one). Let us write $U(n) = b$ if n is such a Gödel number, $U(n) = 0$ otherwise. Then $U(n)$ is also a (primitive) recursive function.

Now suppose $f(a)$ is any numerical function $\mathbb{N}^k \to \mathbb{N}$ which is representable in $\mathscr{L}_0$. Then, for some $e \in \mathbb{N}$,

$$\#f(a) = F_e{}^{\mathsf{f}}(\#a_1, \ldots, \#a_k),$$

hence

$$f(a) = U(\mu_n(S(e, a, n))).$$

This is clearly a recursive function (even in Kleene normal form), hence the first part of the theorem has been proved.

To prove the second part, take $k = 1$ and write

$$f_e(a) \equiv U(\mu_n(S(e, a, n))).$$

To show that $f_e(a)$ is a total function, we must verify that, for any $e \in \mathbb{N}$ and $a \in \mathbb{N}$, there exists $n \in \mathbb{N}$ such that $S(e, a, n)$, that is to say, there exists $b \in \mathbb{N}$ such that the equation $F_e{}^{\mathsf{f}}(\#a) = \#b$ holds in $\mathscr{L}_0$.

In fact, if t is any closed term of type N in $\mathscr{L}_0$, an equation of the form $t = S^b(0)$ holds in $\mathscr{L}_0$. To see this, we cite Theorem 14.7 of Part I, according to which t is bounded. Consequently, there is an irreducible term t', such that the equation $t = t'$ holds in $\mathscr{L}_0$. By Exercise 1 of Section 12 of Part I, $t' \equiv S^b(0)$ for some $b \in \mathbb{N}$.

Then, evidently, $f_0(a), f_1(a), f_2(a), \ldots$ is an enumeration of all representable numerical functions $\mathbb{N} \to \mathbb{N}$. The usual diagonal argument now gives a recursive function $f_a(a) + 1$ which is not representable. For, if there is an $e \in \mathbb{N}$ such that $f_a(a) + 1 = f_e(a)$ for all $a \in \mathbb{N}$, we obtain the contradiction $f_e(e) + 1 = f_e(e)$.

For further details, the reader may consult Thibault (1977, 1982) or the text by Shoenfield (Chapters 6.6 and 8.4).

Corollary 2.8. The set of numerical functions representable in $\mathbf{C}(\mathscr{L}_0)$, the free cartesian closed category with weak natural numbers object, is a proper subset of the set of recursive functions, which properly contains the set of primitive recursive functions.

Let us take another look at Lemma 2.5 and its proof. Suppose that $\mathbf{C}(\mathscr{L})$ contains a (strong) natural numbers object. Then, from (ii) in the proof of Lemma 2.5, we can infer that $\Phi(u) =_{\bar{u}} u$ holds, hence (iii) becomes

$$\Psi(0) = a, \quad \Psi(S(u)) =_{\bar{x}} g^{\int} \langle u, \Psi(u) \rangle .$$

Now look at the proof of Theorem 2.4. In particular, suppose we are given $G \in N^{N^k}$ and $H \in N^{(N \times N) \times N^k}$. The argument in $\mathscr{L}(x)$ then produces a term $\Psi_x(u)$ such that the following equations hold:

$$\Psi_x(0) =_{\bar{x}} G^{\int} x, \quad \Psi_x(S(u)) =_{\bar{x}} H^{\int} \langle \langle u, \Psi_x(u) \rangle, x \rangle .$$

If we wish, we may replace $\Psi_x(u)$ by $F^{\int} \langle u, x \rangle$, where $F \in N^{N \times N}$, and we obtain the equations

$$F^{\int} \langle 0, x \rangle =_{\bar{x}} G^{\int} x, \quad F^{\int} \langle S(u), x \rangle \underset{\{u,x\}}{=} H^{\int} \langle \langle u, F^{\int} \langle u, x \rangle \rangle, x \rangle .$$

Note that this equation holds for a variable u, not just for a numeral. This result allows us to do primitive recursion within $\mathscr{L}$, as we normally do in the language of set theory. (See also Part II, Proposition 3.3.)

For example, we can now define a term $+ \in N^{N \times N}$ such that the following equations hold:

$$+^{\int} \langle 0, x \rangle =_{\bar{x}} x, \quad +^{\int} \langle S(u), x \rangle \underset{\{x,u\}}{=} S(+^{\int} \langle u, x \rangle).$$

Adopting the more common notation $x + u$ for $+^{\int} \langle u, x \rangle$, we obtain the usual Dedekind–Peano axioms for addition:

$$x + 0 \underset{x}{=} x, \quad x + S(u) \underset{\{x,u\}}{=} S(x+u).$$

A similar generalization may be carried out for all primitive recursive functions as well as the Ackermann function. Summarizing this discussion, we obtain the following result.

Proposition 2.9. All primitive recursive functions and the Ackermann function may be defined by their usual free variable equations inside any typed λ-calculus $\mathscr{L}$ for which $\mathbf{C}(\mathscr{L})$ has a strong natural numbers object.

We end this section with another application of arithmetization, which is essentially Gödel's incompleteness theorem for pure typed λ-calculus.

Proposition 2.10. In $\mathscr{L}_0$ there is a closed term F of type N^N such that $F{}^{\varsigma}\#k = 0$ holds for all $k \in \mathbb{N}$, yet $F = \lambda_{x \in N} 0$ does not hold.

Proof (sketched). We consider all terms $\varphi(x, y)$ of type N in the variables x and y of type N. They may be enumerated effectively, say $\varphi_0(x, y)$, $\varphi_1(x, y), \ldots$. Let $R(m, n)$ mean that m is the Gödel number of a proof of the equation $\varphi_n(x, \#n) \underset{x}{=} 0$.

The binary relation $R(m, n)$ and its complement may be shown to be primitive recursive, hence representable in $\mathscr{L}_0$. Therefore, there is a term $\varphi_e(x, y)$ such that

(∗) $\quad \neg R(m, n)$ if and only if $\varphi_e(\#m, \#n) = 0$ holds in $\mathscr{L}_0$.

We shall prove that $\varphi_e(\#m, \#e) = 0$ holds in $\mathscr{L}_0$ for all $k \in \mathbb{N}$, but that $\varphi_e(x, \#e) \underset{x}{=} 0$ does not hold. The proposition then follows if we take $F \equiv \lambda_{x \in N} \varphi_e(x, \#e)$.

Suppose $\varphi_e(x, \#e) \underset{x}{=} 0$ holds in $\mathscr{L}_0$. Let k be the Gödel number of its proof, then $R(k, e)$. Now substituting $\#k$ for x, we obtain the equation $\varphi_e(\#k, \#e) = 0$ in $\mathscr{L}_0$, hence $\neg R(k, e)$, by (∗). We have arrived at a contradiction, so we may conclude that $\varphi_e(x, \#e) \underset{x}{=} 0$ does not hold in $\mathscr{L}_0$. But then no k is the Gödel number of its proof, hence $\forall_{k \in \mathbb{N}} \neg R(k, e)$ and therefore, by (∗), for all $k \in \mathbb{N}$, $\varphi_e(\#k, \#e) = 0$.

Corollary 2.11. In pure typed λ-calculus one cannot infer that $F{}^{\varsigma}x \underset{x}{=} 0$ holds for a variable x of type N from the fact that $F{}^{\varsigma}a = 0$ holds for all closed terms a of type N.

Proof. All closed terms of type N are provably equal (convertible) to terms of the form $\#k \equiv S^k 0$.

Corollary 2.12. In the free cartesian closed category with weak natural numbers object, the terminal object 1 is not a generator.

3 Representing numerical functions in toposes

Which numerical functions are representable in a topos? We can reformulate this question linguistically, since every topos is equivalent to one of the form $T(\mathfrak{L})$, the topos generated by an intuitionistic type theory $\mathfrak{L}$. Translating Definition 2.1 into linguistic terms, we obtain the following (for total functions):

Proposition 3.1. A function $f: \mathbb{N}^k \to \mathbb{N}$ is representable in $T(\mathfrak{L})$ if and only if there is a formula $\varphi(x, y)$ in $\mathfrak{L}$, with x a variable of type N^k and y a variable of type N, such that

(*a*) for all $a \in \mathbb{N}^k$, $\vdash \varphi(\#a, \#f(a))$,

(*b*) $\vdash \forall_{x \in N^k} \exists !_{y \in N} \varphi(x, y)$,

where the trunstile denotes provability in $\mathfrak{L}$.

The proof of this will be left as an exercise.

In the literature (e.g. Mendelson 1974), a formula $\varphi(x, y)$ satisfying (*a*) and (*b*) above is said to *strongly represent f in the language* $\mathfrak{L}$. Our question then becomes: which numerical functions are strongly representable in a type theory?

If $\mathfrak{L}$ is consistent, we can rewrite (*a*) of Proposition 3.1 as

(*a'*) for all $a \in \mathbb{N}^k$ and $b \in \mathbb{N}$, $f(a) = b$ if and only if, $\vdash \varphi(\#a, \#b)$.

The 'only if part' of (*a'*) is clearly equivalent to (*a*). To deduce the 'if part', we use (*b*). Then from $\vdash \varphi(\#a, \#b)$ and (*a*) it follows that $\vdash \#f(a) = \#b$. We need consistency of $\mathfrak{L}$ to infer that $f(a) = b$.

This argument also tells us that if f is strongly represented in a consistent type theory $\mathfrak{L}$ by a formula $\varphi(x, y)$, then, for all $a \in \mathbb{N}^k$, there exists a unique $b \in \mathbb{N}$ such that $\vdash \varphi(\#a, \#b)$ in $\mathfrak{L}$. In the case of pure type theory $\mathscr{L}_0$, we have even a stronger result.

Proposition 3.2. Suppose $\varphi(x, y)$ is a formula of pure type theory $\mathfrak{L}_0$, with x a variable of type N^k and y a variable of type N. If, for every $a \in \mathbb{N}^k$, there exists $b \in \mathbb{N}$ such that $\vdash \varphi(\#a, \#b)$, then there is a recursive function $f: \mathbb{N}^k \to \mathbb{N}$ such that, for all $a \in \mathbb{N}^k$, $\vdash \varphi(\#a, \#f(a))$.

Proof. Let Proof (q, p) be the (primitive) recursive relation asserting that p is the Gödel number of a proof of a closed formula in $\mathfrak{L}_0$ which Gödel number q. Thus

$$\vdash \varphi(\#a, \#b) \text{ if and only if } \mathrm{Proof}(q(a, b), p) \text{ for some } p \in \mathbb{N},$$

where $q(a, b)$ is the Gödel number of the formula $\varphi(\#a, \#b)$. Let $(-)_0$, $(-)_1$: $\mathbb{N}^2 \rightrightarrows \mathbb{N}$ be primitive recursive inverses of a pairing function (see Section 1,

Exercise 5) and suppose that for every $a \in \mathbb{N}^k$ there is a $b \in \mathbb{N}$ such that $\vdash \varphi(\#a, b)$. Then $\mathrm{Proof}(q(a,b),p)$ for some $p \in \mathbb{N}$. Define

$$f(a) \equiv (\mu_k \; \mathrm{Proof}(q(a,(k)_0),(k)_1))_0.$$

In view of the assumption on φ, for every a there is a k such that $\mathrm{Proof}(q(a,(k)_0), (k)_1)$, hence f is a total recursive function. Moreover, $\vdash \varphi(\#a, \#f(a))$, as required.

The crucial fact about $\mathfrak{L}_0$ in the above argument was the existence of a recursive proof predicate. In fact, Proposition 3.2 may be generalized to other 'recursively axiomatizable' type theories, as may be the following result.

Corollary 3.3. If a numerical function is strongly representable in $\mathfrak{L}_0$, then it is recursive.

Proof. Suppose f is strongly represented by φ. By (*a*) of Proposition 3.1, for every $a \in \mathbb{N}^k$ there is a $b \in \mathbb{N}$ such that $\vdash \varphi(\#a, \#b)$. Hence, by Proposition 3.2, there is a recursive function $g\colon \mathbb{N}^k \to \mathbb{N}$ such that $\vdash \varphi(\#a, \#g(a))$. It now follows from (*b*) of Proposition 3.1 that $\vdash \#f(a) = \#g(a)$. Since $\mathfrak{L}_0$ is consistent, we may infer that $f(a) = g(a)$ for all a, hence f is recursive.

Unfortunately, the converse of Corollary 3.3 fails.

Proposition 3.4. Not every recursive function is strongly representable in pure type theory.

Proof. We consider unary functions $f\colon \mathbb{N} \to \mathbb{N}$. We shall diagonalize over the set of $\forall\exists!$-proofs.

Let E be the recursively enumerable set of Gödel numbers of proofs of formulas of the form $\forall_{x \in N} \exists!_{n \in N} \varphi(x, y)$. Thus, for each $e \in E$ there is a formula $\varphi_e(x, y)$ such that

$$\vdash \forall_{x \in N} \exists!_{y \in N} \varphi_e(x, y).$$

By the unique existence property for $\mathfrak{L}_0$ (Part II, Lemma 20.3), for every $a \in \mathbb{N}$ there is a $b \in \mathbb{N}$ such that $\vdash \varphi_e(\#a, \#b)$. Therefore, by Corollary 3.3, there is a recursive function $f_e\colon \mathbb{N} \to \mathbb{N}$ such that $\vdash \varphi_e(\#a, \#f_e(b))$ for all $a \in \mathbb{N}$.

Let $h\colon \mathbb{N} \to \mathbb{N}$ be a recursive function which enumerates E. Thus, $e \in E$ if and only if $e = h(m)$ for some $m \in \mathbb{N}$. Consider the function

$$g(m) = f_{h(m)}(m) + 1.$$

Clearly, g is computable, hence recursive. On the other hand, g is not strongly representable. For, if it were strongly representable by φ then

$\vdash \forall_{x \in N} \exists !_{y \in N} \varphi(x, y)$, hence $\varphi \equiv \varphi_{h(k)}$, where $h(k)$ is the Gödel number of the proof of this theorem. Moreover, since φ strongly represents g,

$$g(m) = f_{h(k)}(m) \text{ for all } m \in \mathbb{N}.$$

But then,

$$f_{h(k)}(k) + 1 = g(k) = f_{h(k)}(k),$$

a contradiction.

Translating Corollary 3.3 and Proposition 3.4 into statements about arrows in $T(\mathfrak{L}_0)$, the free topos, we obtain the following.

Corollary 3.5. Every arrow $f: N^k \to N$ in the free topos $T(\mathfrak{L}_0)$ is sent by the functor $\Gamma \equiv \mathrm{Hom}(1, -): T(\mathfrak{L}_0) \to \mathbf{Sets}$ onto a recursive function $\mathbb{N}^k \cong \Gamma(N^k) \to \Gamma(N) \cong \mathbb{N}$, but not all recursive functions arise in this manner.

Proof. The isomorphism $\Gamma(N) \cong \mathbb{N}$ was established in Part II (Lemma 20.3).

Remark 3.6. If a numerical function is representable in every cartesian closed category, it is representable in every topos. Hence the primitive recursive functions and the Ackermann function (Theorem 2.4) are strongly representable in every type theory (Proposition 3.1). Since type theories are more powerful than typed λ-calculi, one expects more numerical functions to be strongly representable in pure type theory than in pure λ-calculus. (See Fortune *et al.* 1983.)

Actually, the primitive recursive functions and the Ackermann function are more than representable in a topos: their definitions may be carried out in the internal language of the topos, in view of Proposition 2.9.

The situation is radically different in *classical* type theory, in which the Boolean axiom $\beta \equiv \forall_{t \in \Omega}(t \vee \neg t)$ is assumed. As we shall see, every total recursive function then becomes strongly representable. First we shall look at a weak form of representability called 'numeralwise' representability by Kleene.

Definition 3.7. A numerical function $f: \mathbb{N}^k \to \mathbb{N}$ is *numeralwise representable* in a type theory $\mathfrak{L}$ if there is a formula $\varphi(x, y)$, with x of type N^k and y of N such that

(a) for all $a \in \mathbb{N}^k$, $\vdash \varphi(\#a, \#f(a))$

(b') for all $a \in \mathbb{N}^k$, $\vdash \exists !_{y \in N} \varphi(\#a, y)$

Here (b') differs from (b) in the definition of strong representability (Proposition 3.1) in that functionality of φ is proved only for k-tuples of numerals and not for variables of type N^k. For classical type theories, it turns out that numeralwise is not weaker than strong.

Proposition 3.8. (V. Huber-Dyson). A numerical function is numeralwise representable in a classical type theory if and only if it is strongly representable.

Proof. Suppose $f: \mathbb{N}^k \to \mathbb{N}$ is numeralwise representable by $\varphi(x, y)$, we shall find another formula $\psi(x, y)$ which strongly represents f.

First consider the formula

$$\varphi'(x, y) \equiv (\exists_{z\in N}\varphi(x, z) \Rightarrow \varphi(x, y)).$$

By classical logic, we have $\vdash \forall_{x\in N^k}\exists_{y\in N}\varphi'(x, y)$. Now define

$$\psi(x, y) \equiv (\varphi'(x, y) \wedge \forall_{z\in N}(\varphi'(x, z) \Rightarrow y \leqslant z)).$$

(For the meaning of $\leqslant$ see Remark 3.9 below.)
Applying the (classical) least number principle to φ', we deduce that $\vdash \forall_{x\in N^k}\exists!_{y\in N}\,\psi(x, y)$. It remains to prove that, for all $a\in\mathbb{N}^k$, $\vdash \psi(\#a, \#f(a))$.

Now $\vdash \varphi(\#a, \#f(a))$ by (*a*) above, hence clearly $\vdash \varphi'(\#a, \#f(a))$. It remains to prove that $\vdash \forall_{z\in N}(\varphi'(\#a, z) \Rightarrow \#f(a) \leqslant z)$.

We argue informally. Suppose $\varphi'(\#a, z)$. Since $\exists_{y\in N}\varphi(\#a, y)$ by (b'), therefore $\varphi(\#a, z)$. But since $\exists!_{y\in N}\varphi(\#a, y)$ by (b'), $z = \#f(a)$, and so $\#f(a) \leqslant z$.

Remark 3.9. The above proof made use of the symbol $\leqslant$ and its properties in the type theory $\mathfrak{L}$. This is defined easily enough: for terms a, b of type N, let $a \leqslant b$ mean that $\langle a, b\rangle$ belongs to (the intersection of) all $u\in P(N \times N)$ such that

$$\forall_{x\in N}\langle x, x\rangle \in u$$

and

$$\forall_{x\in N}\forall_{y\in N}(\langle x, y\rangle \in u \Rightarrow \langle x, Sy\rangle \in u).$$

However, it may be a little tedious to derive the usual properties of $\leqslant$ from this definition, for example, the transitive law, not to speak of the least number principle (assuming the Boolean axiom of course).

An alternative approach is to introduce function symbols $+$ and $\cdot$ into the type theory $\mathfrak{L}$ to satisfy the usual Peano axioms:

$$x + 0 =_{x} x, \quad x\cdot 0 =_{x} 0,$$
$$x + Sy =_{\{x,y\}} S(x + y), \quad x\cdot Sy =_{\{x,y\}} (x\cdot y) + x,$$

and then refer to the book by Kleene, where $<$ and $\leqslant$ are defined as follows:

$$a < b \equiv \exists_{z\in N}(Sz + a = b),$$
$$a \leqslant b \equiv (a < b \vee a = b).$$

All necessary properties of these symbols are then derived in the book by Kleene, including the least number principle if the Boolean axiom holds.

How do we introduce the symbols $+$ and $\cdot$ in a type theory $\mathfrak{L}$? Since the topos $T(\mathfrak{L})$ generated by $\mathfrak{L}$ is cartesian closed, we may use Proposition 2.9 to find an arrow $+: N \times N \to N$ in $T(\mathfrak{L})$ such that

$$+\langle x, 0\rangle \cdot \underset{x}{=} \cdot x, \quad +\langle x, Su\rangle \cdot \underset{\{x,u\}}{=} \cdot S + \langle x, u\rangle.$$

If we write $a+b$ for $+\langle a, b\rangle$, this shows that the statements

$$\forall_{x\in N}(x+0=x), \quad \forall_{x\in N}\forall_{u\in N}(x+Su=S(x+u))$$

are satisfied in $T(\mathfrak{L})$, that is, provable in $LT(\mathfrak{L})$. (See also Exercise 4.)

For terms a and b of type N in $LT(\mathfrak{L})$, we may now define $a<b$ as above in terms of addition. Now recall from Section 14 of Part II that the translation $\eta_{\mathfrak{L}}: \mathfrak{L} \to LT(\mathfrak{L})$ given by

$$\eta_{\mathfrak{L}}(a) \equiv \mathbf{a}, \quad |\mathbf{a}| \equiv \{\langle *, a\rangle\}$$

is a conservative extension. We may therefore define $a<b$ in $\mathfrak{L}$ to mean $\mathbf{a}<\mathbf{b}$ in $LT(\mathfrak{L})$ and infer all the expected properties of $<$ in $\mathfrak{L}$.

The proof of Proposition 3.8 used classical logic in two places: to show $\vdash \forall_{x\in N^k}\exists_{y\in N}\varphi'(x, y)$ we made use of independence of premisses for $\exists_{z\in N}\varphi(x, z)$ and we applied the least number principle to φ'. Intuitionistically, independence of premisses is known to be applicable only when the hypothesis is a negative formula (see Part II, Section 21). The least number principle is only known to be valid for decidable predicates (Kleene 1952, §40, *149°).

In his famous paper of 1931, Gödel characterized the representable functions of pure classical type theory, in our notation $\mathfrak{L}_0/\beta$.

Theorem 3.10. A numerical function is numeralwise (or strongly) representable in pure classical type theory if and only if it is recursive.

Proof. If f is numeralwise (hence strongly) representable in $\mathfrak{L}_0/\beta$, then it is recursive by the generalization of Corollary 3.3 to 'recursively axiomatizable type theories' (discussed just prior to Corollary 3.3).

In the converse direction, we already know that all primitive recursive functions are strongly representable in $\mathfrak{L}_0$, hence a fortiori in $\mathfrak{L}_0/\beta$. It remains to show that the set of numeralwise representable functions is closed under the minimization scheme.

Suppose $f: \mathbb{N}^k \to \mathbb{N}$ is defined by

$$f(a) \equiv \mu_b(g(a, b)=0),$$

where for every $a \in \mathbb{N}^k$ there is a $b \in \mathbb{N}$ such that $g(a, b)=0$. Suppose furthermore that $g: \mathbb{N}^k \times \mathbb{N} \to \mathbb{N}$ is numeralwise represented by $\varphi(x, y, z)$.

Put

$$\Psi(x, y) \equiv \varphi(x, y, 0) \wedge \forall_{w \in N}(w < y \Rightarrow \neg\varphi(x, w, 0)).$$

A straightforward calculation then shows that Ψ numeralwise represents f.

Theorem 3.10 is even valid for very weak fragments of first-order arithmetic (see the books by Kleene and Shoenfield).

What goes wrong with the diagonal argument which was used to prove that not every recursive function is strongly representable in $\mathfrak{L}_0$ (Proposition 3.4) when $\mathfrak{L}_0$ is replaced by $\mathfrak{L}_0/\beta$? The point is that, having proved classically $\vdash \forall_{x \in N} \exists !_{y \in N} \varphi(x, y)$, we cannot guarantee that φ represents a total function, as the unique existence property fails for $\mathfrak{L}_0/\beta$.

For example, let γ be an undecidable sentence in $\mathfrak{L}_0/\beta$ and consider the formula

$$\varphi(x, y) \equiv (\gamma \wedge y = 0) \vee (\neg\gamma \wedge y = S0)$$

in $\mathfrak{L}_0/\beta$. Classically we have $\vdash \forall_{x \in N} \exists !_{y \in N} \varphi(x, y)$; but, since φ does not contain x, any total function represented by φ would have to be constantly 0 or 1. In either case we would be able to decide γ.

Taking $k = 0$, we see that the above formula $\varphi(x, y)$ defines a 'non-standard numeral' in $T(\mathfrak{L}_0/\beta)$, that is, an arrow $1 \to N$ not of the form S^n0 with $n \in \mathbb{N}$. This shows that the canonical mapping $\S: \mathbb{N} \to \mathrm{Hom}(1, N)$ such that $\S n \equiv S^n 0$ is not surjective. Hence an arrow $g: N^k \to N$ in $T(\mathfrak{L}_0/\beta)$, the free Boolean topos, gives rise to a function

$$\mathbb{N}^k \xrightarrow{\S^k} \mathrm{Hom}(1, N^k) \xrightarrow{\mathrm{Hom}(1, g)} \mathrm{Hom}(1, N)$$

which, in general, yields only a partial function from $\mathbb{N}^k$ to $\mathbb{N}$. If it so happens that g represents a total function $f: \mathbb{N}^k \to \mathbb{N}$, then f must be recursive by Theorem 3.10.

In pure intuitionistic type theory $\mathfrak{L}_0$ all numerals are standard, that is, all terms of type N are provably equal to terms of the form S^n0. This suggests that we may do better in representing partial recursive functions in $\mathfrak{L}_0$ than total ones.

Definition 3.11. A partial function f from $\mathbb{N}^k$ to $\mathbb{N}$ is numeralwise representable in a type theory $\mathfrak{L}$ if there is a formula $\varphi(x, y)$ in $\mathfrak{L}$ such that

(i) $\vdash \forall_{x \in N^k} \forall_{y \in N} \forall_{z \in N}((\varphi(x, y) \wedge \varphi(x, z)) \Rightarrow y = z)$

and, for all $a \in \mathbb{N}^k$, $f(a)$ is defined and equal to b if and only if

(ii) $\vdash \varphi(\#a, \#b)$.

Theorem 3.12. A partial numerical function is numeralwise representable in pure type theory $\mathfrak{L}_0$ if and only if it is a partial recursive function.

Proof. Suppose a partial function f from $\mathbb{N}^k$ to $\mathbb{N}$ is numeralwise representable in $\mathfrak{L}_0$ by a formula $\varphi(x, y)$. As in the proof of Proposition 3.2, we have

$$f(a) = (\mu_k \text{ Proof } (q(a, (k)_0), (k)_1))_0,$$

where $q(a, b)$ is the Gödel number of the formula $\varphi(\#a, \#b)$. Thus f is seen to be a partial recursive function.

In the converse direction, it suffices to show that, if $g(a, n)$ is a numeralwise representable partial recursive function from $\mathbb{N}^k \times \mathbb{N}$ to $\mathbb{N}$, then $f(a) \equiv \mu_n(g(a, n) = 0)$ is also numeralwise representable. In view of Kleene's normal form theorem (Section 1), we may assume that $g(a, n)$ is a total function.

We may write the definition of $f(a)$ in the language of set theory thus:

$$g(a, f(a)) = 0 \wedge \forall_{b \in \mathbb{N}}(b < f(a) \Rightarrow g(a, b) \neq 0).$$

Now g is representable by $\varphi(x, y, z)$, hence

$$g(a, b) = c \Leftrightarrow \vdash \varphi(\#a, \#b, \#c).$$

We claim that the following formula represents f:

$$\psi(x, y) \equiv (\varphi(x, y, 0) \wedge \forall_{z \in N}(z < y \Rightarrow \neg\varphi(x, z, 0))).$$

We must show two things:

(i) $\vdash_x \forall_{y \in N} \forall_{z \in N}((\psi(x, y) \wedge \psi(x, z)) \Rightarrow y = z)$,

(ii) $\vdash \psi(\#a, \#f(a))$ whenever $f(a)$ is defined.

To prove (i), we argue informally as follows. Suppose $\psi(x, y)$ and $\psi(x, z)$. It follows from the second assumption that $\varphi(x, z, 0)$, hence from the first that $\neg(y > z)$. The law of trichotomy then yields $y \leqslant z$. Similarly we obtain $z \leqslant y$, hence $y = z$. This shows (i).

To prove (ii), assume that $f(a)$ is defined. Then $g(a, f(a)) = 0$, hence $\vdash \varphi(\#a, \#f(a), 0)$. It remains to show that also

$$\vdash \forall_{z \in N}(z < \#f(a) \Rightarrow \neg\varphi(\#a, z, 0)).$$

Again, we argue informally as follows. Suppose $z < \#f(a)$. Then it is easily seen that $z = \#b$ for some $b < f(a)$. Therefore $g(a, b) \neq 0$, say $g(a, b) = c + 1$. (Remember that g is a total recursive function.) Therefore $\varphi(\#a, \#b, \#(c + 1))$ and so, $\neg\varphi(\#a, \#b, 0)$, that is, $\neg\varphi(\#a, z, 0)$. (Recall that, if $\varphi(x, z, t)$, t is unique.) This completes the proof of (ii).

For additional results along these lines, see Coste-Roy *et al.* (1980).

Exercises

1. Write out a proof of Proposition 3.1.
2. Using the first definition of $a \leqslant b$ suggested in Remark 3.9, prove some of the usual properties of $\leqslant$.
3. Using the definition of $a < b$ in terms of addition given in Remark 3.9, show that, for each $n \in \mathbb{N}$,

   $\vdash \forall_{x \in N}(x < \#n \Rightarrow (x = \#0 \vee x = \#1 \vee \cdots \vee x = \#(n-1)))$.
4. In a type theory $\mathfrak{L}$, let $\alpha \in P(N \times N \times N)$ be the intersection of all $u \in P(N \times N \times N)$ such that $\forall_{x \in N}\langle 0, x, x\rangle \in u$ and $\forall_{x \in N}\forall_{y \in N}(\langle x, y, z\rangle \in u \Rightarrow \langle x, Sy, Sz\rangle \in u)$. Show that $\vdash \forall_{x \in N}\forall_{y \in N}\exists!_{z \in N}\langle x, y, z\rangle \in \alpha$. Infer that this also justifies the introduction of a function symbol $+$ satisfying the Dedekind–Peano axioms.

4 Representing numerical functions in C-monoids

We recall from Part I that an *extended λ-calculus* is an untyped λ-calculus with surjective pairing, that is, with term-forming operations $\pi(-)$, $\pi'(-)$ and $(-,-)$ satisfying

$$\pi(a, b) = a, \quad \pi'(a, b) = b, \quad (\pi(c), \pi'(c)) = c.$$

It was shown there that the category of extended λ-calculi is isomorphic to a certain equational category of monoids with additional structure, called C-monoids. Moreover, C-monoids are in bijective correspondence with cartesian closed categories which have, up to isomorphism, only two objects 1 and U such that $U^U \cong U \cong U \times U$.

One of the earliest results in the λ-calculus was the identification of recursive functions with λ-definable ones. This result had also been extended to partial functions, and here we shall re-examine the result for partial functions in the context of extended λ-calculi.

In an extended λ-calculus $\mathscr{L}$, natural numbers are not given as in typed λ-calculus but may be defined according to Church as follows. The idea is that 2 is the function which to any function f assigns its iterate $f^2 \equiv f \circ f$. Here $f \circ g$ is defined by

$$f \circ g \equiv \lambda_x(f^{\varsigma}(g^{\varsigma}x)).$$

The reader will recall the underlying implicit ontology according to which all entities are functions. In general, $n^{\varsigma} f \equiv f^n$, which is defined in the usual way by induction. We are thus led to define $0^{\varsigma}x$ as I and $(S^{\varsigma}y)^{\varsigma}x$ as $x \circ (y^{\varsigma}x)$, that is,

$$0 \equiv \lambda_x I, \quad S \equiv \lambda_y \lambda_x(x \circ (y^{\varsigma}x)).$$

As usual,

$$1 \equiv S^{\ulcorner}0, \quad 2 \equiv S^{\ulcorner}1, \ldots$$

and it is easily verified that $1 = I$ holds in $\mathscr{L}$.

From now on we write $\#n \equiv S^{n\ulcorner}0$, to distinguish the natural number n from its representation in $\mathscr{L}$. Note however that $\#0 \equiv 0$. Mimicking the definition in typed λ-calculus, we say:

Definition 4.1. A numerical function $f: \mathbb{N}^k \to \mathbb{N}$ is *represented* by a closed term F of an extended λ-calculus $\mathscr{L}$ if the equation

$$F^{\ulcorner}(\#n_1, \ldots \#n_k) = \#f(n_1, \ldots n_k)$$

holds in $\mathscr{L}$ for all k-tuples $(n_1, \ldots n_k)$ of natural numbers. We shall also say that f is *represented* in a C-monoid $\mathscr{M}$, if $\mathscr{M}$ corresponds to $\mathscr{L}$.

Proposition 4.2 (Church). All primitive recursive functions are representable in any extended λ-calculus, hence in any C-monoid $\mathscr{M}$.

Proof. The zero function is represented by $\lambda_x 0$ and the successor function by S as defined above. We illustrate the representation of the projections by an example: p_2^3 is represented by $\lambda_x(\pi'(\pi(x)))$. If $g(a_1, \ldots, a_k), \ldots, g_m(a_1, \ldots a_k)$ are represented by $G_1, \ldots G_m$ and $h(b_1, \ldots, b_m)$ is represented by H, $h(g(a_1, \ldots, a_k), \ldots, g_m(a_1, \ldots, a_k))$ is represented by $\lambda_x H^{\ulcorner}(G_1{}^{\ulcorner}x, \ldots, G_m{}^{\ulcorner}x)$. It remains to prove that the set of representable functions is closed under primitive recursion.

Suppose $g(a)$ and $h(n, m, a)$ are represented by G and H respectively, where $a \equiv (a_1, \ldots, a_k)$. We claim that $f(n, m)$, defined by

$$f(0, a) = g(a), \quad f(n+1, a) = h(n, f(n, a), a)$$

is also representable. We seek a closed term F such that the equations

$$F^{\ulcorner}(0, \#a) = G^{\ulcorner}\#a, \quad F^{\ulcorner}(\#(n+1), \#a) = H^{\ulcorner}(\#n, F^{\ulcorner}(\#n, \#a), \#a)$$

hold in $\mathscr{L}$. Our argument follows the treatment in Hindley, Lercher and Seldin (1972). Suppose for the moment we can find a term $\Psi(z, u, v)$ in $\mathscr{L}$ such that the following equations hold in $\mathscr{L}$:

$$\Psi(0, u, v) = u, \quad \Psi(S^{\ulcorner}\#n, u, v) = v^{\ulcorner}(\#n, \Psi(\#n, u, v)).$$

Then we define

$$\beta(x) \equiv \lambda_z H^{\ulcorner}(\pi(z), \pi'(z), x), \quad F \equiv \lambda_w \Psi(\pi(w), G^{\ulcorner}\pi'(w), \beta(\pi'(w))).$$

We then calculate

$$F^{\ulcorner}(0, x) = \Psi(0, G^{\ulcorner}x, \beta(x)) = G^{\ulcorner}x,$$
$$F^{\ulcorner}(S^{\ulcorner}\#n, x) = \Psi(S^{\ulcorner}\#n, G^{\ulcorner}x, \beta(x))$$

$$= \beta(x)^{\varsigma}(\#n, \Psi(\#n, G^{\varsigma}x, \beta(x)))$$
$$= H^{\varsigma}(\#n, \Psi(\#n, G^{\varsigma}x, \beta(x)), x)$$
$$= H^{\varsigma}(\#n, F^{\varsigma}(\#n, x), x).$$

Replacing x by $\#a$, we see that F satisfies the claimed equations.

It remains to construct Ψ with the required properties. Let

$$\theta_v \equiv (\lambda_x(S^{\varsigma}\pi(x), v^{\varsigma}x)$$

and write

$$(z^{\varsigma}\theta_v)^{\varsigma}(0, u) \equiv (\Phi(z, u, v), \Psi(z, u, v)),$$

that is, $\Phi(z, u, v) \equiv \pi(\text{LHS})$ and $\Psi(z, u, v) \equiv \pi'(\text{LHS})$. In particular, when z is replaced by $\#n$, we have

$$\theta_v^{n\varsigma}(0, u) = (\Phi(\#n, u, v), \Psi(\#n, u, v)).$$

Now

$$\theta_v^0(0, u) = (0, u),$$

hence

$$\Phi(0, u, v) = 0, \quad \Psi(0, u, v) = u.$$

Moreover,

$$\theta_v^{n+1}(0, u) = \theta_v{}^{\varsigma}\theta_v^n(0, u)$$
$$= \theta_v{}^{\varsigma}(\Phi(\#n, u, v), \Psi(\#n, u, v))$$
$$= (S^{\varsigma}\Phi(\#n, u, v), v^{\varsigma}(\Phi(\#n, u, v), \Psi(\#n, u, v))),$$

and therefore

$$\Phi(S^{\varsigma}\#n, u, v) = S^{\varsigma}\Phi(\#n, u, v).$$
$$\Psi(S^{\varsigma}\#n, u, v) = v^{\varsigma}(\Phi(\#n, u, v), \Psi(\#n, u, v)).$$

By induction on n, we see that

$$\Phi(\#n, u, v) = \#n,$$

and so we obtain

$$\Psi(0, u, v) = u, \quad \Psi(S^{\varsigma}\#n, u, v) = v^{\varsigma}(\#n, \Psi(\#n, u, v))$$

as required.

Actually, not only the primitive recursive functions, but all recursive functions are representable in any extended λ-calculus. However, the proof of this involves partial recursive functions, so we may as well state the result more generally.

Definition 4.3. A partial numerical function f from $\mathbb{N}^k$ to $\mathbb{N}$ is represented by a closed term F of an extended λ-calculus if, for all k-tuples $a_1, \ldots, a_k$ of natural numbers such that $f(a_1, \ldots, a_k)$ is defined,

$$F^{\varsigma}(\#a_1, \ldots, \#a_k) = \#f(a_1, \ldots, a_k).$$

Theorem 4.4 (Church). All partial recursive functions are representable in any extended λ-calculus, hence in any C-monoid $\mathcal{M}$.

Proof. Since we already know that all primitive recursive functions are representable, in view of Kleene's Normal Form Theorem (Section 1), it suffices to show that the minimization scheme when applied to a primitive recursive predicate will yield a representable partial function.

Suppose $g(a, n)$ is a primitive recursive function with $a \in \mathbb{N}^k$, $n \in \mathbb{N}$, and suppose the partial recursive function $f(a)$ is defined by

$$f(a) \equiv \mu_n(g(a, n) = 0).$$

We claim that $f(a)$ is representable, and more generally so is the partial recursive function

$$f(a, b) \equiv \mu_n(n \geqslant b \wedge g(a, n) = 0).$$

of course, $f(a) = f(a, 0)$.

Since g is representable, there is a closed term G such that $\#g(a, n) = (G^{\mathsf{f}}\#a)^{\mathsf{f}}\#n$ holds in $\mathscr{L}$. We seek a closed term F such that $\#f(a, b) = F^{\mathsf{f}}(\#a, \#b)$ whenever $f(a, b)$ is defined.

Suppose for the moment, we can find a term $P(u, v)$ in $\mathscr{L}$ such that

$$P(u, \#n) = \#n \text{ if } u^{\mathsf{f}}\#n = 0,$$
$$P(u, \#n) = P(u, \#n + 1) \text{ if } u^{\mathsf{f}}\#n = \#(m + 1) \text{ for some } m \in \mathbb{N}.$$

Let

$$F^{\mathsf{f}}(x, y) \equiv P^{\mathsf{f}}(G^{\mathsf{f}}x, y),$$

then

$$F^{\mathsf{f}}(\#a, \#b) = \#b \text{ if } (G^{\mathsf{f}}\#a)^{\mathsf{f}}\#b = 0, \text{ i.e. } g(a, b) = 0.$$

and

$$F^{\mathsf{f}}(\#a, \#b) = F^{\mathsf{f}}(\#a, \#(b + 1)) \text{ if } g(a, b) \neq 0.$$

We claim that F represents f. For suppose $f(a, b)$ is defined and $= c$, then $g(a, c) = 0$ and $c \geqslant b$, but $g(a, n) \neq 0$ for any n such that $c > n \geqslant b$. Hence $F^{\mathsf{f}}(\#a, \#c) = \#c$ and $F^{\mathsf{f}}(\#a, \#n) = F^{\mathsf{f}}(\#a, \#n + 1)$ for all $c > n \geqslant b$. If $c = b$, we get $F^{\mathsf{f}}(\#a, \#b) = \#c = \#f(a, b)$. If $c > b$, we get $F^{\mathsf{f}}(\#a, \#b) = F^{\mathsf{f}}(\#a, \#b + 1) = \cdots = F^{\mathsf{f}}(\#a, \#c) = \#c = \#f(a, b)$. Thus F represents f.

It remains to construct the term $P(u, v)$ with the required property. Let us assume for the moment that, for any terms α and β, we have a term $Q(\alpha, \beta)$ such that

$$Q(\alpha, \beta)^{\mathsf{f}}0 = \beta, \quad Q(\alpha, \beta)^{\mathsf{f}}\#(m + 1) = \alpha$$

for all $m \in \mathbb{N}$. We shall specify $\alpha \equiv \alpha_u$ and β presently, but in the meantime we define

$$P(u, v) \equiv (Q(\alpha_u, \beta)^{\text{f}}(u^{\text{f}} v))^{\text{f}}(Q(\alpha_u, \beta), v)$$

If $u^{\text{f}}\#n = 0$, we get

$$P(u, \#n) = \beta^{\text{f}}(Q(\alpha_u, \beta), \#n) = \#n$$

provided we stipulate

$$\beta^{\text{f}}(u, v) = v.$$

If $u^{\text{f}}\#n = \#(m+1)$, we get

$$P(u, \#n) = \alpha_u{}^{\text{f}}(Q(\alpha_u, \beta), \#n).$$

Since we want this to be equal to

$$P(u, \#(n+1)) = (Q(\alpha_u, \beta)^{\text{f}}(u^{\text{f}}\#(n+1)))^{\text{f}}(Q(\alpha_u, \beta), \#(n+1)),$$

we stipulate that

$$\alpha_u{}^{\text{f}}(x, y) \equiv (x^{\text{f}}(u^{\text{f}}(S^{\text{f}} y)))^{\text{f}}(x, S^{\text{f}} y).$$

It remains to find $Q(\alpha, \beta)$ with the desired property. Consider the numerical function $\delta(n)$ given by

$$\delta(0) = 1, \quad \delta(m+1) = 0.$$

This is clearly primitive recursive, hence representable by a closed term Δ. Thus

$$\Delta^{\text{f}}0 = \#1, \quad \Delta^{\text{f}}\#(m+1) = 0.$$

Define

$$R(\alpha, \beta)^{\text{f}} x \equiv (x^{\text{f}}(K^{\text{f}}\beta))^{\text{f}}\alpha,$$

where $K \equiv \lambda_x \lambda_y x$, so $(K^{\text{f}}x)^{\text{f}} y \equiv x$. Then

$$R(\alpha, \beta)^{\text{f}}0 = (0^{\text{f}}(K^{\text{f}}\beta))^{\text{f}}\alpha = I^{\text{f}}\alpha = \alpha,$$
$$R(\alpha, \beta)^{\text{f}}\#1 = (I^{\text{f}}(K^{\text{f}}\beta))^{\text{f}}\alpha = (K^{\text{f}}\beta)^{\text{f}}\alpha = \beta.$$

Now let

$$Q(\alpha, \beta)^{\text{f}} x \equiv R(\alpha, \beta)^{\text{f}}(\Delta^{\text{f}} x).$$

Then

$$\begin{aligned} Q(\alpha, \beta)^{\text{f}}0 &= R(\alpha, \beta)^{\text{f}}(\Delta^{\text{f}}0) \\ &= R(\alpha, \beta)^{\text{f}}\#1 = \beta, \\ Q(\alpha, \beta)^{\text{f}}\#m+1 &= R(\alpha, \beta)^{\text{f}}(\Delta^{\text{f}}\#m+1) \\ &= R(\alpha, \beta)^{\text{f}}0 = \alpha. \end{aligned}$$

This completes the proof.

The above proof, which is presumably of some antiquity, was influenced by that in the book of Hindley, Lercher and Seldin, but we hope that our presentation is a bit more transparent. Even so, we feel that we have not fully exploited the possibilities of simplification due to the presence of surjective pairing.

We cannot expect a converse to Theorem 4.4 unless we restrict attention to *pure extended λ-calculus*, the language corresponding to the free C-monoid generated by the empty set.

Corollary 4.5 (Church). A partial numerical function is representable in pure extended λ-calculus, that is, in the free C-monoid generated by the empty set, if and only if it is a partial recursive function.

The proof in the converse direction is the same as that already used for Theorem 3.12 and Proposition 3.2.

The following exercise, adapted from the book by Barendregt (1981), where further references will be found, were added as an afterthought. They give alternative and easier proofs of propositions 4.3 and 4.4.

Exercises

1. Show that numerals in an extended λ-calculus can also be defined by $\#n \equiv S^{n\,\varsigma}0$, where

   $$0 \equiv (\lambda_x \pi(x), I), \quad S \equiv \lambda_x(\lambda_y \pi'(y), x),$$

   and that π' acts like a predecessor function:

   $$\pi'(S^\varsigma x) = x, \quad \text{but } \pi'(0) = I.$$

   Infer that

   $$\pi(\#n)^\varsigma(a, b) = \begin{cases} a & \text{if } n = 0, \\ b & \text{if } n > 0. \end{cases}$$

2. Suppose the numerical functions $g(a)$ and $h(n, m, a)$ are represented by G and H (as in the proof of Proposition 4.2, but using the numerals of Exercise 1) and that

   $$f(0, a) = g(a), \; f\hat{\ }(n+1, a) = h(n, f(n, a), a).$$

   Show that $f(n, a)$ is represented by F provided F satisfies

   $$F^\varsigma(x, y) \equiv \pi(x)^\varsigma(G^\varsigma y, H^\varsigma(\pi'(x), F^\varsigma(\pi'(x), y), y)).$$

   Such an F may be constructed using the following 'fixpoint operator'

   $$\text{Fix}(z) \equiv R(z)^\varsigma R(z), \quad R(z) \equiv \lambda_x(z^\varsigma(x^\varsigma x)),$$

   which satisfies $z^\varsigma \text{Fix}(z) = \text{Fix}(z)$.

3. Taking

$$K_x \equiv \lambda_y \lambda_z((\pi(G^{\varsigma}(x,z))^{\varsigma}(z, y^{\varsigma}(S^{\varsigma}z))),$$

and $H_x \equiv \text{Fix}(K_x)$, verify that

$$H_x{}^{\varsigma}c = (K_x{}^{\varsigma}H_x)^{\varsigma}c = \begin{cases} c & \text{if } G^{\varsigma}(x,c) = 0 \\ H_x{}^{\varsigma}(S^{\varsigma}c) & \text{if } G^{\varsigma}(x,c) = S^{\varsigma}d_x \end{cases}$$

from some term d_x.

4. Consider the partial function $f(a) \equiv \mu_m(g(a,m) = 0)$, which is defined for a if $g(a,m) = 0$ for some m. Suppose $g(a,m)$ is represented by G in the sense that $\#g(a,m) = G^{\varsigma}(\#a, \#m)$. Show that $F \equiv \lambda_x(H_x{}^{\varsigma}0)$ represents $f(a)$ (see Definition 4.3).

Historical Comments on Part III

Section 1. For the history of the primitive rescursive and recursive functions, along with the Church–Turing thesis, see Kleene (1952) and Mendelson (1974) and their references. Of course the first person to define precisely the abstract notion of calculating machine was Turing. A more recent approach, dealing with calculability on an abacus or register machine, was discovered by Melzak (1961), who influenced Lambek (1961), by Minsky (1961) and again by Shepherdson and Sturgis (1963). See also the paper by Dana Scott (1967).

Section 2. The fact that the numerical functions representable in pure typed λ-calculus are properly contained between the primitive recursive and recursive functions was known to logicians in the 1960s. Our proof follows Shoenfield's text, chapter 8, which in turn appears to be based on earlier work of Tait. This entire line of research stems from Gödel's influential Dialectica paper (1958). Gregorczyk (1964) elucidates many of the finer points of Gödel's remarks. See also Troelstra (1973) and Barendregt (1981, Appendix A), for surveys.

Independently from the logicians, M.-F. Thibault (1977) investigated functions representable in free cartesian closed categories and obtained the results mentioned in the text. Our results in Part I on the equivalence between λ-calculi and cartesian closed categories complete this circle of ideas. Once again this illustrates our main theme: the close connections between natural categorical questions and central results in logic.

The representability of the Ackermann function is in both (Thibault 1977) and (Barendregt 1981, Appendix A), although a related result seems to have been published first by Gregorczyk (1964, Theorem 6.2).

Finally, what are the numerical functions representable in the pure typed λ-calculus? Gödel (1958) remarks that they are the ε_0-recursive functions (see Péter, 1967), while Shoenfield (1967), following Kreisel and Tait, proves they are also the same as the provably recursive functions of classical first-order arithmetic. For a recent survey of these results, see Fortune *et al.* (1983). Note however, that none of these authors use product types with surjective pairing.

Section 3. The representability of functions in formal systems comes from Gödel's famous 1931 paper. Many of the results in this section are standard (see, for example, Kleene, 1952) although the notion of strong representability (which is the natural notion categorically) appears only in Mendelson's book (1964). Huber–Dyson's result (Proposition 3.8) answers a question posed in early editions of Mendelson.

A stronger version of Theorem 3.12 (the numeralwise representability of partial recursive functions) using variables rather than numerals is in Coste-Roy *et al.* (1980), but their proof uses reflection principles (which were also mentioned in connection with the projectivity of N in the free topos).

Section 4. We only represent the partial recursive functions in the untyped λ-calculus in a 'weak' sense. What happens if $f(a_1, \ldots, a_k)$ is undefined? Church (1940) proposed in his version of λI-calculus that $f(a_1, \ldots, a_k)$ undefined should imply that the term $F^f(\#a_1, \ldots, \#a_k)$ has no normal form. However Barendregt (1981) argues persuasively that this was not a good definition and discusses numerous recent results on these matters. For us the issue is further complicated by the fact that the original Church–Rosser theorem *fails* for untyped λ-calculi with surjective pairing, as we remarked in Part I . We feel the final story here has not yet been written.

Bibliography

(LNM refers to Springer Lecture Notes in Mathematics)

W. Ackermann [1928] Zum Hilbertschen Aufbau der reellen Zahlen. *Math. Annalen* **99**, 118–33.

P.H. Aczel [1969] Saturated intuitionistic theories. In H.A. Schmidt, K. Schütte and H.J. Thiele (eds), *Contributions to mathematical logic*, North-Holland Publishing Co., Amsterdam, pp. 1–11.

T. Adachi [1983] A categorical characterization of lambda calculus models. Research Reports on Information Sciences. Tokyo Institute of Technology, No. C–49.

M. Artin, A. Grothendieck and J.L. Verdier [1972] *Théorie des topos et cohomologie étale des schémas* (SGA 4), Springer LNM 269.

A.A. Babaev [1981] Equality of maps and coherence theorem for biclosed categories. Translated from *Zapiski Nauchnykh Seminarov Leningradskogo Otdeleniya Mat. Instituta im. V.A. Steklova AN SSSR* **105**, 1281–5.

R. Balbes and P. Dwinger [1974] *Distributive lattices.* University of Missouri Press, Columbia.

H.P. Barendregt [1977] The type-free lambda calculus. In Barwise, pp. 1091–1132.

[1981] *The lambda calculus, its syntax and semantics.* Studies in Logic vol. 103. North-Holland Publishing Co., Amsterdam.

M. Barr and C. Wells [1985] *Toposes, triples, and theories.* Springer-Verlag.

J. Barwise (ed.) [1977] *Handbook of mathematical logic.* Studies in Logic and the Foundations of Mathematics vol. 90. North-Holland Publishing Co., Amsterdam.

J. Barwise, H.J. Keisler and K. Kunen (eds) [1980] *The Kleene symposium.* North-Holland Publishing Co., Amsterdam.

M.J. Beeson [1982] Problematic principles in constructive mathematics. In D. van Dalen, D. Lascar and J. Smiley (eds), *Logic Colloquium '80*, North-Holland Publishing Co., Amsterdam, pp. 11–55.

J. Bénabou [1963] Catégories avec multiplication, *C.R. Acad. Sci. Paris* **256**, 1887–90.

G. Birkhoff and J. Lipson [1970] Heterogeneous algebras. *J. Comb. Theory* **8**, 115–33.

E. Bishop [1967] *Foundations of constructive analysis.* McGraw-Hill, New York.

[1980] Mathematics as a numerical language. In Kino *et al.*, pp. 53–71

A. Boileau [1975] Types vs topos. Thesis, Université de Montréal.

A. Boileau and A. Joyal [1981] La logique des topos. *J. Symbolic Logic* **46**, 6–16.

N. Bourbaki [1948] *Algèbre multilinéaire.* Hermann, Paris.

A. Burroni [1980] Sur une utilisation des graphes dans le langage de la logique (Colloque d'Amiens), Manuscript.

[1981] Algèbres graphiques, 3ème colloque sur les catégories. *Cahiers de top. et géom. diff.* **23**, 249–65.

C.C. Chang and H.J. Keisler [1973] *Model theory.* North-Holland, Amsterdam and New York.

A. Church [1936] An unsolvable problem of elementary number theory. *Amer. J. Math.* **58**, 345–63.

[1937] Combinatory logic as a semigroup (abstract). *Bull. Amer. Math. Soc.* **43**, 333.

[1940] A foundation for the simple theory of types. *J. Symbolic Logic* **5**, 56–68.

[1941] The calculi of lambda conversion. *Ann. Math. Studies* **6**, Princeton University Press.

[1956] *Introduction to mathematical logic.* Princeton University Press.

S.D. Comer [1971] Representations of algebras by sections over Boolean spaces. *Pacific J. Math.* **38**, 29–38.

[1972] A sheaf-theoretic duality theory for cylindrical algebras. *Trans. Amer. Math. Soc.* **169**, 75–87.

M. Coste (see also M.-F. Coste-Roy) [1972] Language interne d'un topos. Séminaire de M. Bénabou, Université Paris-Nord.

[1973] Logique du premier ordre dans les topos élémentaires. Séminaire de M. Bénabou, Université Paris-Nord.

[1974] Logique d'ordre supérieur dans les topos élémentaires. Séminaire de M. Bénabou, Université Paris–Nord.

M.-F. Coste-Roy, M. Coste and L. Mahé [1980] Contribution to the study of the natural number object in elementary topoi. *J. Pure Appl. Algebra* **17**, 35–68.

P.-L. Curien [1983] Combinateurs catégoriques, algorithmes séquentiels et programmation applicative. Thèse de Doctorat d'état, Université Paris VII.

[1984] Logique combinatoire catégorique, preprint.

H.B. Curry [1930] Grundlagen der kombinatorischen Logik. *Amer. J. Math.* **28**, 789–834.

[1932] Some additions to the theory of combinators. *Amer. J. Math.* **54**, 551–8.

[1942] The inconsistency of certain formal logics. *J. Symbolic Logic* **7**, 115–17.

[1968] Recent advances in combinatory logic. *Bull. Soc. Math. Belg.* **20**, 288–98.

H.B. Curry and R. Feys [1958] *Combinatory logic*, vol. I. North-Holland Publishing Co., Amsterdam.

H.B. Curry, J.R. Hindley and J.P. Seldin [1972] *Combinatory logic*, vol. II. North-Holland Publishing Co., Amsterdam.

A. Day [1975] Filter monads, continuous lattices and closure systems. *Canad. J. Math.* **27**, 50–9.

R. Diaconescu [1975] Axiom of choice and complementation. *Proc. Amer. Math. Soc.* **51**, 176–8.

J. Dieudonné – see A. Grothendieck

M. Dummett [1977] *Elements of intuitionism.* Clarendon Press, Oxford.

P. Dwinger – see R. Balbes

H. Egli [1973] An analysis of Scott's λ-calculus models. *Cornell University Tech. Report* (TR 73–191).

S. Eilenberg, D.K. Harrison, S. MacLane and H. Röhrl (eds) [1966] *Proceedings of the Conference on Categorical Algebra, La Jolla 1965.* Springer-Verlag.

S. Eilenberg and G.M. Kelly [1966] Closed categories. In Eilenberg *et al.*, pp. 421–562.

S. Eilenberg and S. MacLane [1945] General theory of natural equivalences. *Trans. Amer. Math. Soc.* **58**, 231–94.

S. Eilenberg and J.C. Moore [1965] Adjoint functors and triples. *Ill. J. Math.* **9**, 381–98.

S. Feferman [1977] Theories of finite type related to mathematical practice. In Barwise, pp. 913–71.

J.E. Fenstad (ed.) [1971] *Proceedings of the Second Scandinavian Logic Symposium*, Studies in Logic vol. 63. North-Holland Publishing Co., Amsterdam.

R. Feys – see H.B. Curry

G.D. Findlay and J. Lambek [1955] Calculus of bimodules, unpublished manuscript.

F.B. Fitch [1952] *Symbolic logic.* Ronald, New York.

I. Fleischer [1972] *An adult's guide to Scottery.* Unpublished manuscript of lectures from University of Waterloo, 1972.

S. Fortune, D. Leivant and M. O'Donnell [1983] The expressiveness of simple and second-order type structures. *J. Assoc. for Computing Machinery* **30**, 151–85.

M.P. Fourman [1974] Connections between category theory and logic. Thesis, Oxford University.

[1977] The logic of topoi. In Barwise, pp. 1053–90.

M.P. Fourman, C. Mulvey and D.S. Scott (eds) [1979] *Applications of sheaves.* Proceedings L.M.S. Durham Symposium 1977, Springer LNM 753.

M.P. Fourman and A. Ščedrov [1982] The "world's simplest axiom of choice" fails. *Manuscripta Math.* **38**, 325–32.

M.P. Fourman and D.S. Scott [1979] Sheaves and logic. In Fourman *et al.*, pp. 302–401.

P. Freyd [1964] *Abelian categories: An introduction to the theory of functors.* Harper and Row, New York.

[1972] Aspects of topoi. *Bull. Austral. Math. Soc.* **7**, 1–76 and 467–80.

[1978] On proving that 1 is an indecomposable projective in various free categories, manuscript.

H. Friedman [1973] Some application of Kleene's methods for intuitionistic systems. In Mathias and Rogers, pp. 113–70.

H. Friedman and A. Ščedrov [1984] Set existence property for intuitionistic theories with dependent choice, manuscript.

R.O. Gandy [1980*a*] An early proof of normalization by A.M. Turing. In Seldin and Hindley, pp. 453–6.

[1980*b*] Proofs of strong normalization. In Seldin and Hindley, pp. 457–78.

G. Gierz, K.H. Hofmann, K. Keimel, J.D. Lawson, M. Mislove and D.S. Scott [1980] *A compendium of continuous lattices.* Springer-Verlag.

J.Y. Girard [1971] Une extension de l'interprétation de Gödel à l'analyse, et son application à l'élimination des coupures dans l'analyse et la théorie des types. In Fenstad, pp. 63–92.

[1972] Interprétation fonctionnelle et élimination des coupures de l'arithmétique d'ordre supérieur. Thèse de Doctorat d'état, Université Paris VII.

K. Gödel [1931] Über formal unentscheidbare Sätze der Principia mathematica und verwandter Systeme I. *Monatsh. Math. Phys.* **38**, 173–98. Translated in van Heijenoort.

[1958] Über eine bisher noch nicht benützte Erweiterung des finiten Standpunktes. *Dialectica* **12**, 280–7.

R. Goldblatt [1979] *Topoi: The categorial analysis of logic.* Studies in Logic and the Foundations of Mathematics vol. 98. North-Holland Publishing Co., Amsterdam.

J.W. Gray [1964] Sheaves with values in a category. *Topology* **3**, 1–18.

A. Grothendieck and J. Dieudonné [1960] *Eléments de géométrie algébrique, tome I: le langage des schémas.* I.H.E.S. Publ. Math. 4, Paris (Second edition published 1971 by Springer-Verlag in *Die Grundlehren der math. Wissenschaffen*, Band 166).

A. Grothendieck and J.L. Verdier (see also M. Artin) [1972] Exposé IV: Topos. In M. Artin *et al.*, pp. 229–515.

A. Grzegorczyk [1964] Recursive objects in all finite types. *Fund. Math.* **54**, 73–93.

R. Guitart [1973] Les monades involutives en théorie élémentaire des ensembles. *C.R.A.S., Paris* **277**, 935–7.

W.S. Hatcher [1968] *Foundations of mathematics.* W.B. Saunders Co., Philadelphia.

[1982] *The logical foundations of mathematics.* Pergamon Press.

S. Hayashi [1983] Adjunction of semifunctors: categorical structures in non-extensional lambda calculus, preprint.

L.A. Henkin [1949] The completeness of the first-order functional calculus. *J. Symbolic Logic* **14**, 159–66.

[1950] Completeness in the theory of types. *J. Symbolic Logic* **15**, 81–91. Reprinted in J. Hintikka, *The philosophy of mathematics*, Oxford University Press, 1969.

[1963] A theory of propositional types. *Fund. Math.* **52**, 323–33.
A. Heyting [1966] *Intuitionism* (Second revised edition). North-Holland Publishing Co., Amsterdam.
D. Higgs [1973] A category approach to Boolean-valued set theory, manuscript (see also Higgs 1984).
[1974] Some examples of cartesian closed categories (= CC), manuscript.
[1984] Injectivity in the topos of complete Heyting algebra valued sets. *Canad. J. Math.* **36**, 550–68.
J.R. Hindley, B. Lercher and J.P. Seldin (see also H.B. Curry) [1972] *Combinatory logic.* Cambridge University Press.
M. Hochster [1969] Prime ideal structure in commutative rings. *Trans. Amer. Math. Soc.* **142**, 43–60.
K.H. Hofmann – see G. Giertz
W.A. Howard [1970] Assignment of ordinals to terms for primitive recursive functionals of finite type. In A. Kino *et al.*, pp. 443–58.
[1980] The formulae-as-types notion of construction. In Seldin and Hindley, pp. 479–90.
V. Huber-Dyson [1965] Strong representability of number-theoretic functions. *Hughes Aircraft Report*, 1–5.
J.M.E. Hyland, P.T. Johnstone and A.M. Pitts [1980] Tripos Theory. *Math. Proc. Cambridge Philos. Soc.* **88**, 205–52.
N. Jacobson [1974] *Basic algebra I.* W.H. Freeman and Co., San Francisco.
P.T. Johnstone – see also J.M.E. Hyland
[1977] *Topos theory.* L.M.S. Mathematical Monographs no. 10. Academic Press, London.
[1979*a*] Automorphisms of Ω. *Algebra Universalis* **9**, 1–7.
[1979*b*] Conditions related to De Morgan's law. In Fourman *et al.*, pp. 479–91.
[1981] Factorization theorems for geometric morphisms I. *Cahiers top. et géom. diff.* **22**, 3–17.
[1982] *Stone spaces.* Cambridge studies in advanced mathematics 3. Cambridge University Press.
A. Joyal – see A. Boileau
K. Keimel – see G. Gierz
H.J. Keisler – see C.C. Chang
G.M. Kelly and S. MacLane (see also S. Eilenberg) [1971] Coherence in closed categories, *J. Pure Appl. Algebra* **1**, 97–140 and **2**, 219.
A. Kino, J. Myhill and R.E. Vesley (eds) [1970] *Intuitionism and proof theory.* North-Holland Publishing Co., Amsterdam.
S.C. Kleene [1936] λ-definability and recursiveness. *Duke Math. J.* **2**, 340–53.
[1952] *Introduction to metamathematics.* Van Nostrand, New York and Toronto.
S.C. Kleene and J.B. Rosser [1935] The inconsistency of certain formal logics. *Ann. Math.* **36**, 630–6.
S.C. Kleene and R.E. Vesley [1965] *The foundations of intuitionistic mathematics.* North-Holland Publishing Co., Amsterdam.
H. Kleisli [1965] Every standard construction is induced by a pair of adjoint functors. *Proc. Amer. Math. Soc.* **16**, 544–6.
A. Kock [1970] On a theorem of Läuchli concerning proof bundles. Aarhus University, manuscript.
A. Kock, P. Lecouturier and C.J. Mikkelsen [1975] Some topos-theoretic concepts of finiteness. In Lawvere *et al.*, pp. 209–83.
A. Kock and G.E. Reyes [1977] Doctrines in categorical logic. In Barewise, pp. 282–313.
A. Kock and G.C. Wraith [1971] *Elementary toposes.* Lecture Notes Series no. 30. Aarhus University.

C.P.J. Koymans [1984] Models of the lambda calculus. Thesis, Rijksuniversiteit Utrecht.

S.A. Kripke [1965] Semantical analysis of Intuitionistic Logic I. In J.N. Crossley and M.A.E. Dummett (eds), *Formal systems and recursive functions*, North-Holland Publishing Co., Amsterdam.

K. Kuratowski [1933] *Topologie*, Vol. I. Monographie Matematyczne tom 3, PWN-Polish Scientific Publishers, Warszawa, Z 8–132.

J. Lambek (see also G.D. Findlay) [1958] The mathematics of sentence structure. *Amer. Math. Monthly* **65**, 154–69.

[1961*a*] How to program an infinite abacus. *Canad. Math. Bull.* **4**, 295–302.

[1961*b*] On the calculus of syntactic types. *Amer. Math. Soc. Proc. Symposia Appl. Math.* **12**, 166–78.

[1968] Deductive systems and categories I. *J. Math. Systems Theory* **2**, 278–318.

[1969] Deductive systems and categories II, Springer LNM 86, pp. 76–122.

[1970] Subequalizers. *Canad. Math. Bull.* **13**, 337–49.

[1971] On the representation of modules by sheaves of factor modules. *Canad. Math. Bull.* **14**, 359–68.

[1972] Deductive systems and categories III, Springer LNM 274, pp. 57–82.

[1974] Functional completeness of cartesian categories. *Ann. Math. Logic* **6**, 259–92.

[1979] Review of Strooker 1978. *Bull. Amer. Math. Soc.* **1**, 919–28.

[1980*a*] From types to sets. *Advances in Math.* **36**, 113–64.

[1980*b*] From λ-calculus to Cartesian closed categories. In Seldin and Hindley, pp. 375–402.

[1981] The influence of Heraclitus on modern mathematics. In J. Agassi and R.S. Cohen (eds), *Scientific philosophy today*, Reidel Publishing Co., pp. 111–14.

[1982] Toposes are monadic over categories, Springer LNM 962, pp. 153–66.

J. Lambek and I. Moerdijk [1982*a*] Two sheaf representations of elementary toposes. In Troelstra and van Dalen, pp. 275–95.

[1982*b*] A Henkin–Kripke completeness theorem for intuitionistic type theory, unpublished manuscript.

J. Lambek and B.A. Rattray [1975*a*] Localization and duality in additive categories. *Houston J. Math.* **1**, 87–100.

[1975*b*] Localization and sheaf reflectors. *Trans. Amer. Math. Soc.* **210**, 279–93.

[1978] Functional completeness and Stone duality. In *Studies in foundations and combinatorics*, Advances in Math. Suppl. Studies no. 1, pp. 1–9.

[1979] A general Stone–Gelfand duality. *Trans. Amer. Math. Soc.* **248**, 1–35.

J. Lambek and P.J. Scott [1980] Intuitionist type theory and the free topos. *J. Pure Appl. Algebra* **19**, 215–57.

[1981*a*] Algebraic aspects of topos theory. *Cahiers top. et géom. diff.* **22**, 129–40.

[1981*b*] Intuitionist type theory and foundations. *J. Philos. Logic* **7**, 101–15.

[1981*c*] Independence of premisses and the free topos. *Proc. Symp. Constructive Math.*, Springer LNM 873, pp. 191–207.

[1983] New proofs of some intuitionistic principles. *Zeit. f. Math. Logik und Grundlagen d. Math.* **29**, 493–504.

[1984] Aspects of higher order categorical logic. In J.W. Gray (ed.) Mathematical applications of category theory, *Contemporary Mathematics* **30**, 145–74.

J.D. Lawson – see G. Gierz

F.W. Lawvere [1963] Functorial semantics of algebraic theories. *Proc. Nat. Acad. Sci. U.S.A.* **50**, 869–72.

[1964] An elementary theory of the category of sets. *Proc. Nat. Acad. Sci. U.S.A.* **52**, 1506–11.

[1966] The category of categories as a foundation for mathematics. In Eilenberg *et al.*, pp. 1–20.

[1967] Category-valued higher order logic. Presented at UCLA 1967 Set Theory Symposium, unpublished manuscript.

[1969*a*] Adjointness in foundations. *Dialectica* **23**, 281–96.

[1969*b*] Diagonal arguments and cartesian closed categories. In *Category theory, Homology theory and their applications II*, Springer LNM 92, pp. 134–45.

[1970] Equality in hyperdoctrines and comprehension schema as an adjoint functor. In A. Heller (ed.), *Proc., New York Symposium on Applications of Categorical Algebra*, Amer. Math. Soc., Providence R.I., pp. 1–14.

[1971] Quantifiers and sheaves. In *Actes du Congrès Intern. des Math., Nice 1970*, tome I, Gauthier-Villars, Paris, pp. 329–34.

[1972] *Introduction to toposes, algebraic geometry and logic*, Springer LNM 274, pp. 1–12.

[1975*a*] Continuously variable sets: Algebraic Geometry = Geometric Logic. In Rose and Shepherdson, pp. 135–56.

[1975*b*] Introduction to part I. In Lawvere *et al.*, pp. 3–14.

[1976] Variable quantities and variable structures in topoi. In A. Heller and M. Tierney (eds), *Algebra, topology and category theory*: a collection of papers in honor of Samuel Eilenberg, Academic Press, pp. 101–31.

F.W. Lawvere, C. Maurer and G.C. Wraith (eds) [1975] *Model theory and topoi*, Springer LNM 445.

P. Lecouturier – see A. Kock

B. Lercher – see J.R. Hindley

F.E.J. Linton [1969] An outline of functorial semantics. In B. Eckmann (ed.), *Seminar on triples and categorical homology theory*, Springer LNM 80, pp. 7–52.

J. Lipson – see G. Birkhoff

S. MacLane (see also S. Eilenberg, G.M. Kelly) [1963] Natural associativity and commutativity. *Rice University Studies* **49**, no. 4, 28–46.

[1971] *Categories for the working mathematician.* Graduate Texts in Mathematics 5. Springer-Verlag.

[1975] Sets, topoi and internal logic in categories. In Rose and Shepherdson, pp. 119–34.

[1982] Why commutative diagrams coincide with equivalent proofs. *Contemporary Mathematics* **13**, 387–401.

L. Mahé – see M.-F. Coste-Roy

M. Makkai [1980] *N* is projective in the initial topos, manuscript.

M. Makkai and G.E. Reyes [1976] Model-theoretic methods in the theory of topoi and related categories. *Bull. Acad. Pol. Sci.* **27**, 379–92.

[1977] *First order categorical logic*, Springer LNM 611.

E.G. Manes [1976] *Algebraic theories.* Graduate Texts in Mathematics 26. Springer-Verlag.

C.R. Mann [1975] The connection between equivalence of proofs and cartesian closed categories. *Proc. London Math. Soc.*, **31**, 289–310.

P. Martin-Löf [1974] An intuitionistic theory of types: predicative part. In Rose and Shepherdson, pp. 73–118.

A.R.D. Mathias and H. Rogers (eds) [1973] *Cambridge Summer School in Math. Logic, Proc. 1971*, Springer LNM 337.

Z.A. Melzak [1961] An informal arithmetical introduction to computability and computation. *Canad. Math. Bull.* **4**, 279–93.

E. Mendelson [1974] *Introduction to mathematical logic.* Van Nostrand, Princeton.

K. Menger [1949] Are variables necessary in calculus? *Amer. Math. Monthly* **56**, 609–20.

A.R. Meyer [1982] What is a model of the Lambda Calculus. *Information and Control* **52**, 87–122.

C.J. Mikkelsen (see also A. Kock) [1976] *Lattice-theoretic and logical aspects of elementary topoi.* Aarhus Universitet Various Publications Series 25.

G.E. Minc (G.E. Mints) [1975] Theory of proofs (arithmetic and analysis). Translated from *Itogi Nauki i Tekhniki* (*Algebra, topologiya, Geometriya*), **13**, 5–49.

[1977] Closed categories and the theory of proofs. Translated from *Zapiski Nauchnykh Seminarov Leningradskogo Otdeleniya Mat. Instituta im. V.A. Steklova AN SSSR* **68**, 83–114.

[1979?] Proof theory and category theory, manuscript.

M.L. Minsky [1961] Recursive unsolvability of Post's problem of "tag" and other topics in the theory of Turing machines. *Ann. Math.* **74**, 437–55.

W. Mitchell [1972] Boolean topoi and the theory of sets. *J. Pure Appl. Algebra* **2**, 261–74.

M. Mislove – see G. Gierz

I. Moerdijk (see also J. Lambek) [1982] Glueing topoi and higher order disjunction and existence. In Troelstra and van Dalen, pp. 359–75.

J.C. Moore – see S. Eilenberg

A. Mostowski [1951] On the rules of proof in the pure functional calculus of the first order. *J. Symbolic Logic* **16**, 107–11.

C. Mulvey (see also M.P. Fourman) [1974] Intuitionistic algebra and representations of rings. *Memoirs Amer. Math. Soc.* **148**, 3–57.

J. Myhill [1973] Some properties of intuitionistic Zermelo–Fränkel set theory. In Mathias and Rogers, pp. 206–31.

A. Obtułowicz [1979] On the consistent Church algebraic theories, algebraic theories of type $\lambda - \beta\eta$, and nontrivial models of the type-free lambda calculus, manuscript.

A. Obtułowicz and A. Wiweger [1982] Categorical, functional and algebraic aspects of the type-free Lambda Calculus. *Universal algebra and applications.* Banach Center Publications vol. 9, PWN-Polish Scientific Publishers, Warszawa.

G. Osius [1974*a*] Categorical set theory: a characterisation of the category of sets. *J. Pure Appl. Algebra* **4**, 79–119.

[1974*b*] The internal and external aspects of logic and set theory in elementary topoi. *Cahiers top. et géom. diff.* **15**, 157–80.

[1975*a*] Logical and set-theoretical tools in elementary topoi. In Lawvere *et al.*, pp. 297–346.

[1975*b*] A note on Kripke–Joyal semantics for the internal language of topoi.

R. Paré [1974] Co-limits in topoi. *Bull. Amer. Math. Soc.* **80**, 556–61.

B. Pareigis [1970] *Categories and functions.* Academic Press, New York.

R. Péter [1967] *Recursive functions.* Academic Press, New York and London.

R.S. Pierce [1967] Modules over commutative regular rings. *Memoirs Amer. Math. Soc.* no. 70.

A.M. Pitts – see J.M.E. Hyland

G.D Plotkin and M.B. Smyth – see also M.B. Smyth

[1977] The category-theoretic solution of recursive domain equations (extended abstract). *18th Annual Symposium on Foundations of Computer Science* (*Providence, R.I.*), pp. 13–29. IEEE Comput. Sci., Long Beach, Calif.

G. Pottinger [1977] Normalization as a homomorphic image of cut-elimination. *Ann. Math. Logic* **12**, 323–57.

[1978] Proofs of the normalization and Church–Rosser Theorems for the typed λ-calculus. *Notre Dame J. Formal Logic* **19**, 445–51.

[1981] The Church–Rosser Theorem for the typed λ-calculus with surjective pairing. *Notre Dame J. Formal Logic* **22**, 264–8.

D. Prawitz [1965] *Natural deduction.* Almquist and Wiksell, Stockholm.

[1971] Ideas and results in proof theory. In Fenstad, pp. 235–307.

W.V. Quine [1960] Variables explained away. *Proc. Amer. Philos. Soc.* **104**, 343–7.
H. Rasiowa and R. Sikorski [1963] *The Mathematics of metamathematics.* Monographie Matematyczne tom 41, PWN-Polish Scientific Publishers, Warszawa.
B.A. Rattray – see J. Lambek
G.E. Reyes (see also A. Kock, M. Makkai) [1974] From sheaves to logic. In A. Daigneault (ed.), *Studies in Algebraic Logic*, M.A.A. Studies in Math., vol. 9, Math. Assoc. of America, pp. 143–204.
A. Robinson [1966] *Non-standard analysis.* North-Holland Publishing Co., Amsterdam.
H. Rogers – see A.R.D. Mathias
H.E. Rose and J.C. Shepherdson (eds) [1974] *Proceedings of the A.S.L. Logic Colloquium, Bristol 1973.* Studies in Logic and the Foundations of Mathematics vol. 28. North-Holland Publishing Co., Amsterdam.
P.C. Rosenbloom [1950] *The elements of mathematical logic.* Dover, New York.
J.B. Rosser (see also S.C. Kleene) [1935] A mathematical logic without variables. *Ann. Math.* **36**, 127–50; *Duke Math. J.* **1**, 328–55.
[1942] New sets of postulates for combinatory logics. *J. Symbolic Logic* **7**, 18–27.
B. Russell [1908] Mathematical logic based on the theory of types. *Amer. J. Math.* **30**, 222–63. Reprinted in van Heijenoort.
B. Russell and A.N. Whitehead [1910–13] *Principia mathematica* I–III. Cambridge University Press.
L.E. Sanchis [1967] Functionals defined by recursion. *Notre Dame J. of Formal Logic* **8**, 161–74.
B. Scarpellini [171] A model for bar recursion of higher types. *Comp. Math.* **23**, 123–53.
A. Ščedrov (see also M.P. Fourman, H. Friedman, P.J. Scott) [1984*a*] Forcing and classifying topoi. *Memoirs Amer. Math. Soc.* **48**, no. 295, Providence, R.I.
[1984*b*] On some non-classical extensions of second-order intuitionistic propositional calculus. *Ann. Pure Appl. Logic*, **27**, 155–64.
A. Ščedrov and P.J. Scott [1982] A note on the Friedman slash and Freyd covers. In Troelstra and van Dalen, pp. 443–52.
D.J. Schlomiuk [1977] Logique des topos. Séminaire de Mathématiques supérieures no. 53. Les Presses de l'Université de Montréal.
M. Schönfinkel [1924] Über die Bausteine der mathematischen Logik. *Math. Ann.* **92**, 305–16. Translated in van Heijenoort.
H. Schubert [1972] *Categories.* Springer-Verlag.
D.S. Scott (see also M.P. Fourman, G. Gierz) [1967] Some definitional suggestions for automata theory. *J. of Computer and Systems Sciences* **1**, 187–212.
[1968] Extending the topological interpretation to intuitionistic analysis I. *Compositio Math.* **20**, 194–210.
[1970*a*] Extending the topological interpretation to intuitionistic analysis II. In Kino *et al.* pp. 235–56.
[1970*b*] Constructive validity, Springer LNM 125, pp. 237–75.
[1972] Continuous lattices, Springer LNM 274, pp. 97–136.
[1973] Models for various type-free calculi. In P. Suppes *et al.* (eds), *Logic, methodology and philosophy of Science* IV, North-Holland Publishing Co., Amsterdam, pp. 157–87.
[1976] Data types as lattices. *SIAM J. Computing* **5**, 522–87.
[1977] Logic and programming languages. *Comm. Assoc. for Comp. Mach.* **20**, 634–41.
[1979] Identity and existence in intuitionistic logic. In Fourman *et al.*, pp. 660–96.
[1980*a*] Lambda Calculus: some models, some philosophy. In Barwise *et al.*, pp. 381–421.
[1980*b*] Relating theories of the lambda calculus. In Seldin and Hindley, pp. 403–50.
P.J. Scott (see also J. Lambek, A. Ščedrov) [1978] The "dialectica" interpretation and categories. *Zeitschr. f. Math. Logik und Grundlagen d. Math.* **24**, 553–75.

J. Seebach, L. Seebach and L. Steen [1970] What is a sheaf? *Amer. Math. Monthly* **77**, 681–703.

R.A.G. Seely [1977] Hyperdoctrines and natural deduction. Thesis, U. of Cambridge.
[1983] Hyperdoctrines, natural deduction and the Beck condition. *Zeitschr. f. Math. Logik und Grundlagen d. Math.* **29**, 505–42.
[1984] Locally cartesian closed categories and type theory. *Math. Proc. Camb. Philos. Soc.* **95**, 33–48.

J.P. Seldin (see also H.B. Curry, J.R. Hindley) [1980] Curry's program. In Seldin and Hindley, pp. 3–34.

J.P. Seldin and J.R. Hindley (eds) [1980] *To H.B. Curry: essays on combinatory logic, lambda calculus and formalism.* Academic Press, London.

J.C. Shepherdson and H.E. Sturgis [1963] Computability of recursive functions. *J. Assoc. Comput. Mach.* **10**, 217–55.

J.R. Shoenfield [1967] *Mathematical logic.* Addison-Wesley, Reading, Mass.

R. Sikorski – see H. Rasiowa

C. Smorynski [1973] Applications of Kripke models. In Troelstra, pp. 324–91.

M.B. Smyth and G.D. Plotkin (see also G.D. Plotkin) [1982] The category-theoretic solution of recursive domain equations. *SIAM Journal on Computing*, **11**, 761–783.

S.V. Solov'ov [1981] The category of finite sets and Cartesian closed categories. Translated from *Zapiski Nauchnykh Seminarov Leningradskogo Otdeleniya Mat. Instituta im. V.A. Steklova An SSSR* **105**, 1387–1400.

C. Spector [1962] Provably recursive functionals of analysis: a consistency proof of analysis by an extension of principles formulated in current intuitionistic mathematics. In J.C.E. Dekker (ed.), *Recursive function theory.* Proceedings of Symposia in Pure Math., vol. 5, Amer. Math. Soc., Providence, R.I., pp. 1–27.

J. Staples [1973] Combinator realizability of constructive finite type analysis. In Mathias and Rogers, pp. 253–73.

L. Steen – see J. Seebach S.

S. Stenlund [1972] *Combinators, λ-terms and proof theory.* D. Reidel, Dordrecht, Holland.

J.R. Strooker [1978] *Introduction to categories, homological algebra and sheaf cohomology.* Cambridge University Press.

H.F. Sturgis – see J.C. Sheperdson

M.E. Szabo (ed.) [1969] *The collected papers of Gerhard Gentzen.* Studies in Logic and the Foundations of Mathematics. North-Holland Publishing Co., Amsterdam.

M.E. Szabo [1974*a*] A categorical characterization of Boolean algebras. *Algebra Universalis*, 109–11.
[1974*b*] A categorical equivalence of proofs. *Notre Dame J. Formal Logic* **15**, 177–91.
[1975] A counter-example to coherence in cartesian closed categories. *Canad. Math. Bull.* **18**, 111–14.
[1978] *Algebra of proofs.* Studies in Logic and the Foundations of Mathematics vol. 88. North-Holland Publishing Co., Amsterdam.

W.W. Tait [1967] Intentional interpretation of functionals of finite type I. *J. Symbolic Logic* **32**, 198–212.

A. Tarski (A. Tajtelbaum) [1923] Sur le terme primitif de la logistique. *Fund. Math.* **4**, 196–200.

M.-F. Thibault [1977] Représentations des fonctions récursives dans les catégories. Thesis, McGill University, Montreal.
[1982] Prerecursive categories. *J. Pure Appl. Algebra* **24**, 79–93.

M. Tierney [1972] Sheaf theory and the continuum hypothesis. In F.W. Lawvere (ed.), *Toposes, algebraic geometry and logic*, Springer LNM 274, pp. 13–42.

A.S. Troelstra (ed.) [1973] *Metamathematical investigations of intuitionistic arithmetic and analysis*, Springer LNM 344.

A.S. Troelstra [1971] Notions of realizability for intuitionistic arithmetic and intuitionistic arithmetic in all finite types. In Fenstad, pp. 369–405.

[1973] Notes in intuitionistic second order arithmetic, *Cambridge Summer School in Math. Logic*, Springer LNM 337, pp. 171–203.

A.S. Troelstra and D. van Dalen (eds) [1982] *The L.E.J. Brouwer Centenary Symposium.* Studies in Logic and the Foundations of Mathematics vol. 110. North-Holland Publishing Co., Amsterdam.

A. Turing [1973] On computable numbers, with an application to the Entscheidungsproblem. *Proc. London Math. Soc.* **42**, 230–65; **43**, 544–6.

D. van Dalen, D. Lascar and J. Smiley (eds) [1982] *Logic Colloquium '80.* North-Holland Publishing Co., Amsterdam.

J. van Heijenoort (ed.) [1967] *From Frege to Gödel.* Harvard University Press.

J.L. Verdier – see M. Artin, A. Grothendieck

R.E. Vesley – see S.C. Kleene

H. Volger [1975*a*] Completeness theorem for logical categories. In Lawvere *et al.*, pp. 51–86.

[1975*b*] Logical categories, semantical categories and topoi. In Lawvere *et al.*, pp. 87–100.

R. Voreadou [1977] Coherence and non-commutative diagrams in closed categories. *Memoirs Amer. Math. Soc.* **182**, Providence, R.I.

R.C. de Vrijer [1982] Strong normalization in $N\text{-}HA_p^\omega$, manuscript.

A. Wiweger – see A. Obtułowicz

C. Wells – see M. Barr

A.N. Whitehead – see B. Russell

A. Wolf [1974] Sheaf representation of arithmetical algebras. In Recent advances in the representation theory of rings and C-algebras by continuous sections. *Memoirs Amer. Math. Soc.* **148**, 87–93.

G.C. Wraith (see also A. Kock) [1974] Artin glueing. *J. Pure Appl. Algebra* **4**, 345–8.

J. Zucker [1974] The correspondence between cut-elimination and normalization. II. *Ann. Math. Logic* **7**, 113–55.

Author index*

* This index lists all names mentioned in the text but does not refer to the bibliography.

Subject index